Power System Wide-Area
Stability Analysis and Control

Power System Wide-Area Stability Analysis and Control

Jing Ma

State Key Laboratory of Alternate Electrical Power System with Renewable Energy Sources
North China Electric Power University
Beijing, PR China

WILEY

SCIENCE PRESS
Beijing

This edition first published 2018 by John Wiley & Sons Singapore Pte. Ltd under exclusive licence granted by Science Press for all media and languages (excluding simplified and traditional Chinese) throughout the world (excluding Mainland China), and with non-exclusive license for electronic versions in Mainland China.

Registered Offices
John Wiley & Sons, Inc., 111 River Street, Hoboken, NJ 07030, USA

John Wiley & Sons Singapore Pte. Ltd, 1 Fusionopolis Walk, #07-01 Solaris South Tower, Singapore 138628

Editorial Office
1 Fusionopolis Walk, #07-01 Solaris South Tower, Singapore 138628

For details of our global editorial offices, customer services, and more information about Wiley products visit us at www.wiley.com.

Wiley also publishes its books in a variety of electronic formats and by print-on-demand. Some content that appears in standard print versions of this book may not be available in other formats.

Library of Congress Cataloging-in-Publication Data
Names: Ma, Jing (Electrical engineer), author.
Title: Power system wide-area stability analysis and control / Dr. Ma Jing, State Key Laboratory of Alternate Electrical Power System with Renewable Energy Sources, North China Electric Power University, Beijing, China.
Description: First edition. | Singapore ; Hoboken, NJ, USA : Wiley, [2018] | Includes bibliographical references and index. |
Identifiers: LCCN 2017013166 (print) | LCCN 2017039657 (ebook) | ISBN 9781119304876 (pdf) | ISBN 9781119304869 (epub) | ISBN 9781119304845 (cloth)
Subjects: LCSH: Electric power system stability. | Electric power system–Control.
Classification: LCC TK1010 (ebook) | LCC TK1010 .M328 2017 (print) | DDC 621.319–dc23
LC record available at https://lccn.loc.gov/2017013166

Cover design by Wiley
Cover images: (Powerline) © Trudy Fletcher / EyeEm/Gettyimages; (Wind Turbines) © Nicolas Balcazar / EyeEm/Gettyimages

Set in 10/12pt Warnock by SPi Global, Pondicherry, India

Printed in Singapore by C.O.S. Printers Pte Ltd

10 9 8 7 6 5 4 3 2 1

Contents

About the Author

Professor Jing Ma has been working in this area since 2003. His research mainly concentrates in the area of power system protection and control. A very systematic research and practice on power system wide-area stability analysis and control, especially the approaches on the small-signal stability analysis and studies on the novel principle of wide-area robust and adaptive control, have been carried out over more than 12 years. He was the first to apply Guardian Map Theory, Perturbation Theory, and the Markov Model to fully consider the characteristics of large time-delay and high uncertainties into the power system stability analysis process. He also invented a variety of robust and adaptive control strategies using the Federated Kalman Filter, Dual Youla Parameterization and Classification, and the Regression Tree to establish a wide-area control system with high accuracy and efficiency. A series of papers was published in the authoritative journals such as *IEEE Transactions on Power Systems* and *IEEE Transactions on Power Delivery*. The work has been widely acknowledged and cited by international peers. Part of his research results was used in many practical engineering projects, which highly accelerate the application and spread of wide-area control technology. In recent years, he has published more than 30 papers (indexed by SCI), authorized more than 20 patents, and undertaken many major projects in China. For instance, he guided two projects of the National Natural Science Foundation of China to study the Stability and Coordinated Control for a Complicated Power System. He set up an advanced wide-area protection and control platform for stability analysis of a power system, and pioneered to design and realize the corresponding control techniques. He was also responsible for several projects from governments and enterprises on the study of wide-area stability analysis and control. He was also a major member of the National Key Research and Development Plan of China on the study of the Wide-Area Protection and Control for Complicated Power Systems. He cooperated with the China Electric Power Research Institute to guide the exploitation of the integrated protection and control system of substation areas and wide areas. He has been teaching the courses of *Power System Automation* for years. Many materials of the book have been taught to students and other professionals.

Preface

Power system stability has been a key issue in the "three defense lines" of power system security defense in China, and has drawn the continuous attention of many researchers. With the large-scale integration of renewable energy sources and the wide application of power electronic devices, the stability problem caused by grid interconnection has become ever prominent, greatly endangering the safe and stable operation of a power system, even causing blackouts, which result in immeasurable losses to the society and the economy. Therefore, the stability of the power grid under new circumstances and trends requires the in-depth exploration and research of all electric power practitioners.

Challenges usually come with innovation and opportunities. With the development and wide application of PMU technology, wide-area information from WAMS has provided abundant synchronous and dynamic information for the stability analysis and control of a large complex power grid. It is an issue of great theoretical value and practical significance in the smart grid research to combine PMU technology with stability control, so that the safety of the power grid can be guaranteed to most users. Currently, many problems in wide-area stability control are yet to be solved. For example, it is the basis and premise of wide-area stability control to conduct system stability identification, instability mode identification, and stability region calculation in a complex time-varying grid with time delay and uncertainties. Meanwhile, it is the goal and task of wide-area stability control to design reasonable and effective robust and adaptive control strategies to solve the problems of lack-of-control, over-control, and out-of-control in the power grid. By summarizing the existing research achievements and learning from the advantages and disadvantages of traditional stability control, this book conducts a prospective discussion on the key techniques in wide-area stability control, such as wide-area signal monitoring, wide-area stability analysis, and wide-area control strategies, etc. The discussion is centered on research achievements of the author, and is presented in great breadth and depth, endeavoring to point out the evolving direction of wide-area stability control.

The author strives to make the basic theories simple and the formula derivations precise and complete. On this basis, through a large number of case studies, rigorous verification that fits the engineering practice is carried out concerning the wide-area stability control schemes introduced in the book. It should be noted that, since wide-area stability control is still developing, the contents in this book may not be the final solutions. For the issues that have not gained unified understanding, the author has put forward distinct viewpoints in the book. It is the author's sincere hope that readers will be inspired by this book to pursue better achievements in their research. Due to limited space, references are provided for the issues that cannot be discussed in detail, so that readers could study them in depth.

This book is readable by graduate students in universities, technical personnel in research institutes, and professionals with certain theoretical knowledge and practical experience in their research on wide-area stability control and relevant technological innovations. Mistakes are inevitable in the book and any criticism from readers will be welcome.

This work is supported by National Natural Science Foundation of China (No. 51277193), the Chinese University Scientific Fund Project (No. 2014ZZD02), Beijing Metropolis Beijing Nova program (Z141101001814012), the Excellent Talents in Beijing City (2013B009005000001), and the Fund of Fok Ying Tung Education Foundation (141057).

Beijing, China *Jing Ma*
2016

1

Basic Theories of Power System Security Defense

1.1 Introduction

A power system is a large-scale system with wide geographical distribution, large numbers of components, and fast dynamic response. Disturbance on one single component may quickly spread to the whole system. The most important task in power system design and operation is to analyze the transient and dynamic behaviors of the power system under different levels of disturbance and then determine the appropriate control strategies and corresponding measures.

Ever since the 1960s, large area blackouts have occurred from time to time, causing huge economic losses. The "8–14" blackout in the eastern North American power grid that occurred in 2003 has inspired a worldwide wave of research on the prevention of large power grid blackouts. The 2012 India blackout triggered the largest scale of blackout in human history. With the integration of large-scale renewable energy sources, the power grid operating mode has become changeable. The integration of distributed generation and micronetworks and self-healing control has caused the distribution network to be changeable, even in configuration. Besides, the application of power electronic devices introduces a large number of nonlinear controlled components to the grid, which causes power grid stable operation and control to be increasingly difficult. Therefore, there is an urgent need to study the security defense of a large power grid.

Power system personnel in China have conducted a lot of research to ensure the safe and stable operation of the power system, and have put forward "three defense lines" in a power system to deal with serious faults. The first defense line ensures that the system has a certain degree of safety margin in normal operating condition, does not lose the power source and load, and maintains stable operation when nonserious faults occur. The second defense line ensures that system stability is not destroyed and faults do not expand when relatively serious faults occur. The third defense line ensures that the system does not collapse and that large-area blackout does not occur in the case of extremely serious faults.

This chapter is an overview of the basic theories of power system security defense. First, the basic requirements on power system reliability and the definition and classification of power system stability in China and abroad are introduced. The different types of disturbances that a power system may encounter and their impacts on the system are introduced. In order to ensure the

Power System Wide-Area Stability Analysis and Control, First Edition. Jing Ma.

safe and stable operation of a power system, different kinds of protection and control measures should be taken concerning different types of disturbances, including prevention control, emergency control, splitting control, and restoration control.

1.2 Power System Reliability and Stability

1.2.1 Reliability of Power System

The basic function of a power system is to provide all users with a continuous power supply that is in accordance with relevant regulations in power quality (voltage and frequency). Power system reliability is a measure of the capability of the power system to provide users with the required quantity of power of acceptable quality standard continuously, including two aspects: system adequacy and security [1].

1.2.1.1 Adequacy

Adequacy (also known as static reliability) refers to the capability of a power system to provide users with the required quality and quantity of power when the power system is in steady-state operation and, within the allowed ranges of system component rated capacity, bus voltage, and system frequency, to consider the planned outage and reasonable unplanned outage of system components to the user and provides all of the required electric power and the ability described in reference [1]. Detailed indexes to characterize adequacy are as follows:

1) *LOLD* (loss of load probability) refers to the probability that the system cannot meet load demand in a given time interval, that is,

$$LOLD = \sum_{i \in S} P_i \tag{1.1}$$

   where P_i is the probability of the system being at state i. S is the complete set of system states in which the system cannot meet load demand in the given time interval.
2) *LOLE* (loss of load expectation) refers to the expected number of hours or days when the system cannot meet load demand in a given time interval, that is,

$$LOLE = \sum_{i \in S} P_i T \tag{1.2}$$

   where P_i is the probability of the system being at state i, S is the complete set of system states in which the system cannot meet load demand in the given time interval, and T is the number of hours or days in the given time interval.
3) *LOLF* (loss of load frequency) refers to the number of times when the system cannot meet load demand in a given time interval, that is,

$$LOLE = \sum_{i \in S} F_i \tag{1.3}$$

where F_i is the probability of the system being at state i and S is the complete set of system states in which the system cannot meet load demand in the given time interval.

4) *LOLD* (loss of load duration) refers to the average time duration when the system cannot meet load demand in a given time interval, that is,

$$LOLE = \frac{LOLE}{LOLF} \tag{1.4}$$

where *LOLE* is loss of load expectation and *LOLF* is loss of load frequency.

5) *EDNS* (expected demand not supplied) refers to the expected reduction of load demand power due to generation capacity shortage or power grid constraints in a given time interval, that is,

$$ENDS = \sum_{i \in S} C_i P_i \tag{1.5}$$

where P_i is the probability of the system being at state i, C_i is the reduced load power at state i, and S is the complete set of system states in which the system cannot meet load demand in the given time interval.

6) *EENS* (expected energy not supplied) refers to the expected reduction of load demand energy due to generation capacity shortage or power grid constraints in a given time interval, that is,

$$EENS = \sum_{i \in S} C_i F_i D_i = \sum_{i \in S} C_i P_i T \tag{1.6}$$

where P_i is the probability of the system being at state i, F_i is the probability of the system being at state i, D_i is the time duration (in days) at state i, C_i is the reduced load power at state i, S is the complete set of system states in which the system cannot meet load demand in the given time interval, and T is the number of hours in the given time interval.

1.2.1.2 Security

Security (also known as dynamic reliability) refers to the capability of a power system to endure emergent disturbances such as a short-circuit fault or unexpected withdrawal of system components. For safe operation of the power system, the following constraints should be satisfied:

1) Load constraint. For a system containing n nodes, the following power balance equations should be satisfied, that is, the load constraints:

$$\begin{aligned} & P_i - U_i \sum_{j \in i} U_j \left(G_{ij} \cos\theta_{ij} + B_{ij} \sin\theta_{ij} \right) = 0 \\ & Q_i - U_i \sum_{j \in i} U_j \left(G_{ij} \sin\theta_{ij} - B_{ij} \cos\theta_{ij} \right) = 0 \\ & \theta_{ij} = \theta_i - \theta_j \left(i = 1,2,\ldots,n \right) \end{aligned} \tag{1.7}$$

where θ_i and θ_j are the phase angles of voltages at node i and node j, respectively, U_i and U_j are the amplitudes of voltages at node i and node j, P_i and Q_i

are the active and reactive power injections to node i, and G_{ij} and B_{ij} are corresponding elements in the node admittance matrix.

2) Operation constraint. The node voltage amplitude U, phase angle difference θ, branch power flow S, and generator power P and Q should be within certain ranges:

$$\begin{aligned} U_i^l \le U_i \le U_i^u \\ |\theta_{ij}| \le \theta_{i,j}^m \\ |S_{ij}| \le S_{i,j}^m \\ P_i^l \le P_i \le P_i^u \\ Q_i^l \le Q_i \le Q_i^u \end{aligned} \tag{1.8}$$

where the symbols in the upper right corner, l, u, and m, represent the lower limit, upper limit, and maximum value respectively.

The operation constraints are inequalities, which could be integrated into

$$\mathbf{U}^l \le \mathbf{U} \le \mathbf{U}^u \tag{1.9}$$

where $\mathbf{U}$ represents the column vector of state variables.

1.2.2 Stability of Power System

The modern power system is a large and complex dynamic system, the basic requirement of which is security and stability. The high-dimensional characteristics of models, the uncertainty of system operation mode, the strong nonlinearity of components and the randomness of disturbance all make the mechanism of the power system stability very complicated. With the interconnection of large-scale power grids, the wide application of flexible AC transmission technology such as HVDC and FACTS, and the gradual increase of renewable energy integration, the analysis of power system dynamic mechanism, and power system stability analysis and control have become more and more difficult.

Power system stability can be summarized as the capability of a system to maintain at the equilibrium state under given initial conditions or to restore to an allowed equilibrium state after disturbance occurs. Through classification and definition, a general understanding of power system stability can be gained, including the characteristics of different types of stability, the causes, and the relationship between them. In the 1960s and before, it was customary to divide power system stability into static stability and dynamic stability. In 1981, the Institute of Electrical and Electronic Engineers (IEEE) proposed a new classification and definition of power system stability at the winter session of the IEEE power engineering seminar (PES) [2–5]:

1) Static stability/small disturbance stability. For a certain steady-state operation state, if the system is static-stable, then when a small disturbance occurs the system will be restored to an operation state the same or close to the predisturbance operation state.

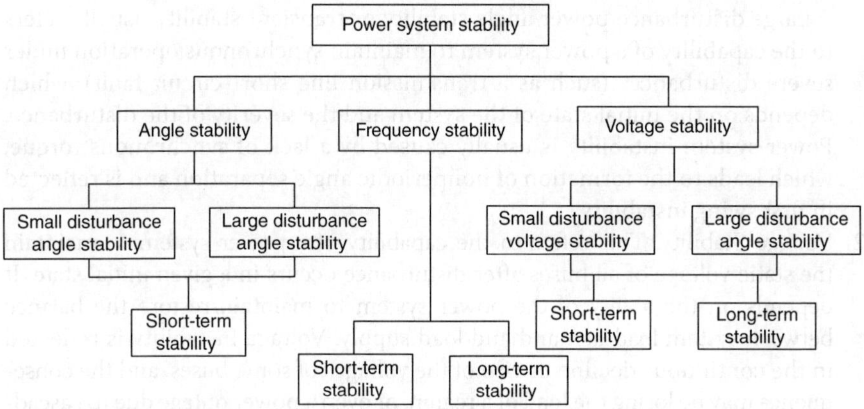

Figure 1.1 Classification of power system stability by IEEE/CIGRE.

2) Transient stability/large disturbance stability. For a certain steady-state operation state and a certain disturbance, if the system is transient stable, then when such disturbance occurs, the system could be restored to an acceptable steady-state operation state.

In August 2004, IEEE published the latest definition and classification of power system stability formulated by the CIGRE 38th Committee and the IEEE Dynamic Behavior Committee joint group. In the new classification method, power system stability is divided into three different forms of stability – power angle stability, voltage stability, and frequency stability, according to the reflection of power system instability characteristics in the three operation variables. Each form of stability is then divided into small disturbance stability and large disturbance stability. The connection between short-term stability/long-term stability and the above forms of stability is established. The classification of power system stability is shown in Figure 1.1 [6,7].

According to the size of disturbance, the physical properties that lead to instability mode, flow, and time span, etc. IEEE/CIGRE divides power system stability into power angle stability, frequency stability and voltage stability:

1) Power angle stability. This refers to the capability of synchronous generators in an interconnected system to maintain synchronous operation after disturbance occurs. It depends on the ability of the electromagnetic torque and mechanical torque of each synchronous generator to maintain/restore balance.

   Small disturbance power angle stability is the capability of a power system to maintain synchronous operation under small disturbances, which is determined by the initial operating state of the system. Small disturbance problems usually have to do with lack of oscillation damping. Local problems usually involve rotor angle oscillation between certain generator and other parts of the system. Global problems are caused by the interaction between large generator clusters, which involve oscillation between the generator cluster in one region and the generator cluster in another region; thus they have complex characteristics obviously different from those of local oscillation.

Large disturbance power angle stability or transient stability usually refers to the capability of a power system to maintain synchronous operation under severe disturbances (such as a transmission line short-circuit fault), which depends on the initial state of the system and the severity of the disturbance. Power system instability is usually caused by a lack of synchronous torque, which leads to the formation of nonperiodic angle separation and is reflected in first swing instability.

2) Voltage Stability. This refers to the capability of a power system to maintain the static voltage of all buses after disturbance occurs in a given initial state. It depends on the ability of the power system to maintain/restore the balance between system load demand and load supply. Voltage instability is reflected in the continuous decline or rise of the voltages of some buses, and the consequence may be losing the load of a region, or even a power outage due to cascading failure caused by protection tripping.

   Voltage stability can also be divided into large disturbance voltage stability and small disturbance voltage stability, which describe the capability of a power system to maintain the static voltage under large disturbances (such as system fault or loss of generator or line) and small disturbances (such as slight variations of load), respectively. The forms of voltage stability are affected by the interaction between system and load characteristics, discrete and continuous control, and protection.

3) Frequency stability. This refers to the capability of a system to maintain the frequency within the allowed limit by adjusting the output of the system hot spares or automatically removing part of the load when serious disturbances (such as generator tripping, system splitting, loss of a large load, etc.) occur and result in active power imbalance. Frequency instability is reflected in the tripping of protection devices of the generator or load due to continuous declining or swing of frequency.

   At the same time, based on long-time design, operation, and research experience, power system personnel in China proposed a definition and classification of power system stability suitable for the actual power system in China, as shown in Figure 1.2.

The overall framework and main concept of two definition and classification methods are basically the same. A comparison between the two methods is shown in Table 1.1. The main difference is as follows. For power angle stability, in addition to the three short-term stability processes IEEE/CIGRE proposes, that is, nonperiodic instability, periodic instability, and transient instability, DL755-2001 gives the definition of large disturbance dynamic stability, which refers to the capability of a power system to maintain relatively long-term power angle stability under the effects of slow automatic regulation and control devices when a large disturbance occurs. For voltage stability, IEEE/CIGRE considers that the small disturbance voltage stability includes a short-term process and a long-term process, but DL755-2001 considers that a small disturbance voltage stability mainly refers to static voltage stability and does not include a long-term process such as a chain reaction.

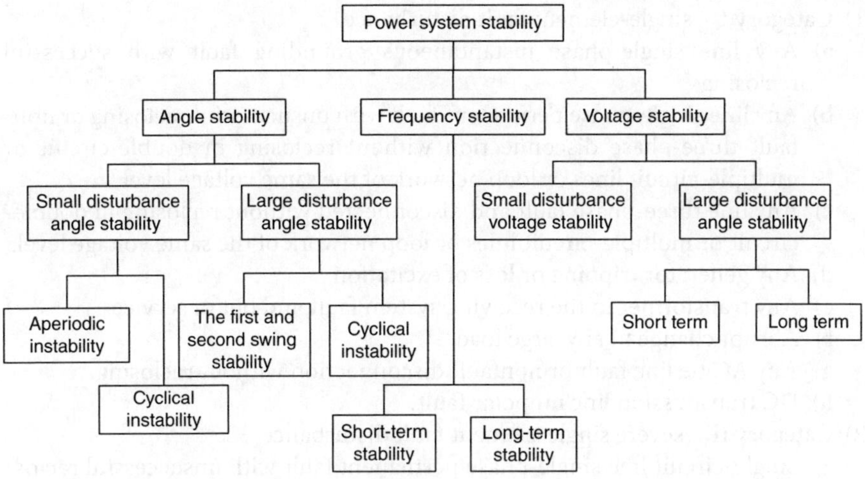

Figure 1.2 Classification of power system stability by industry standard DL755-2001.

Table 1.1 Comparison between two stability classification standards.

Classification standard	**IEEE/CIGRE**		**Industry standard DL755-2001**	
Power angle stability	Small disturbance power angle stability	Short-term process	Steady-state stability Small disturbance dynamic stability	
	Large disturbance power angle stability	Short-term process	Dynamic stability Large disturbance dynamic stability	1st/2nd swing process Short-term/long-term process
Voltage stability	Small disturbance voltage stability	Short-term/long-term process	Steady-state voltage stability	
	Large disturbance voltage stability	Short-term/long-term process	Large disturbance voltage stability	Short-term/long-term process
Frequency stability	Short-term/long-term process		Short-term/long-term process	

1.3 Three Defense Lines in the Power System

1.3.1 Classification of Disturbance in the Power System

"Power system security and stability guidelines" DL 755-2001 classifies large disturbances into three categories according to the severity degree [8], and puts forward corresponding security and stability requirements:

1) Category I – single-element fault disturbance
   a) Any line single-phase instantaneous grounding fault with successful reclosing.
   b) Any line single-phase permanent fault with unsuccessful reclosing or nonfault three-phase disconnection without reclosing in double-circuit or multiple-circuit lines or loop network of the same voltage level.
   c) Any line three-phase fault and disconnected without reclosing in double-circuit or multiple-circuit lines or loop network of the same voltage level.
   d) Any generator tripping or loss of excitation.
   e) Any transformer in the receiving system fault and out of service.
   f) Abrupt change of any large load.
   g) Any AC tie line fault or nonfault disconnection without reclosing.
   h) DC transmission line unipolar fault.
2) Category II – severe single-element fault disturbance
   a) Single-circuit line single-phase permanent fault with unsuccessful reclosing or nonfault three-phase disconnection without reclosing.
   b) Single-phase grounding fault with unsuccessful reclosing occurring simultaneously on two different-name phases of double-circuit lines on the same tower and double-circuit lines three-phase tripped off.
   c) DC transmission line bipolar fault.
   d) Any bus fault.
3) Category III – multiple severe fault disturbances
   a) Switch refusing to operate in the case of a fault.
   b) Relay protection/automatic device maloperation or refusing to operate in the case of a fault.
   c) Failure of an automatic regulation device.
   d) Multiple faults.
   e) Loss of a large-capacity power plant.
   f) Other accidental factors.

1.3.2 Power System Operation State

A power system is a nonlinear dynamic system with structure, parameter, and dynamic uncertainties. The operating conditions of a power system are generally described with three groups of equations: one group of differential equations describing the dynamic behavior of the power system components and control devices, and the other two groups of algebraic equations forming the equality and inequality constraints of power system operation, respectively.

The equality constraints represent the balance between system total power generation and total load capacity; the inequality constraints represent certain system variables, such as voltage and current, that must not exceed the maximum limits of physical devices. According to whether the constraints are satisfied, system operation can be divided into five states, as shown in Figure 1.3 [9].

The power system operation state can be divided into a normal state and an abnormal state. The normal state can be further divided into the safety state and the alert state and the abnormal state can be further divided into the emergency

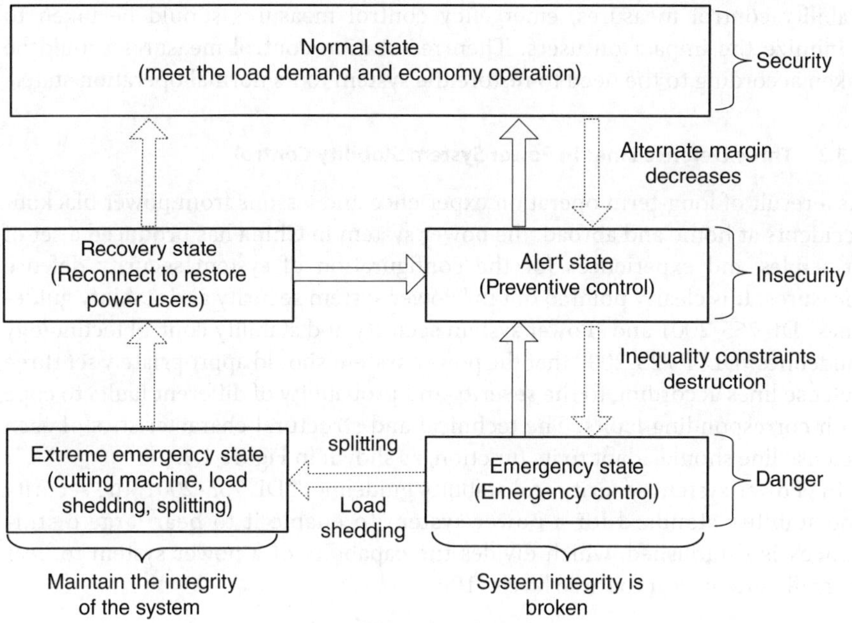

Figure 1.3 Conversion relationship between power system operation states.

state and the restoration state. The operation of a power system includes all these states and the mutual transfer between them.

The safety state refers to when system frequency, node voltages, and the load of each component are all within the allowed ranges, and a general small disturbance does not cause the system to break from the normal operation state.

The alert state refers to when the overall system is still within the safe operation range, but the operating parameters of certain components or regions are close to the threshold values; thus any new disturbance will cause the system to enter the emergency state. For a power system in the alert state, preventive control measures should be taken for the system to remain in the safe state.

The emergency state refers to when a disturbance occurs in the power system in the normal operation state (including load fluctuation and various kinds of fault), and the power balance between the power source and load is destroyed, causing system frequency and node voltage to deviate more than the allowed values, or the load of components to exceed the limit of safe operation. For a power system in the emergency state, all kinds of corrective and stability control measures should be taken to restore the system to the normal operation state.

The restoration state refers to when a power system is split into a number of local systems and part of the systems cannot ensure normal power supply to users, although the other systems could maintain the normal operation state; or when the system is not split but cannot ensure normal power supply to all users and part of the load has been cut off. If a power system in the emergency state cannot be restored to the normal operation state by taking corrective and stability control measures, emergency control measures should be taken to minimize the impact on users. Then restoration control measures should be taken according to the need to restore the system to its normal operation state.

1.3.3 Three Defense Lines in Power System Stability Control

As a result of long-term operation experience and lessons from power blackout accidents at home and abroad, the power system in China has produced a set of principles and experiences for the configuration of system security defense measures. It is clearly pointed out in "Power system security and stability guidelines" DL 755-2001 and "Power system security and stability control technology guidelines" DL/T 723-2000 that the power system should appropriately set three defense lines according to the severity and probability of different faults to cope with corresponding faults. The technical and structural characteristics of each defense line should adapt to its function, as shown in Figure 1.4.

In "Power system security and stability guidelines" DL 755-2001, the security and stability standard for a power system to enable it to bear large disturbances is established, which divides the capability of a power system to bear large disturbances into three levels [10]:

1) Security and stability standard level I. When large disturbance category I occurs in a power system in the normal operation mode, protection, switches, and reclosing all operate correctly. The power system should maintain stable operation and a normal power supply without taking stability

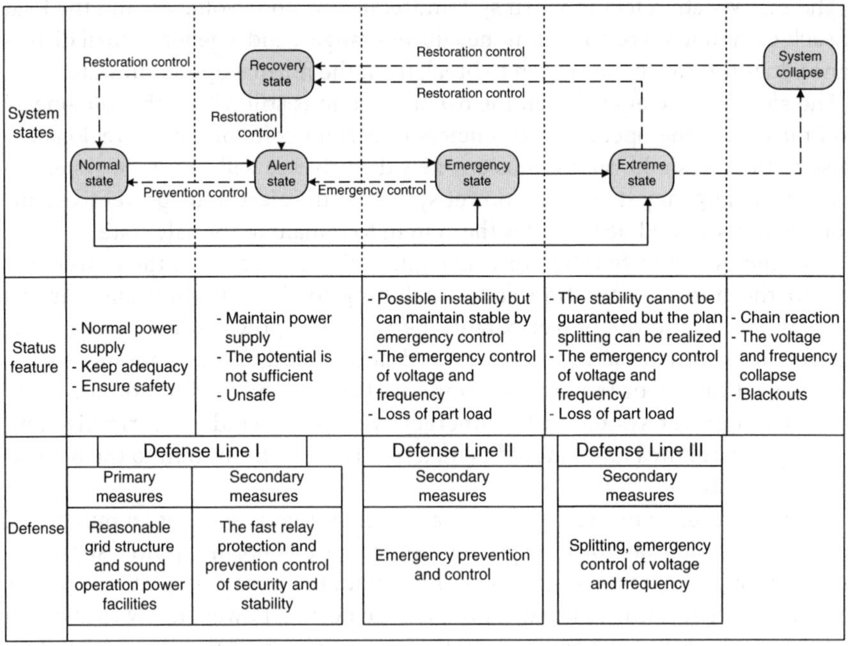

Figure 1.4 Three defense lines in power system safety and stability control.

control measures. Other components should remain within the accident overload capacity and no cascading tripping should occur.

2) Security and stability standard level II. When large disturbance category II occurs in a power system in the normal operation mode, protection, switches, and reclosing all operate correctly. The power system should maintain stable operation, and if necessary stability control measures such as generator tripping and load shedding could be taken.
3) Security and stability standard level III. When large disturbance category III occurs in a power system in the normal operation mode and system stability is destroyed, measures must be taken to prevent the system from collapsing, to avoid long-time blackout and disastrous power outage for most important users (including the station service power supply), and to minimize the loss of load. The power system should be restored to the normal operation state as soon as possible.

Correspondingly, in order to ensure the safe and stable operation of a power system, a complete secondary defense system should also include three defense lines:

1) Defense line I. Defense line I ensures that the power system could maintain normal operation and meet the security requirements under large disturbance category I. The measures taken include a primary system facility and a secondary protection device. The primary system should establish a reasonable grid structure, be equipped with a complete power facility, and arrange reasonable operation modes. The secondary system should be equipped with relay protection devices with good performance that could isolate faults immediately and restrict the development of faults.

   Power system security and stability control defense line I is to clear fault components immediately and prevent faults from expanding. It mainly consists of relay protection devices with good performance, and it is required to clear fault components fast, accurately, and reliably, so that fault expansion could be effectively prevented and the impact of the fault could be limited to the minimum range.
2) Defense line II. Defense line II ensures that the power system could meet the security requirements under large disturbance category II. Emergency control measures such as generator tripping, centralized load shedding, interconnected system splitting, HVDC power emergency modulation, and series compensation, etc., are taken to prevent stability destruction and the parameter seriously exceeding the limit.

   Power system security and stability control defense line II is to take the necessary security and stability control measures such as generator tripping, load shedding, system splitting, and DC modulation, etc., to prevent the system from losing stability. It mainly consists of power system security and stability control devices, and is required to operate accurately and reliably to ensure that the power system could maintain stable operation.
3) Defense line III. Defense line III ensures that the power system could meet the security requirements under large disturbance category III. Emergency control measures such as system splitting, resynchronization, frequency, and

voltage emergency control, etc., are taken to prevent the fault from expanding and the system from collapsing.

Power system security and stability control defense line III is to take measures such as system splitting, generator tripping, load shedding, and emergency control measures by dispatchers to avoid blackout after system loses stability. It mainly consists of automatic devices such as loss of synchronization splitting, high-frequency generator tripping, low-frequency load shedding, low-voltage load shedding, and emergency control measures taken by dispatchers, and is required to prevent blackouts effectively.

1.3.4 Functions of Defense System

In order to ensure the safe and stable operation of a power system, a reasonable grid structure should be established first, equipped with a relay protection system with good performance. If the power grid cannot maintain security by its own structure and protection devices alone, then security and stability control devices and corresponding measures should be deployed according to specific circumstances of the grid in order to form a complete power grid security defense system, so that all kinds of disturbance and fault could be dealt with and the safe and stable operation of the power grid could be guaranteed. According to the operation state of the power grid, stability control can be divided into five categories: prevention control, emergency control, out-of-synchronization control, postsplitting control, and restoration control, as shown in Figure 1.5:

1) Preventive control. According to the power system safe and stable operation guidelines, a power system in the normal operation state should be able to withstand faults required by standard level I and remain stable without loss of load, which means that the system in the normal operation state should have a certain security and stability margin. Control measures to improve the adequacy and security of a power system and help restore the system to the normal operation state are called prevention control. Prevention control in cooperation with relay protection constitutes defense line I to ensure the security and stability of the power system. Prevention control includes: (1) power angle and power flow control to maintain the stability margin and avoid overload; (2) frequency control to maintain system frequency in the allowed range and appropriately use the power reserve; (3) voltage control to maintain bus voltage in the allowed range and appropriately allocate the reactive power; (4) control measures to improve the damping characteristics to maintain system damping at an appropriate level.

   Prevention control measures include generator power prevention control, generator excitation supplementary control, parallel and series capacitor compensation control, HVDC power modulation, and other flexible AC transmission (FACTS) control, etc. It should be noted that control measures to improve the damping characteristics such as PSS and FACTS, which work both in the normal operation state and after fault occurs, also belong to prevention control.

   Prevention control usually applies the continuous control mode, so that the operating indexes of system and relevant parameters can gradually draw close

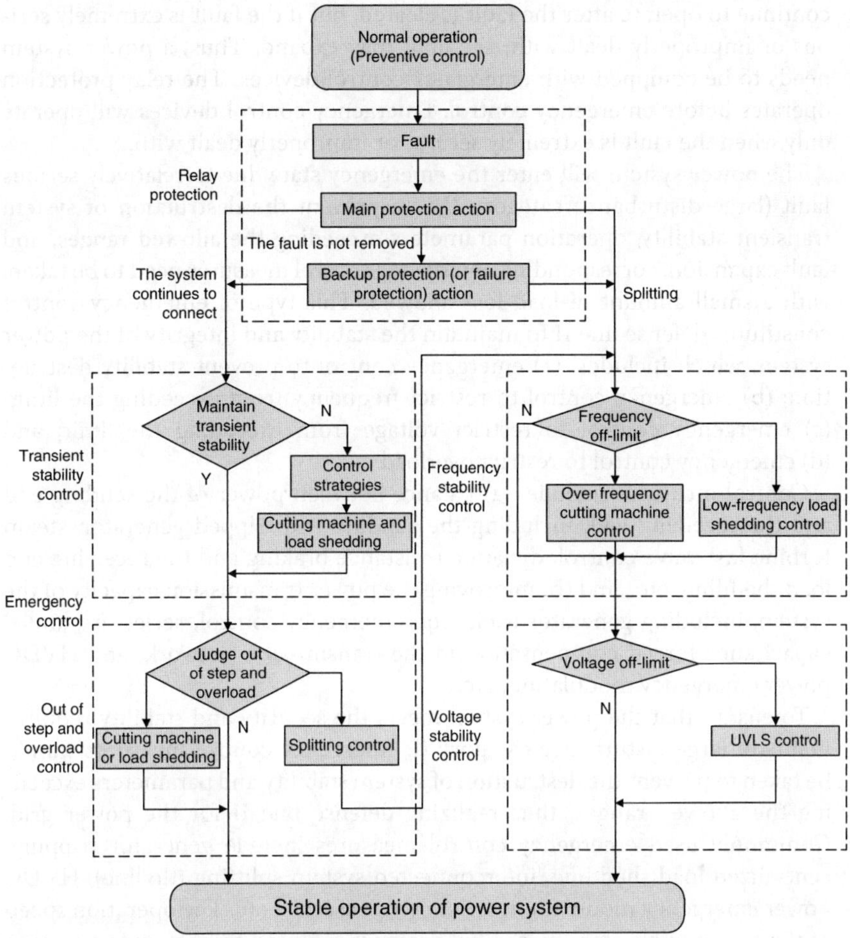

Figure 1.5 Power system safety and stability control.

to the predetermined values. To get the system out of a potential insecure state, the control time should not exceed 10 min, and to eliminate the impact of large disturbance, the control time should not be more than 30 min.

Prevention control is not only for power system operation safety but also for the economic operation of a system. Many factors in a wide range need to be considered. At present the general practice is to centralize relevant information widely distributed in different areas to the system control center (dispatch center) in order to make the control decision. Then the control center carries out control directly (for example, the dispatch of AGC directly controls the generators) or transmit control strategies to different control points for local control (for example, the dispatch of AGC issues control commands to power plants).

2) Emergency control. When a fault such as a short circuit occurs in a power system, relay protection is the first to act and clear the fault, as an important measure to ensure the effectiveness of defense line I. Usually, the system could

continue to operate after the fault is cleared, but if the fault is extremely serious or improperly dealt with, the fault may expand. Thus, a power system needs to be equipped with emergency control devices. The relay protection operates before emergency control. Emergency control devices will operate only when the fault is extremely serious or improperly dealt with.

The power system will enter the emergency state due to relatively serious fault (large disturbance category II). To prevent the destruction of system transient stability, operation parameters exceeding the allowed ranges, and fault expansion, corresponding emergency control measures need to be taken, with a small amount of load loss allowed. This type of emergency control constitutes defense line II to maintain the stability and integrity of the power system, which includes: (a) emergency control to prevent stability destruction; (b) emergency control to restrict frequency from exceeding the limit; (c) emergency control to restrict voltage from exceeding the limit; and (d) emergency control to restrict overload.

Control measures include: (a) balance between power of the sending end and the receiving end, including the sending end tripped generator, steam turbine fast valve control, dynamic resistance braking and the receiving end load shedding, etc., and (b) improving the power transmission capacity of the system, including generator excitation emergency control, series or parallel capacitance forced compensation in the transmission network, and HVDC power emergency modulation, etc.

To ensure that the power system meets the security and stability requirements of large disturbance category II, emergency control measures should be taken to prevent the destruction of system stability and parameters exceeding the allowed ranges, thus realizing defense line II for the power grid. Commonly used emergency control measures include generator tripping, centralized load shedding, interconnected system splitting (tie line), HVDC power emergency modulation, series compensation, etc. The operation speed of the power angle stability control device is required to be very fast (within 50 ms) and the operation speed of the overload control device is required to be relatively slow (from a few seconds to tens of seconds).

To ensure that the power system meets the security and stability requirements of large disturbance category III, emergency control to prevent fault expansion and system collapse should be available, such as system out-of-synchronization splitting (or resynchronization when possible), frequency, and voltage emergency control, etc. Meanwhile, measures should be taken to avoid maloperation of the line and generator protection and cascading tripping of the line and generator in the case of oscillation, in order to realize defense line III for the power grid. The out-of-synchronization splitting device operates according to the set number of oscillation cycles, for example, a 500 kV out-of-synchronization splitting device usually operates at 1 ~ 2 oscillation cycles. The operation time of the emergency control device for voltage and frequency stability is 0.1 ~ 0.5 s (the delay time is usually set to be 0.2 s).

3) Out-of-step and overload control. The cause for power grid out-of-synchronization may be static stability damage caused by transmission line power exceeding the limit value, or power system transient stability damage

caused by a power grid short-circuit fault, cutting of large capacity power generation, transmission or transformation devices, and an instantaneous large mutation of load, or a sudden open loop of the loop system (or parallel double-circuit lines), which cause the connection impedance between two parts of the system to increase suddenly, damaging system stability and resulting in loss of synchronization. Currently, power grid out-of-synchronization control mainly relies on out-of-synchronization splitting devices, which is an important part of defense line III of the power grid. The core technology of the out-of-synchronization splitting device includes a sound out-of-synchronization criterion, the cooperation between operation of splitting devices at different installation sites, and locking measures to prevent maloperation in various cases.

4) Splitting. When power outage occurs in a power system due to serious disturbance or fault expansion, to restore the system to normal operation and power supply, regional systems should be equipped with necessary black start measures and restoration control measures (including automatic control and manual control). Automatic restoration control includes power source automatic fast start and paralleling, transmission line automatic recharge, automatic restoration of the split part of the system, and restoration of power supply to users, etc.

1.4 Summary

This chapter introduces the requirements of a power system on reliability, describes the classification of power system stability in different periods and its adaptability, and illustrates the focus of this book – power angle stability analysis. On this basis, three defense lines to ensure the safe and stable operation of the power system are introduced. First, a reasonable grid structure should be formed, equipped with a relay protection system with good performance. Then, according to specific circumstances of the power grid, security and stability control devices and corresponding control measures should be deployed to form a complete power grid security defense system that could deal with all kinds of disturbance and fault. Only by building a strong power grid structure, deploying the optimal automatic control system, arranging for a reasonable and safe operation mode, and further strengthening the three defense lines in power system security and stability control can the safe and stable operation of an extra-large-scale AC/DC hybrid power system be guaranteed.

References

1 DL/T 861–2004 (2004) *Basic Vocabulary of Electric Power Reliability*, China Electric Power Press, Beijing.
2 Byerly, R.T. and Kimbark, E.W. (1974) *Stability of Large Electric Power System*, IEEE Press, New York.
3 Anderson, P.M. (1977) Reliability criteria for system dynamic performance. *IEEE Transactions on Power Apparatus and Systems*, **96**(6), 1815–1818.

4 IEEE Task Force on Terms and Definitions (1982) Proposal terms and definitions for power system stability. *IEEE Transactions on Power Apparatus and Systems*, **101**(7), 1894–1898.
5 Ames, I. (1985) Proceedings of the International Symposium on Power System Stability.
6 Kundur, P., Paserba, J., Ajjarapu, V., *et al.* (2004) Definition and classification of power system stability. *IEEE Transactions on Power Systems*, **19**(3), 1387–1401.
7 Kundur, P., Paserba, J., Vittal, V., and Andersson, G. (2006) Closure of definition and classification of power system stability. *IEEE Transactions on Power Systems*, **21**(1), 446–458.
8 DL 755–2001 (2001) *Guide on Security and Stability for Power System*, China Electric Power Press, Beijing.
9 DL/T 723–2000 (2000) *Technical Guide for Electric Power System Security and Stability Control*, China Electric Power Press, Beijing.
10 State Grid Electric Power Dispatching and Communication Center (2012) *Relay Protection Training Materials of State Grid Corporation of China*, China Electric Power Press, Beijing.

2

Power System Analysis and Control Theory

2.1 Introduction

A power system is a complex dynamic system that is strongly stochastic and time varying with a time delay, but the difficulty of its stability control is rarely seen all over the world. Meanwhile, with the introduction of fast, high-gain excitation technology, the fast increase of electrical load, and the economic and environmental constraints, the power system stability problem has become increasingly prominent. In this case, it is a challenging issue to ensure the security and stability of complex large power grids.

Power system dynamic analysis and automatic control theory is the basis of power system security and stability control. With the continuous expansion of power system scale, system structure becoming increasingly complex, and the components constantly being updated, the requirement on system dynamic analysis and stability control has also become higher. At the same time, with the continuous progress of computational tools and mathematical methods, the theory and method of power system stability analysis and control are also developing. To enable readers to understand better the new wide area stability analysis and control theory proposed later in this book, this chapter will give a brief introduction to the basics of power system dynamic analysis and automatic control theory.

This chapter mainly includes the following three aspects: (1) a mathematical model of the power system, including that of a synchronous generator, excitation system, the prime mover and speed governor, and load; (2) power system stability analysis methods, including the time-domain simulation method, eigenvalue analysis method, and transient energy function method; (3) automatic control theory, including classical control theory, modern control theory, large system theory, and intelligent control theory.

2.2 Mathematical Model of Power System

2.2.1 Mathematical Model of Synchronous Generator

2.2.1.1 Park Model

A synchronous generator is a component that could realize conversion between electrical energy and mechanical energy, and its dynamic mathematical model is

Power System Wide-Area Stability Analysis and Control, First Edition. Jing Ma.

the basis of study on power system dynamic behavior. To facilitate analysis and calculation, the following assumptions need to be made concerning the actual three-phase synchronous generator [1,2]:

1) Neglecting the effects of magnetic hysteresis and magnetic saturation.
2) Stator three-phase windings are exactly the same in structure and 120° apart in spatial location, and the spatial magnetomotive forces generated are all sine distribution.
3) The stator/rotor slot does not affect the stator/rotor inductance.

The generator that satisfies the above assumptions is called an ideal generator, the schematic diagram of which is shown in Figure 2.1. By writing the electromagnetic equations of the coupled circuit in Figure 2.1 and transformation from the *abc-dq0* coordinate system, the Park model of synchronous generator can be obtained, shown in the following equations (2.1) to (2.5):

Voltage equation:

$$\begin{bmatrix} u_d \\ u_q \\ u_0 \\ u_f \\ 0 \\ u_g \\ 0 \end{bmatrix} = \begin{bmatrix} \dot{\varphi}_d \\ \dot{\varphi}_q \\ \dot{\varphi}_0 \\ \dot{\varphi}_f \\ \dot{\varphi}_D \\ \dot{\varphi}_g \\ \dot{\varphi}_Q \end{bmatrix} + \begin{bmatrix} -\omega\varphi_q \\ -\omega\varphi_d \\ 0 \\ 0 \\ 0 \\ 0 \\ 0 \end{bmatrix} + \begin{bmatrix} -r_a i_d \\ -r_a i_q \\ -r_a i_0 \\ r_f i_f \\ r_D i_D \\ r_g i_g \\ r_Q i_Q \end{bmatrix} \tag{2.1}$$

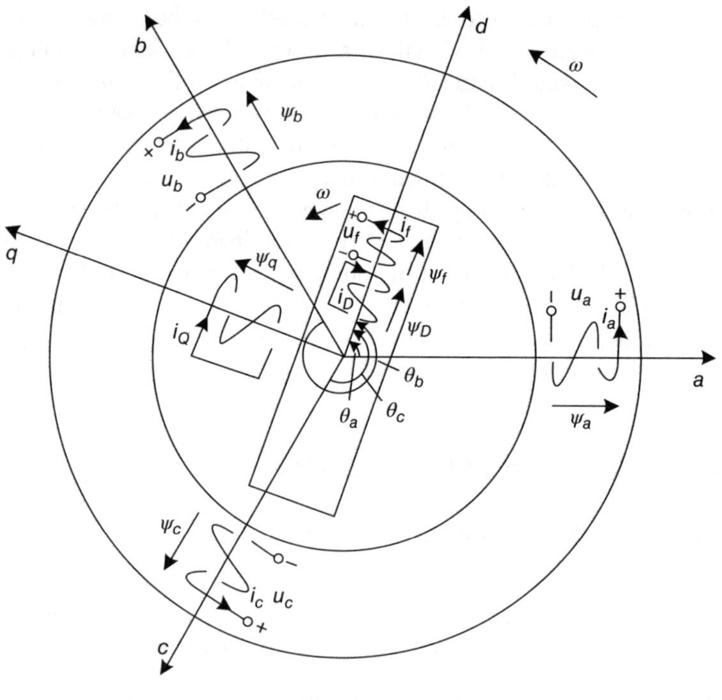

Figure 2.1 Schematic diagram of the ideal generator.

where u, i, and φ represent the voltage, current, and flux of each winding, r_a represents the resistance of each stator winding, r_f, r_D, r_g, and r_Q represent the resistance of winding f, D, g, and Q, respectively, and ω is the synchronous rotating angular speed. Subscripts d and q represent d-axis winding and q-axis winding, subscript f represents the excitation winding on the d-axis, subscript g represents the transient damping winding on the q-axis, and subscripts D and Q represent the subtransient damping windings on the d-axis and q-axis, respectively.

Flux equation:

$$\begin{bmatrix} \varphi_d \\ \varphi_q \\ \varphi_0 \\ \varphi_f \\ \varphi_D \\ \varphi_g \\ \varphi_Q \end{bmatrix} = \begin{bmatrix} X_d & 0 & 0 & X_{ad} & X_{ad} & 0 & 0 \\ 0 & X_q & 0 & 0 & 0 & X_{aq} & X_{aq} \\ 0 & 0 & X_0 & 0 & 0 & 0 & 0 \\ X_{ad} & 0 & 0 & X_f & X_{ad} & 0 & 0 \\ X_{ad} & 0 & 0 & X_{ad} & X_D & 0 & 0 \\ 0 & X_{aq} & 0 & 0 & 0 & X_g & X_{aq} \\ 0 & X_{aq} & 0 & 0 & 0 & X_{aq} & X_Q \end{bmatrix} \begin{bmatrix} -i_d \\ -i_q \\ -i_0 \\ i_f \\ i_D \\ i_g \\ i_Q \end{bmatrix} \tag{2.2}$$

where X_d, X_q, X_f, X_D, X_g, and X_Q represent the self-inductive reactance of d-axis winding, q-axis winding, f excitation winding, and D, g, and O damping windings; X_d and X_q are also called synchronous reactance, X_{ad} is the synchronous reactance of the d-axis armature reaction, which represents the mutual inductive reactance between winding d, f, and D, and X_{aq} is the synchronous reactance of the q-axis armature reaction, which represents the mutual inductive reactance between windings q, g, and Q.

Electromagnetic power equation:

$$\begin{aligned} P_e &= \frac{3}{2}\left(u_d i_d + u_q i_q\right) + 3u_0 i_0 \\ &= \frac{3}{2}\left(i_d \dot{\varphi}_d + i_q \dot{\varphi}_q + 2i_0 \dot{\varphi}_0\right) + \frac{3}{2}\omega\left(\varphi_d i_q - \varphi_q i_d\right) - \frac{3}{2}\left(i_d^2 + i_q^2 + 2i_0^2\right) r_a \end{aligned} \tag{2.3}$$

Electromagnetic torque equation:

$$T_e = \varphi_d i_q - \varphi_q i_d \tag{2.4}$$

Rotor motion equation:

$$\begin{cases} T_J \dot{\omega} = T_m - T_e \\ \dot{\delta} = \omega - 1 \end{cases} \tag{2.5}$$

where T_J is the inertia time constant, T_m and T_e are the mechanical torque and electromagnetic torque, respectively, δ is the angle (in radians) by which the q-axis is leading the x-axis of the synchronous coordinate system. The other variables are all in per-unit values.

2.2.1.2 Practical Models

For synchronous generator equations in the synchronous rotating coordinate system, if the electromagnetic transient processes of winding d, q, f, D, g, and Q and the rotor mechanical dynamic process are taken into account, then the eighth-order model of generator can be obtained. For a large-scale multimachine system, if dynamic equations of the excitation system, speed governor and prime mover are also considered, "dimension disaster" will result, which will bring great difficulties to the analysis and calculation. Therefore, in engineering practice, the model of synchronous generator is usually simplified to different extents [3]:

1) In electromechanical transient analysis, it is generally considered that transient processes related to the grid attenuate fast and thus the transient process of the stator flux can be neglected, that is, in the stator voltage equation, $\dot{\varphi}_d = 0$ and $\dot{\varphi}_q = 0$.
2) Neglecting the effects of speed variation on stator voltage, that is, in the stator voltage equation, assume that $\omega \approx 1$ (p.u.).

Meanwhile, to facilitate analysis, rotor-side variables (current, flux, etc.) are usually converted to practical stator-side variables:

1) Fifth-order practical model. For hydraulic turbines, the fifth-order practical model is usually used, which takes into account the transient processes of excitation winding and damping winding and the rotor mechanical dynamic process. This model is often used in the comprehensive stability program, which is suitable for the description of a salient pole synchronous generator.

   Stator voltage equation:

$$\begin{cases} u_d = E''_d + X''_q i_q - r_a i_d \\ u_q = E''_q - X''_d i_d - r_a i_q \end{cases} \tag{2.6}$$

   Excitation winding voltage equation:

$$T'_{d0}\dot{E}'_q = E_f - \frac{X_d - X_l}{X'_d - X_l}E'_q + \frac{X_d - X'_d}{X'_d - X_l}E''_q - \frac{(X_d - X'_d)(X''_d - X_l)}{X'_d - X_l}i_d \tag{2.7}$$

   Damping winding voltage equation:

$$\begin{cases} T''_{d0}\dot{E}''_q = \dfrac{X''_d - X_l}{X'_d - X_l}T''_{d0}pE'_q - E''_q + E'_q - (X'_d - X''_d)i_d \\ T''_{q0}\dot{E}''_d = \dfrac{X''_q - X_l}{X'_q - X_l}T''_{q0}pE'_d - E''_d + E'_d + (X'_q - X''_q)i_q \end{cases} \tag{2.8}$$

   Rotor motion equation:

$$\begin{cases} \dot{\delta} = \omega - 1 \\ T_J\dot{\omega} = T_m - \left[E''_q i_q + E''_d i_d - (X''_d - X''_q)i_d i_q\right] \end{cases} \tag{2.9}$$

   where δ and ω are the angle and rotating speed of synchronous generator, T_m is the mechanical torque, T_J is the inertia time constant, E_f is the excitation emf, x_l is the leakage reactance of the synchronous generator, E' and E'' are the transient emf and subtransient emf of the synchronous generator, T'_0 and T''_0 are the open-loop transient and subtransient time constants of the synchronous generator, x, x', and x'' are the steady-state, transient, and subtransient reactance of the synchronous generator, i is the output current of the synchronous generator, and subscripts d and q represent the direct axis and quadrature axis of the synchronous rotating coordinate system.

2) Sixth-order practical model. For steam turbines with solid rotor, the transient process of rotor q-axis is described by two damping windings with different time constants, that is, winding g, which reflects the transient process, and winding Q, which reflects the subtransient process. Compared with the fifth-order model, the q-axis is added with winding g, and thus this model is more accurate. This model is often used in the BPA program, which is suitable for the description of a nonsalient pole synchronous generator.

   Stator voltage equation:

$$\begin{cases} u_d = E_d'' + X_q'' i_q - r_a i_d \\ u_q = E_q'' - X_d'' i_d - r_a i_q \end{cases} \tag{2.10}$$

   Excitation winding voltage equation:

$$T_{d0}' \dot{E}_q' = E_f - \frac{X_d - X_l}{X_d' - X_l} E_q' + \frac{X_d - X_d'}{X_d' - X_l} E_q'' - \frac{(X_d - X_d')(X_d'' - X_l)}{X_d' - X_l} i_d \tag{2.11}$$

   Damping winding voltage equation:

$$\begin{cases} T_{q0}' \dot{E}_d' = -\dfrac{X_q - X_l}{X_q' - X_l} E_d' + \dfrac{X_q - X_q'}{X_q' - X_l} E_d'' + \dfrac{(X_q - X_q')(X_q'' - X_l)}{X_q' - X_l} i_q \\ T_{d0}'' \dot{E}_q'' = \dfrac{X_d'' - X_l}{X_d' - X_l} T_{d0}'' p E_q' - E_q'' + E_q' - (X_d' - X_d'') i_d \\ T_{q0}'' \dot{E}_d'' = \dfrac{X_q'' - X_l}{X_q' - X_l} T_{q0}'' p E_d' - E_d'' + E_d' + (X_q' - X_q'') i_q \end{cases} \tag{2.12}$$

   Rotor motion equation:

$$\begin{cases} \dot{\delta} = \omega - 1 \\ T_J \dot{\omega} = T_m - \left[E_q'' i_q + E_d'' i_d - (X_d'' - X_q'') i_d i_q \right] \end{cases} \tag{2.13}$$

   The meaning of each variable is similar to that in the fifth-order practical model.

3) Classical secondorder model. For a generator far from where the disturbance occurs, the transient process of transient emf can be neglected, as well as the transient salient pole effect, that is it could be assumed that $X_d' = X_q'$. Thus the classical second-order model, which could directly use the xy coordinate system when integrated to the grid, can be gained.

   Stator voltage equation:

$$\begin{cases} u_d = E_d' + X_q' i_q - r_a i_d \\ u_q = E_q' - X_d' i_d - r_a i_q \end{cases} \tag{2.14}$$

Rotor motion equation:

$$\begin{cases} \dot{\delta}' = \omega - 1 \\ T_J \dot{\omega} = T_m - \mathrm{Re}\left(\hat{E}' \overset{*}{I}\right) - D(\omega - 1) \end{cases} \tag{2.15}$$

where D is the damping coefficient of the synchronous generator. "Re" means extracting the real part of…; "*" represents the conjugate complex of …; and $\hat{E}' = E'\angle\delta'$.

2.2.2 Mathematical Model of Excitation System

The basic function of the excitation system is to provide a synchronous generator magnetic field winding with a DC current. At the same time, by regulating the voltage to control the magnetic field current, the excitation system also has a part to play in system control and protection [4].

According to different excitation power sources, the excitation system can be divided into the following three types:

1) DC excitation system, which provides a synchronous generator with excitation power through the DC exciter via the slip ring.
2) AC excitation system, which provides a synchronous generator with excitation power through the AC exciter via the controllable or uncontrollable current rectifier. The rectifier can be stationary or rotating.
3) Stationary excitation system, which provides a synchronous generator with excitation power through the transformer (connected to the generator or grid) via the controllable rectifier, including self-shunt excitation, AC-side series self-compound excitation, and excitation with a constant voltage source.

In a real power system, there are various types of excitation systems, especially voltage regulators. This book mainly introduces a typical excitation system of a large-capacity synchronous generator. As shown in Figure 2.2, the excitation system includes the exciter, voltage regulator, terminal voltage sensor, and excitation system stabilizer.

Considering the influence of exciter saturation, the simplified mathematical model of the exciter is shown in Figure 2.3. S_K is the exciter saturation coefficient. The bigger S_K is, the more saturated the exciter is. T_L is the no-load time constant of the excitation circuit. K_L is the self-shunt excitation coefficient, which

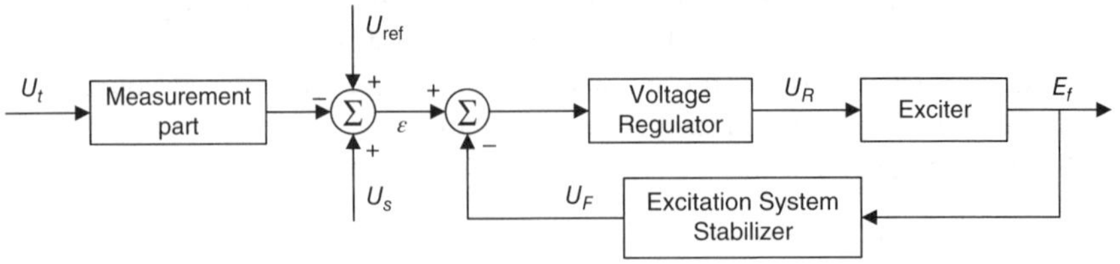

Figure 2.2 Block diagram of a typical excitation system.

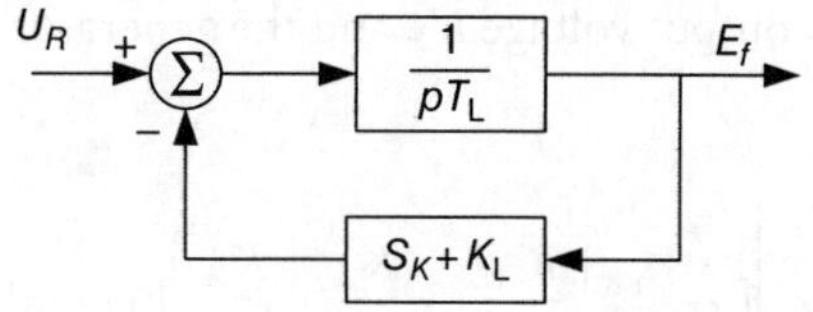

Figure 2.3 Simplified mathematical model of the exciter.

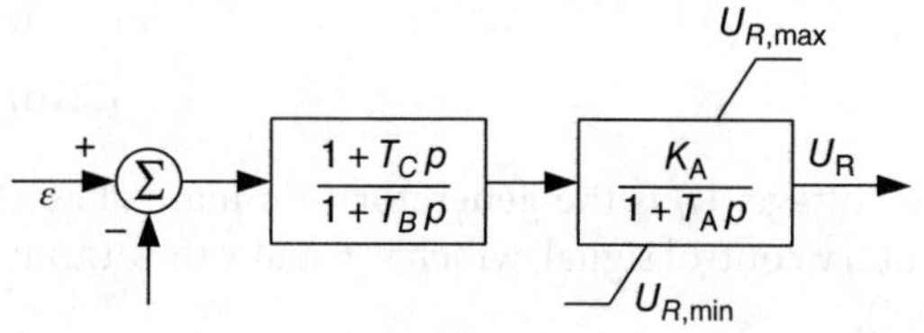

Figure 2.4 Mathematical model of the voltage regulator.

represents the ratio of separately excited ampere turns in the exciter to the total ampere turns. For a separately excited DC exciter and separately excited AC exciter, $K_L = 1$. For a stationary excitation system, there is no exciter.

The voltage regulator is usually represented by a lead-lag link, which reflects the phase characteristics of the regulator and an inertia amplification link, the transfer function shown in Figure 2.4. K_A is the magnification times of the inertia amplification link and T_A is the time constant of the link. T_B and T_C are time constants of the lead-lag link, the values of which are usually small and negligible.

The terminal voltage sensor is the measuring link of an excitation system, which can be expressed by an inertia link with time constant T_R. Since T_R is very small, it can usually be neglected.

In order to improve the dynamic characteristics of an excitation system and ensure its stable operation, a negative feedback link is introduced as the excitation system stabilizer. As shown in Figure 2.5, the excitation system stabilizer usually adopts the soft feedback link, where K_F is the amplification times and T_F is the time constant. In steady-state operation, the output voltage of the negative feedback link $U_F = 0$, which does not affect the static characteristics of the excitation system.

When the time constant of the terminal voltage sensor and the lead-lag link of the voltage regulator are neglected, the typical excitation system could be expressed by a third-order model with the power system regulator output voltage

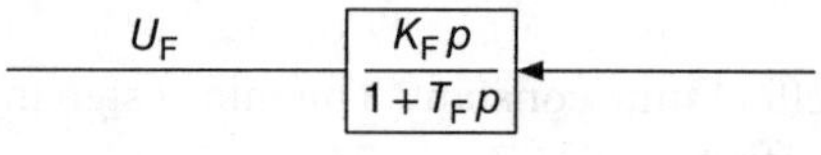

Figure 2.5 Mathematical model of the excitation system stabilizer.

U_R, the excitation system negative feedback output voltage U_F, and the generator excitation emf E_f as state variables:

$$\begin{bmatrix} \dot{U}_R \\ \dot{E}_f \\ \dot{U}_F \end{bmatrix} = \begin{bmatrix} -\dfrac{1}{T_A} & 0 & -\dfrac{K_A}{T_A} \\ \dfrac{1}{T_L} & -\dfrac{(K_L+S_E)}{T_L} & 0 \\ \dfrac{K_F}{T_F T_L} & -\dfrac{K_F(K_L+S_E)}{T_F T_L} & -\dfrac{1}{T_F} \end{bmatrix} \begin{bmatrix} U_R \\ E_f \\ U_F \end{bmatrix} + \begin{bmatrix} \dfrac{K_A}{T_A} & -\dfrac{K_A}{T_A} & \dfrac{K_A}{T_A} \\ 0 & 0 & 0 \\ 0 & 0 & 0 \end{bmatrix} \begin{bmatrix} U_{REF} \\ U_t \\ U_S \end{bmatrix} \tag{2.16}$$

where U_{REF} is the given reference voltage, U_t is the generator terminal voltage, and U_S is the excitation supplementary control signal, which is usually the output of the power system stabilizer (PSS).

2.2.3 Mathematical Model of Prime Mover and Speed Governor

The prime mover is the device that provides the generator with mechanical power and the speed governor (usually installed on the prime mover) is the device that controls the output power of the prime mover to keep the grid frequency stable. In addition to wind, solar, and tidal power generation, the prime mover used for large-scale power generation can be divided into the hydraulic turbine and the steam turbine. Simplified models commonly used in power system analysis are introduced below [5,6].

2.2.3.1 Hydraulic Turbine and Its Speed Governor

Hydraulic Turbine The dynamic characteristics of the hydraulic turbine are mainly affected by the dynamic characteristics of water flow in the water diversion pipeline. Due to the inertia of water, variation of water in the diversion pipeline lags behind variation of the water guide vane opening μ. When the water guide vane opening suddenly increases, the water flow speed at different points in the diversion pipeline cannot increase immediately; thus the inlet pressure of the hydraulic turbine decreases instead of increasing in a short period of time, and so does the output mechanical power P_m. This phenomenon is usually referred to as the "water hammer effect." The simplified mathematical model of the hydraulic turbine mainly considers the water hammer effect of the rigid water diversion pipeline. Its transfer function is shown in Figure 2.6,

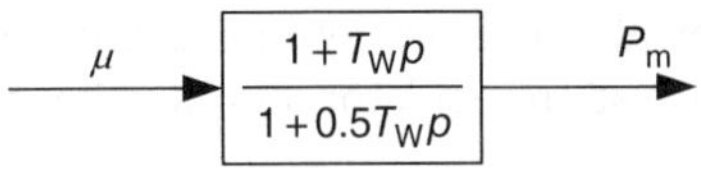

Figure 2.6 Simplified model of hydraulic turbine.

where T_w is the equivalent water hammer effect time constant. The minus sign in the numerator reflects the water hammer effect.

The above models are classical models commonly used in power system stability analysis and are applicable when the load varies modestly. When the load varies significantly, nonlinear mathematical models of the hydraulic turbine should be considered.

Speed Governor of Hydraulic Turbine The speed governors of the hydraulic turbine can be divided into two categories: mechanical speed governor and electrical hydraulic speed governor. Although the structures of the two categories are different, the principles are similar. Thus they could be described with the same mathematical model. The simplified model is shown in Figure 2.7,

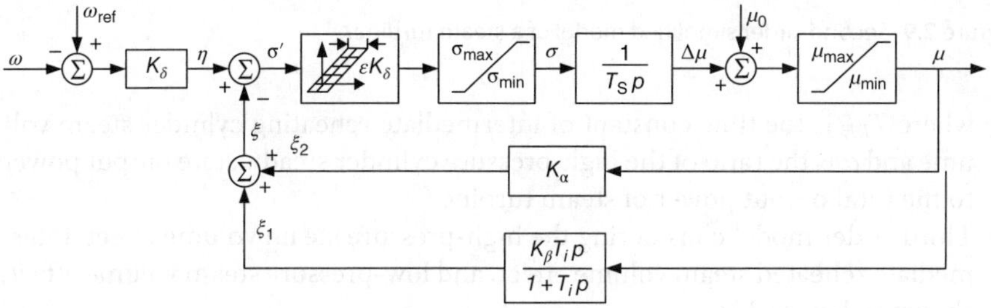

Figure 2.7 Simplified model of the hydraulic turbine speed governor.

where ω and ω_{REF} are the actual rotating speed and reference rotating speed of the hydraulic turbine, μ is the water guide vane opening, K_δ is the derivative of the adjustment coefficient, T_S is the time constant of the servomotor, K_α is the magnification times of hard feedback, and K_β and T_β are the magnification times and time constant of the soft feedback.

2.2.3.2 Steam Turbine and Its Speed Governor

Steam Turbine The dynamic characteristics of steam turbine are mainly affected by the dynamic characteristics of the steam between the valve and nozzle. Due to the inertia of steam, variation of the steam between the valve and nozzle lags behind variation of the steam turbine valve opening μ. When the valve opening suddenly increases, the pressure of the steam between the valve and nozzle cannot change immediately; thus the output mechanical power P_m will not change immediately. This phenomenon is usually referred to as the "steam volume effect." The simplified mathematical model of a steam turbine mainly considers the steam volume effect caused by the inertia of steam between the valve and nozzle, which can usually be expressed with a first-order inertia link. Three commonly used dynamic models of a steam turbine are introduced below:

1) Firstorder model only considering the high-pressure steam volume effect, shown in Figure 2.8,

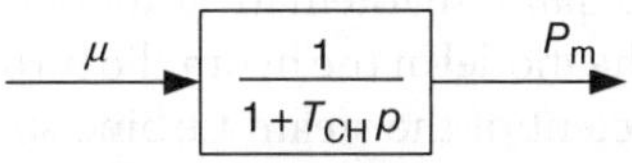

Figure 2.8 Firstorder simplified model of a steam turbine.

where T_{CH} is the time constant of the high-pressure cylinder steam volume.

2) Second-order model considering the high-pressure steam volume effect and intermediate reheated steam volume effect, shown in Figure 2.9,

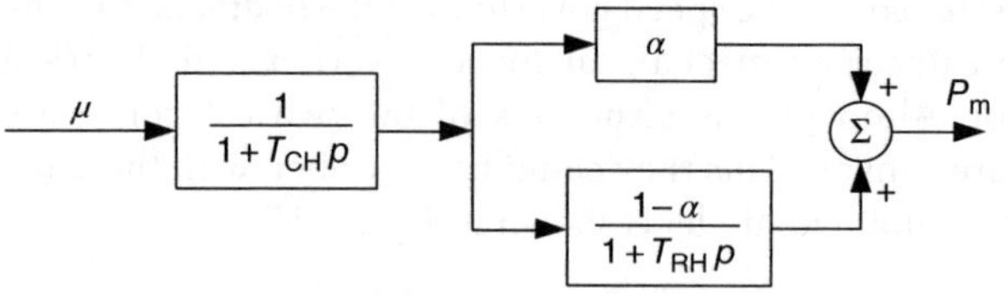

Figure 2.9 Second-order simplified model of a steam turbine.

where T_{RH} is the time constant of intermediate reheating cylinder steam volume and α is the ratio of the high-pressure cylinder steady-state output power to the total output power of steam turbine.

3) Third-order model considering the high-pressure steam volume effect, intermediate reheated steam volume effect, and low-pressure steam volume effect, shown in Figure 2.10,

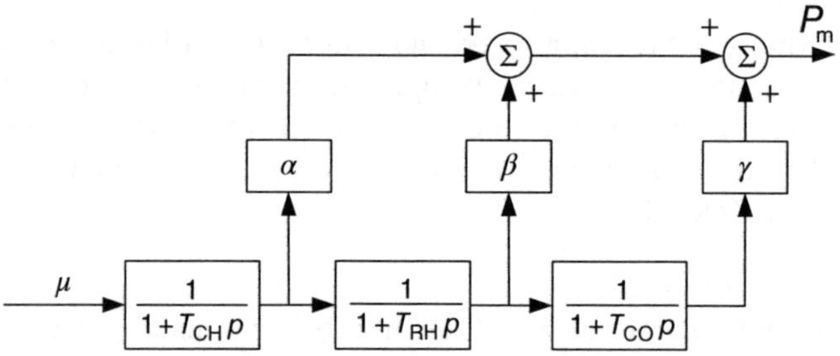

Figure 2.10 Third-order simplified model of a steam turbine.

where T_{CO} is the time constant of the low-pressure cylinder steam volume, α, β, and γ are, respectively, the ratio of the high-pressure cylinder, intermediate reheating cylinder, and low-pressure cylinder steady-state output power to the total output power of steam turbine, and $\alpha + \beta + \gamma = 1$.

Speed Governor of Steam Turbine The speed governors of a steam turbine can be divided into the hydraulic pressure speed governor and power-frequency electrohydraulic speed governor. The former can be further divided into a rotating damping speed governor and high-speed spring-piece speed governor. For hydraulic pressure speed governors, the basic principles of two categories are the same; thus they can be described using the same mathematical model. As shown in Figure 2.11, the main difference from the model of the hydraulic turbine speed governor is that the hard feedback coefficient of the steam turbine speed governor is 1 and there is no soft feedback link.

The power-frequency electrohydraulic speed governor is mainly applicable to the intermediate reheating steam turbine. On the basis of the hydraulic pressure

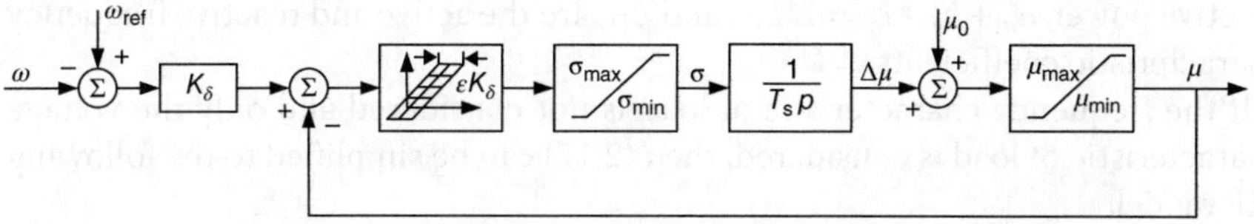

Figure 2.11 Simplified model of a steam turbine hydraulic pressure speed governor.

speed governor, a power measuring link is introduced for output power feedback, thus improving the power-frequency regulation characteristics. In addition, the proportion-integration-differential regulator is used to maintain the static characteristics and to eliminate the influence of the intermediate reheated steam volume effect. The mathematical model is shown in Figure 2.12,

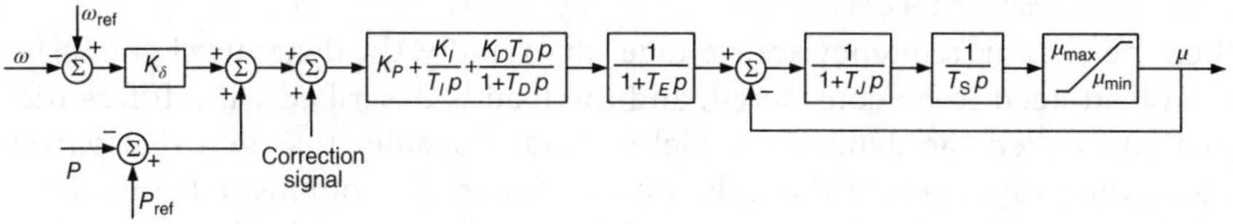

Figure 2.12 Mathematical model of a steam turbine power-frequency electrohydraulic speed governor.

where K_δ is the derivative of the adjustment coefficient, K_P, K_I, and K_D are the magnification times of the proportion, integration, and differential links, T_I and T_D are time constants of the integration and differential links, T_E is the time constant of the electrohydraulic converter, T_J is the time constant of the relay, and T_S is the time constant of the oil servomotor.

2.2.4 Mathematical Model of Load

2.2.4.1 Static Model of Load

The static model of load is used to reflect the variation characteristics of active power and reactive power when the system frequency and voltage vary slowly, and can be expressed with algebraic equations or curves. In the power system analysis, the static load can be expressed with the following polynomial:

$$\begin{cases} P = P_0\left[a_p\left(\dfrac{U}{U_0}\right)^2 + b_p\left(\dfrac{U}{U_0}\right) + c_p\right]\left(1 + L_{Dp}\Delta f/f_0\right) \\ Q = Q_0\left[a_q\left(\dfrac{U}{U_0}\right)^2 + b_q\left(\dfrac{U}{U_0}\right) + c_q\right]\left(1 + L_{Dq}\Delta f/f_0\right) \end{cases} \tag{2.17}$$

where a_p, b_p, and c_p are the ratio of the active power of constant impedance load, constant current load, and constant power load to the total active power, $a_p + b_p + c_p = 1$, a_q, b_q, and c_q are the ratio of the reactive power of constant impedance load, constant current load, and constant power load to the total reactive power, $a_q + b_q + c_q = 1$, L_{Dp} and L_{Dq} are the active and reactive frequency characteristic coefficients.

If the frequency characteristic of load is not considered and only the voltage characteristic of load is considered, then (2.17) can be simplified to the following ZIP model.

$$\begin{cases} P = P_0\left[a_p\left(\dfrac{U}{U_0}\right)^2 + b_p\left(\dfrac{U}{U_0}\right) + c_p\right] \\ Q = Q_0\left[a_q\left(\dfrac{U}{U_0}\right)^2 + b_q\left(\dfrac{U}{U_0}\right) + c_q\right] \end{cases} \tag{2.18}$$

2.2.4.2 Dynamic Model of Load

When the system frequency and voltage vary rapidly, the dynamic characteristics of load need to be considered, and the load is described with differential equations, called the dynamic model of load. Consider that in a real power system, the proportion of the induction motor in the composite load is relatively big; thus the induction motor model is usually used as the dynamic model of load.

In a power system stability analysis, the dynamic models of load commonly used are those considering the mechanical transient process of the induction motor and the electromechanical transient process. Usually the dynamic model that considers the electromagnetic transient process is only used in a power system electromagnetic transient analysis. Therefore, this book only introduces the first two types of load dynamic models [7]:

1) Dynamic model of a load considering the mechanical transient process of the induction motor. The dynamic model of a load that only considers the mechanical process can be expressed with the rotor motion equation of induction motor:

$$T_J\dot{s} = T_m - T_e \tag{2.19}$$

$$T_m = k\left[\alpha + \beta(1-s)^p\right] \tag{2.20}$$

$$T_e \approx \frac{2T_{e,\max}}{\dfrac{s}{s_{cr}} + \dfrac{s_{cr}}{s}}\left(\frac{U_L}{U_R}\right)^2 \tag{2.21}$$

where s is the slip, T_m and T_e are the mechanical torque and electromagnetic torque of the induction motor, T_J is the inertia time constant, k is the load coefficient, which is dependent on the steady-state mechanical torque and slip, α is the constant part of torque, β is the slip-related part of torque, $\alpha + \beta = 1$, p is an index related to the mechanical characteristics of the motor load. $T_{e,\max}$ is the maximum electromagnetic torque, and s_{cr} is the critical slip, that is, the torque and slip when the motor bus voltage U_L equals the rated voltage U_R.

2) Dynamic model of a load considering the electromechanical transient process of the induction motor. The induction motor considering the electromechanical transient process can be regarded as a special case of a synchronous motor with the same d-axis and q-axis parameters, asynchronous rotating speed, and an excitation voltage of 0. Neglecting the transient of the stator winding and using the common xy coordinate system, the model of the induction motor considering the electromechanical transient process can be expressed as

$$\begin{cases} \hat{U} = \hat{E}' + (r_s + jX')\hat{I} \\ \dot{\hat{E}}' = -js\hat{E}' - \left[\hat{E}' - j(X - X')\hat{I}\right]/T' \\ T_J\dot{s} = T_m - T_e \end{cases} \tag{2.22}$$

where $T_e = \mathrm{Re}\left(\hat{E}'\overset{*}{I}\right)$, where Re means extracting the real part of ...; ^ represents the complex form of ... in the xy coordinate system; * represents the conjugate complex of ..., X and X' are the steady-state and transient reactance of the induction motor, r_s is the stator resistance, and T' is the stator open-loop transient time constant. The other parameters are the same as those in the introduction of the load dynamic model considering the mechanical transient process.

2.3 Power System Stability Analysis Method

Concerning the stability analysis method, a set of mature theory based on the differential equation and the Lyapunov stability analysis method has been formed in China and abroad, including a time-domain simulation method, an eigenvlaue analysis method for small-signal stability analysis, and a transient energy function method for transient stability analysis. A brief introduction to these methods follows.

2.3.1 Time-Domain Simulation Method

In a time-domain simulation method, by establishing the differential equations that describe the dynamic characteristics of the power system components and the steady-state equations that describe network topology, and through numerical integration calculation, the trajectory of each variable with time can be gained and thus the dynamic response of a system under disturbance can be analyzed. The equations are shown as follows:

$$\begin{cases} \dot{\boldsymbol{x}} = \boldsymbol{f}(\boldsymbol{x},\boldsymbol{y}) \\ \boldsymbol{0} = \boldsymbol{g}(\boldsymbol{x},\boldsymbol{y}) \end{cases} \tag{2.23}$$

The numerical integration methods to solve the differential equations can be divided into two categories: the explicit method and the implicit method. The representative algorithms in the two categories are listed below [8]:

1) Runge Kutta method for explicit integration. Suppose the time step $\Delta t = h$ and the state variable $\boldsymbol{x}$, and the differential function $\boldsymbol{f}$ from t_0 to $t_k = t_0 + kh$ are both given, and $\boldsymbol{x}_{k+1}$ can be calculated according to (2.24), then it is explicit numerical integration:

$$\boldsymbol{x}_{k+1} = \boldsymbol{F}\left(\boldsymbol{x}_0, \cdots, \boldsymbol{x}_k, \boldsymbol{f}_0, \cdots, \boldsymbol{f}_k\right) \tag{2.24}$$

The Runge Kutta method is widely applied to solving explicit numerical integration. The calculation of each step is

$$\boldsymbol{x}_{k+1} = \boldsymbol{x}_k + \frac{1}{6}\left(\boldsymbol{c}_1 + 2\boldsymbol{c}_2 + 2\boldsymbol{c}_3 + \boldsymbol{c}_4\right) \tag{2.25}$$

where

$$\boldsymbol{c}_1 = h\boldsymbol{f}\left(\boldsymbol{x}_k, t_k\right), \boldsymbol{c}_2 = h\boldsymbol{f}\left(\boldsymbol{x}_k + \frac{\boldsymbol{c}_1}{2}, t_k + \frac{h}{2}\right), \boldsymbol{c}_3 = h\boldsymbol{f}\left(\boldsymbol{x}_k + \frac{\boldsymbol{c}_2}{2}, t_k + \frac{h}{2}\right),$$
$$\boldsymbol{c}_4 = h\boldsymbol{f}\left(\boldsymbol{x}_k + \boldsymbol{c}_3, t_k + h\right)$$

2) Trapezoid method for implicit integration. Suppose the time step $\Delta t = h$ and the state variable $\boldsymbol{x}$ and differential function $\boldsymbol{f}$ from t_0 to $t_k = t_0 + kh$ are both given, and $\boldsymbol{x}_{k+1}$ can be calculated according to (2.26), then it is implicit numerical integration:

$$\boldsymbol{x}_{k+1} = \boldsymbol{F}\left(\boldsymbol{x}_0, \cdots, \boldsymbol{x}_k, \boldsymbol{f}_0, \cdots, \boldsymbol{f}_k, \boldsymbol{f}_{k+1}\right) \tag{2.26}$$

The trapezoid method is used to solve implicit numerical integration. The calculation of each step is

$$\boldsymbol{x}_{k+1} = \boldsymbol{x}_k + \frac{h}{2}\left[\boldsymbol{f}\left(\boldsymbol{x}_k\right) + \boldsymbol{f}\left(\boldsymbol{x}_{k+1}\right)\right] \tag{2.27}$$

When saturation is not considered, $\boldsymbol{f}(\boldsymbol{x})$ is a linear function at $\boldsymbol{x}_k$ and can be expressed as $\boldsymbol{f}(\boldsymbol{x}) = \boldsymbol{A}\boldsymbol{x}$, which applied to (2.27) yields

$$\boldsymbol{x}_{k+1} = \left[1 - \frac{h}{2}\boldsymbol{A}\right]^{-1}\left[1 + \frac{h}{2}\boldsymbol{A}\right]\boldsymbol{x}_k \tag{2.28}$$

The advantage of the time-domain simulation method is that it can deal with all kinds of detailed models, unaffected by system scale and control mode, etc. However, the time-domain simulation method is concerned with the dynamic behavior of a system in a specific operating condition under specific disturbance and thus it is incapable of determining system stability range and conditions. Besides, for some specific types of fault, fault locations, and system operating conditions, the time domain method is likely to miss some oscillation modes that have a great impact on the safe and stable operation of a system. In this case, the analysis results will not be reliable.

2.3.2 Eigenvalue Analysis Method

In the eigenvalue analysis method, by establishing the dynamic model of system and through linearization of the system at the steady-state operating point, the linear equations that describe the system dynamic can be utilized and then the state matrix of system can analyze the oscillation characteristics of system [9].

In a small-signal stability analysis, usually the differential equation of system as shown in (2.23) is linearized at the steady-state operating point $\boldsymbol{x}_{(0)}$:

$$\Delta\dot{\boldsymbol{x}} = \boldsymbol{A}\Delta\boldsymbol{x} \tag{2.29}$$

where

$$\boldsymbol{A} = \begin{bmatrix} \frac{\partial f_1}{\partial x_1} & \cdots & \frac{\partial f_1}{\partial x_n} \\ \vdots & \ddots & \vdots \\ \frac{\partial f_n}{\partial x_1} & \cdots & \frac{\partial f_n}{\partial x_n} \end{bmatrix}_{\boldsymbol{x}=\boldsymbol{x}_{(0)}}$$

Matrix $\boldsymbol{A}$ is called the state matrix of the system or the coefficient matrix, the eigenvalues of which reflect the small-signal stability of the system. Suppose the eigenvalue of matrix $\boldsymbol{A}$ is λ; if c is a real number, then it corresponds to a nonoscillatory mode. If λ is a negative real number, then it corresponds to an attenuating mode. The bigger the absolute value of λ, the faster is the attenuation. If λ is a positive real number, then it corresponds to a nonperiodic unstable mode. If λ is a complex number, it always appears in the form of conjugate complex roots, that is,

$$\lambda = \sigma \pm j\omega \tag{2.30}$$

Each pair of conjugate complex roots corresponds to an oscillatory mode. The real part σ reflects the damping of the oscillation mode and the imaginary part ω reflects the oscillation frequency. The negative real part corresponds to attenuating oscillation and the positive real part corresponds to increasing oscillation. The oscillation frequency (Hz) is

$$f = \frac{\omega}{2\pi} \tag{2.31}$$

Define the damping ratio as

$$\xi = \frac{-\sigma}{\sqrt{\sigma^2 + \omega^2}} \tag{2.32}$$

The damping ratio determines the attenuation rate and attenuation characteristics of the oscillation mode.

In order to measure the correlation between each oscillation mode and the state variables, the left eigenvector $\boldsymbol{u}_i^T$ and right eigenvector $\boldsymbol{v}_i$ of the state matrix $\boldsymbol{A}$ are combined to form the participation matrix $\boldsymbol{P}$ of state matrix $\boldsymbol{A}$:

$$\boldsymbol{P} = \begin{array}{c} \\ \Delta x_1 \\ \\ \Delta x_k \\ \\ \Delta x_n \end{array} \begin{array}{c} \begin{matrix} \lambda_1 & & \lambda_i & & \lambda_n \end{matrix} \\ \begin{bmatrix} u_{11}v_{11} & \cdots & u_{1i}v_{1i} & \cdots & u_{1n}v_{1n} \\ \vdots & \ddots & \vdots & \ddots & \vdots \\ u_{k1}v_{k1} & \cdots & u_{ki}v_{ki} & \cdots & u_{kn}v_{kn} \\ \vdots & \ddots & \vdots & \ddots & \vdots \\ u_{n1}v_{n1} & \cdots & u_{ni}v_{ni} & \cdots & u_{nn}v_{nn} \end{bmatrix} \end{array} \tag{2.33}$$

Generally, the element in matrix $\boldsymbol{P}$, $p_{ki} = u_{ki}v_{ki}$, is called the participation factor, which describes the mutual participation between the ith mode and the kth state variable. The ith column of matrix $\boldsymbol{P}$ is the participation vector of the ith mode. The product of corresponding elements in the left and right eigenvectors causes p_{ki} to be dimensionless. In fact, the participation factor p_{ki} equals the sensitivity of eigenvalue λ_i to the diagonal element a_{kk} of the state matrix $\boldsymbol{A}$, that is $p_{ki} = \partial\lambda_i / \partial a_{kk}$.

From the calculation perspective, eigenvalue analysis methods can be divided into the whole eigenvalue method and the partial eigenvalue method. Representative of the whole eigenvalue method is the QR algorithm, which has the advantages of fast convergence and strong robustness, and could calculate all oscillation modes of the system without missing any. However, when the system is large with a high-order model, "dimension disaster" will result using this algorithm. The partial eigenvalue method mainly analyzes the eigenvalues concerned and obtains eigenvalues of the original system through model reduction methods, including the AESOPS method, the SMA method, the S matrix method, the STEP method, etc. Compared with the traditional method, to some extent the "dimension disaster" problem can be solved using this method, and the computational efficiency can be greatly improved. However, in the iteration process, loss of eigenvalue or convergence to a nonelectromechanical oscillation mode may result.

In general, the eigenvalue analysis method has the following advantages:

1) According to the state matrix, all the eigenvalues of the system can be obtained and thus all of the oscillation modes of the system can be mastered, including the weak damping modes concerned.
2) Through the eigenvectors, the phase relationship between generators in different oscillation modes and the participation degree of each generator in oscillation can be analyzed.
3) By analyzing the sensitivity of the eigenvalue, the influence of system parameters on the eigenvalues can be analyzed quantitatively, providing the basis for placement of the controller and setting of control parameters.

However, since the equilibrium point eigenvlaue analysis method only considers small-signal disturbance around a system operating point, the time-varying and nonlinear factors caused by large disturbance are neglected, which may lead to a relatively large error.

2.3.3 Transient Energy Function Method

The transient energy function method, also known as the Lyapunov direct method, identifies the stability of a system according to the generalized energy function $V(\boldsymbol{x})$ and the derivative of $V(\boldsymbol{x})$ with respect to time $\dot{V}(\boldsymbol{x})$. The direct method is proposed according to the Lyapunov definition of stability [10]. The Lyapunov definition of system asymptotic stability is as follows.

For an n-dimensional autonomous system $\dot{\boldsymbol{x}} = \boldsymbol{f}(\boldsymbol{x})$, when it is not affected by external disturbance, the equilibrium point is the origin, that is $\boldsymbol{f}(\boldsymbol{0}) = \boldsymbol{0}$. If the system is stable and the movement starts from near the origin and converges to the origin when $t \to \infty$, that is $\lim_{t\to\infty} \|\boldsymbol{x}(t)\| = 0$, then the system is asymptotically stable.

According to the above definition, for a dynamic system with $\boldsymbol{x}$ as the state variable, if there exists a scalar function $V(\boldsymbol{x})$ that satisfies

1) $V(\boldsymbol{x})|_{x=0} = 0, V(\boldsymbol{x})|_{x\neq 0} > 0$
2) $\dot{V}(\boldsymbol{x})|_{x=0} = 0, \dot{V}(\boldsymbol{x})|_{x\neq 0} < 0$

Then $V(\boldsymbol{x})$ is a Lyapunov function and the system is asymptotically stable. In a power system, the V function usually takes the form of energy functions.

According to the definition of asymptotic stability, the stability region of the system can also be searched in the following way. Near the origin, determine a $V(\boldsymbol{x})$ for all the state variables that satisfies $V(\boldsymbol{x}) < 0$. Suppose $V_{\min}$ is the minimum value of $V(\boldsymbol{x})$ on the plane of $\dot{V}(\boldsymbol{x}) = 0$. Then the region determined according to $V(\boldsymbol{x}) < V_{\min}$ is contained in the asymptotic stability region and thus the estimated stability region can be obtained.

However, it should be noted that the stability region obtained using the above method is affected by the selected V function. Different V functions will lead to different stability regions, and all of these stability regions are included in the accurate asymptotic stability region. Thus this method is conservative.

Based on the above analysis, the energy function method is unaffected by the degree of system disturbance and is applicable to any size of disturbance. Since the method does not observe the stability problem from the time-domain perspective, there is no need to calculate the trajectory of the whole system. Thus system stability can be identified quickly. However, since the Lyapunov function does not provide the general method to form the V function, practical application of this method is difficult. In addition, the Lyapunov direct method provides the sufficient condition for system stability, but not the necessary condition. Therefore, the stability analysis result and stability region obtained using this method are conservative.

2.4 Automatic Control Theory

Automatic control refers to when no human is directly involved, using an additional device (controller) to control a certain state or parameter of the controlled object (the controlled variable), in order to make it operate automatically according to predetermined rules. As the core of automatic control science, automatic control theory includes the establishment of the mathematical model of the control system, the analysis and identification of system dynamic characteristics, and the design of the control system, etc. Automatic control theory has experienced the stages of classical control theory and modern control theory, and is now developing towards intelligent control theory and large system theory.

2.4.1 Classical Control Theory

Classical control theory, also known as the frequency-domain control theory, has developed as a mature theory as early as the 1950s. Classical control theory mainly studies a linear time-invariant system. By applying Laplace transformation to the linear differential equations with constant coefficients that describe the controlled system and controller, the transfer function and frequency characteristics of each link are established. On this basis, the root-locus method and Bode diagram method can be used for analysis.

According to the difference in control modes and strategies, control systems can be divided into an open-loop control system and a closed-loop control system. An open-loop control system is a simple control system, where only forward control exists between the controller and the controlled object, and the output of the system does not have any influence on the control effect. In a closed-loop control system, the output of the system directly affects the controller. The system output is fed back to the control end of the system and is compared with the reference input signal, and the resulting deviation signal is used for control. Thus closed-loop control is also called feedback control. Classical control theory is mainly used to solve the problem of controller analysis and design in the feedback control system. The control structure is shown in Figure 2.13.

Although classical control theory has great practical value, it also has obvious limitations, which are mainly reflected in the following two aspects:

1) The transfer function and frequency characteristics only describe the external characteristics of the system, which only represent the relationship between the input and output of the system, with no representation of internal variables. Therefore, classical control theory based on the transfer function and frequency characteristics cannot fully reflect the internal state of the system.
2) Analysis methods based on classical control theory are essentially frequency-domain methods. Thus the use of integral transformation is inevitable.

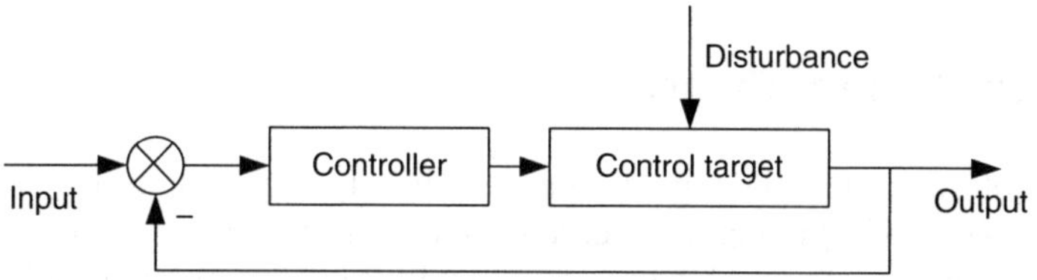

Figure 2.13 Simplified diagram of a feedback control system.

Therefore, in principle they are only suitable for solving the problems of a single-input single-output linear system with constant coefficients. For a multi-input multi-output system, especially for a nonlinear time-varying system, classical control theory has great limitations.

2.4.2 Modern Control Theory

In order to overcome the limitations of classical control theory, modern control theory gradually developed in the early 1960s and so far has been widely applied all over the world. It mainly includes three categories: the multivariable linear system theory, the optimal estimation and system identification theory, and the optimal control theory.

2.4.2.1 Multivariable Linear System Theory

Multivariable linear system theory is the most basic and relatively mature branch of modern control theory, which is focused on the study of linear system state control and observation, and the basic analysis method is the state space method.

Modern control theory uses the state space method for modeling. Define the minimum set of variables that could uniquely determine the dynamic characteristics of a system as the state vector of the system. The n-dimensional Euclidean space formed with each state variable as one axis is defined as the state space. In modern control theory, the system motion equations are expressed with vectors and matrices, etc.; thus various linear algebra methods can be used to analyze the characteristics of a dynamic system, such as the controllability, stability, parameter identification, and comprehensive check, etc. In addition, modern control theory is not based on the transfer function and frequency characteristics of a system, but is based on the state space equation, and thus could solve the problems of a multi-input multi-output system.

2.4.2.2 Optimal Estimation and System Identification Theory

All automatic control systems work under the influence of stochastic disturbance, and the measurements of all control systems have different degrees of stochastic errors. The existence of these stochastic factors causes the state and output of a system to manifest some stochastic process with statistical characteristics. Usually the estimation theory in mathematical statistics needs to be used to estimate the system state and parameter of the given mathematical model through the input and output values of the measurement system. The optimal estimation method includes the Calman filter, optimal linear smoothing, etc.

The estimation and control of the actual process are both based on a correctly described mathematical model. For systems with a complex mechanism and the mathematical model of which cannot be established according to the physical theorem, the mathematical model needs to be constructed with experimental data. System identification is an important method to establish the dynamic model of such system. According to the test data of system input and output, select from a set of given models one that is equivalent to the essential characteristics of the studied system and then determine the structure and parameters of the selected model. System identification methods include the least-square identification method, maximum likelihood method, and stochastic approximation method, etc.

2.4.2.3 Optimal Control Theory

Optimal control theory is the theoretical basis of an optimal control system design. It refers to using the analytical design method to get the optimal control rule according to the requirement of control effect and ensure that the required performance index is optimal (minimum or maximum). Currently, the main methods used in the integrated optimal control system are the variational method, maximum value principle, and dynamic planning, etc.

2.4.3 Large System Theory and Intelligent Control Theory

Since the 1970s, control theory has developed towards "large system theory" and "intelligent control theory." The former is the development of control theory in breadth and the latter in depth.

2.4.3.1 Large System Theory

Large system theory is the technical and theoretical basis of using control and information to study the structural scheme, the decomposition method and coordination, etc., in the overall design of various large systems. The characteristics of large system theory are: large scale, complex structure (with many links and levels or complicated relationship), diverse objectives, and numerous influencing factors, often accompanied by stochastic problems.

The original control theories are all based on centralized control, that is, it is taken that the information of the whole system could be concentrated to a certain location to be processed, and then control signals could be sent to different parts of the system. However, for large systems, since the information is difficult to be concentrated, centralized control is no longer applicable. To make up for the deficiency of the original control theories, the hierarchical control theory and decentralized control theory have been proposed.

Hierarchical Control Theory Hierarchical control theory is based on the decomposition and coordination of a large system. Decomposition is to divide a large system into a group of coupled subsystems and to carry out control and optimization design on the subsystem level. Then a coordinating body (coordinator) is set in the upper level. By coordinating certain variables, the relationship between the subsystems can be constantly adjusted to meet the correlation constraints, so that the global optimal solution of the large system can be obtained and the optimal control effect can be achieved. The information exchange between the centralized controller and the controlled objects is significantly reduced; thus system reliability is greatly improved. A power grid dispatching automation system is a type of hierarchical control system.

Decentralized Control Theory Decentralized control is also based on dividing the large system into several subsystems that are controlled separately. It is different from the hierarchical control system in that the local controller of each subsystem does not consider or rarely acquires information from other subsystems. However, the correlation between subsystems needs to be considered. Take power system stability control, for example, after the stability of subsystems is identified, the conditions that the correlation needs to meet to ensure the stability of the large system and the influence of structural disturbance on system stability should be studied.

The main difference between decentralized control and centralized control is the information structure. In a decentralized control system, each controller can only acquire and use part of the system information, unlike in the centralized control system where each controller can acquire and use all the information of the system. This kind of information structure is called a nonclassical information structure. For a nonclassical information structure, the performance index can only reach suboptimal, but the reliability is often greatly improved.

2.4.3.2 Intelligent Control Theory

The concept of intelligent control was proposed by a Chinese American, Professor K.S. Fu of Purdue University in the United States, in the early 1970s. It is based on the interdisciplinary study of subjects such as artificial intelligence, control theory, operational research, and information theory, etc. An intelligent control system is an engineering control and information processing system with some human-simulating intelligence through studying and simulating the human intelligent activity and the rules of its control and information transfer process. The expert system and artificial neural network are two active branches of intelligent control system.

Expert System An expert system (ES) refers to the computer model system using the experience, knowledge, and reasoning methods of experts. The basic structure includes a human–computer interface, knowledge acquisition, database, knowledge base, inference engine, and interpretation module, etc.

An expert control system refers to the control system built by applying the concept and technology of an expert system and simulating the knowledge and experience of human experts. Although an expert control system is based on an expert system, there are some obvious differences between them. On the one hand, in an expert system, the operation staff are part of the system and play the role of "computer expert" through a man–machine dialogue, while an expert control system could make decisions on control objects independently and automatically without the participation of operation staff. On the other hand, an expert system usually works in the off-line mode, while an expert control system needs to be online not only to acquire the dynamic information but also to conduct real-time control on the system.

Artificial Neural Network An artificial neural network (ANN) has been a research focus in the field of artificial intelligence since the 1980s. From the perspective of information processing, the neural network of a human brain is abstracted, a simple model is set up, and then different networks are formed according to different connection modes. In engineering and academic circles, it is often referred to as a neural network.

Traditional artificial intelligence (including ES) simulates the logical thinking of a human while an ANN mainly simulates the image thinking of a human. It has the following advantages: first, it is adaptive and self-learning; second, due to the structure of a neural network and its multi-input multi-output characteristics, it is easily applicable to the control of a multi-variable system and is more economical compared with other approximation methods.

With the research on the application of an artificial neural network ever deepening, many new neural network models have been proposed and there are now nearly a hundred kinds of neural network models. In the intelligent control field, the most widely applied models are the BP network, self-organizing neural network, Hopfield network, dynamic recurrent network, associative memory network, and Adaline network, etc.

2.5 Summary

This chapter briefly introduces the basic theories and methods of power system stability analysis and control. First, the mathematical models frequently used for power system stability analysis are introduced, including the mathematical models of the synchronous generator, excitation system, prime mover, speed governor, and load. On this basis, analysis methods based on the differential equation and Lyapunov stability theory are introduced, including the time-domain simulation method, the eigenvalue analysis method, and the transient energy function method. Finally, according to the three stages of development of automatic control theory, the classical control theory, modern control theory, large system theory, and intelligent control theory are introduced, respectively, laying the foundation for the proposal of a new-generation wide-area stability analysis and control theories.

References

1 Kundur, P. (1994) *Power System Stability and Control*, McGraw-Hill, New York.
2 Concordia, C. (1951) *Synchronous Machine*, John Wiley & Sons, Inc., New York.
3 Ni, Y., Chen S., and Zhang B. (2001) *Power System Dynamics Theory and Analysis*, Tsinghua University Press, Beijing.
4 IEEE Committee Report (1981) Excitation system models for power system stability studies. *IEEE Transactions on Power Apparatus and Systems*, **100**(2), 494–509.
5 IEEE Working Group Report (1992) Hydraulic turbine and turbine control models for system dynamic studies. *IEEE Transactions on Power Systems*, 7(1), 167–179.
6 IEEE Working Group Report (1991) Dynamic models for fossil fueled steam units in power system studies. *IEEE Transactions on Power Systems*, **6**(2), 753–761.
7 Concordia, C. and Ihara, S. (1982) Load representation in power system stability studies. *IEEE Transactions on Power Apparatus and Systems*, **101**(4), 969–977.

8 Stagg, G.W. and El-Abiad, A.H. (1968) *Computer Methods in Power System Analysis*, McGraw-Hill, New York.

9 Rogers, G. (2000) *Power System Oscillations*, Kluwer Academic Publishers, Dordrecht.

10 Mihalic, R. and Gabrijel, U. (2004) A structure-preserving energy function for a static series synchronous compensator. *IEEE Transactions on Power Systems*, **19**(3), 1501–1507.

3

Wide-Area Information Monitoring

3.1 Introduction

The development and wide application of the WAMS (wide-area measurement system) has provided abundant reliable system information for the stability analysis and control of a large-scale complex power grid. Researchers are making great efforts to study how to effectively monitor and use wide-area information, so that the accuracy and effectiveness of system analysis results could be improved, as well as the control processes and effects concerning disturbances. For wide-area information monitoring, the following two problems need to be solved: (1) among various wide-area signals, how can the optimal wide area feedback signal or combination of feedback signals be chosen in order to improve the control effects of controllers and (2) in a modern power system with multiple types of controllers, how can the optimal controller or combination of controllers be chosen according to wide-area information in order to realize the optimal system control.

In the optimal selection of wide-area signals, this chapter introduces a wide-area signal optimal selection method based on contribution factors that quantify the damping effects of different feedback signals on the closed-loop eigenvalue, so that different supplementary controllers could be coordinated to adapt to different control laws. In the optimal selection of wide-area controllers, a wide-area controller optimal selection scheme based on the eigenvalue transfer factor theory is introduced, which realizes unified quantification of the control effects of different controllers on different oscillation modes. Research shows that the control effect of using the optimal controller is equal to that of using all the controllers.

3.2 Test System

3.2.1 Four-Generator Two-Area System

Take the IEEE four-machine 11-bus system for example. As shown in Figure 3.1, the system contains two areas – area 1 and area 2. G1 and G2 are in area 1 and G3 and G4 are in area 2. The connection lines between two areas – line 7–8 and line 8–9, are both double circuit lines.

Power System Wide-Area Stability Analysis and Control, First Edition. Jing Ma.

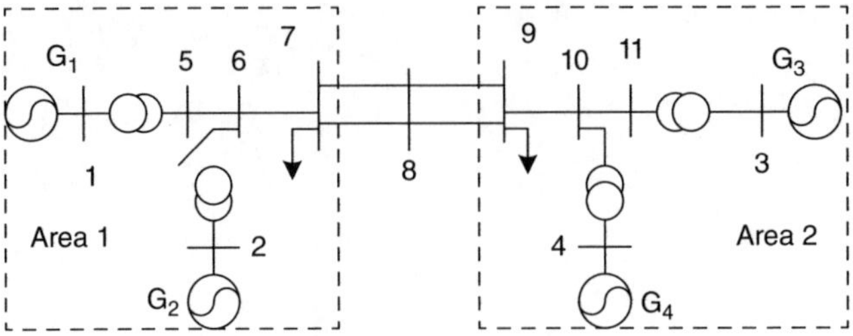

Figure 3.1 Test system of a four-machine 13-bus system.

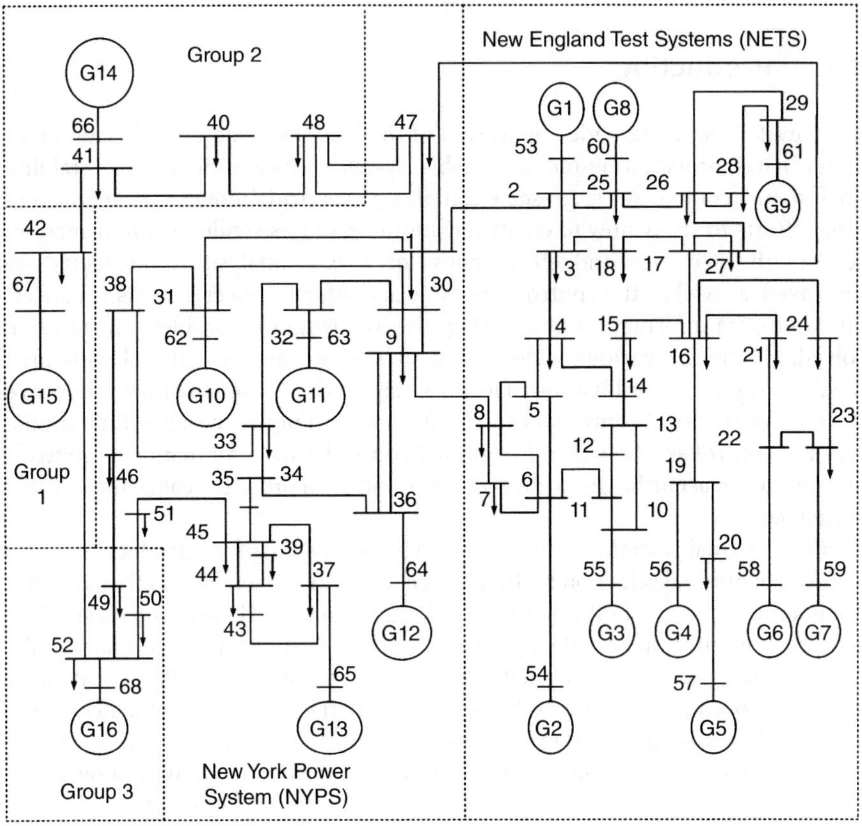

Figure 3.2 The 16-machine, 68 bus test system with tie lines.

3.2.2 Sixteen-Generator System

A 16-machine system with 86 transmission lines and 68 buses study system is shown in Figure 3.2. This is a model of the New England and New York interconnected power systems. The first nine machines (G1–G9) are a simple representation of the New England Test System (NETS) generation. Machines G10–G13 represent the New York Power System (NYPS) generation.

Group 4 contains machines that are a simplified representation of the New England system. Group 5 represents the New York system. The lines 1–2 and 8–9 are double circuits. The thick lines indicate the tie lines, which are the root source of the low-frequency interarea oscillations. The tie lines connecting NETS and NYPS carry 720 MW. Group 3 exports 1163 MW to NYPS and imports 73 MW from Group 1 while 266 MW flows from Group 2 to NYPS and 75.8 MW from Group 2 to Group 1. The detailed description of the study system can be found in reference [1].

3.2.3 Western Electricity Coordinating Council

The 127-bus equivalent of the Western Electricity Coordinating Council (WECC) system is introduced in this section. The system is a multiarea interconnected large-scale power system, which has a generating capacity of 61.4 GW with a load of 60.8 GW, as shown in Figure 3.3. The system model includes 29 generators, 127 buses, and 215 lines.

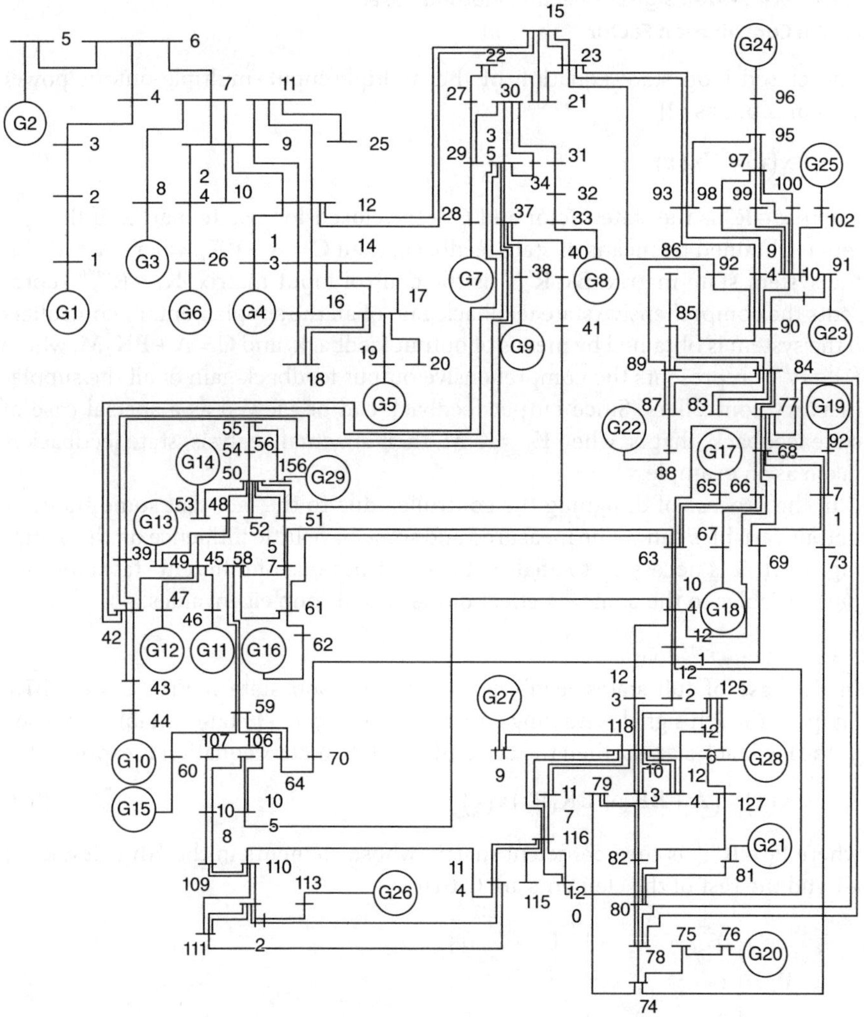

Figure 3.3 WECC system.

3.3 Optimal Selection of Wide-Area Signal

The common wide-area signal selection methods could hardly guarantee the control effect of a closed-loop system. In addition, signals selected by these methods could not adapt to different types of controllers and meanwhile have poor robustness against system uncertainty. Addressing these problems, an optimal signal selection method based on the contribution factor is proposed in this section to improve the damping effect of feedback signals. First, the damping effect of different feedback signals on eigenvalues of a closed-loop is quantified by the interval extension theory and perturbation theory. On this basis, the concepts of a contribution factor and a contribution factor sum are proposed. Then optimal signals are chosen due to the contribution factor sum, which indicate the overall damping effect.

3.3.1 Wide-Area Signal Selection Method Based on the Contribution Factor

The closed-loop state equation of the multiple-input–multiple-output power system is put as [2]

$$\dot{\mathbf{x}}(\mathbf{t}) = \mathbf{Cx}(\mathbf{t}) \tag{3.1}$$

where $\mathbf{x} \in \mathbf{R}^n$ is the state vector and $\mathbf{C}$ is the closed-loop state matrix. If the system is obtained by means of state feedback, then $\mathbf{C} = \mathbf{A} + \mathbf{BK}_1$, where $\mathbf{A} \in \mathbf{R}^{n\times n}$ is the system state matrix, $\mathbf{B} \in \mathbf{R}^{n\times m}$ is the control input matrix, $\mathbf{K}_1 \in \mathbf{R}^{m\times n}$ represents the comprehensive state feedback gain of all the supplementary controllers if the system is obtained by means of output feedback, and $\mathbf{C} = \mathbf{A} + \mathbf{BK}_2\mathbf{M}$, where $\mathbf{K}_2 \in \mathbf{R}^{m\times p}$ represents the comprehensive output feedback gain of all the supplementary controllers. Since output feedback can be viewed as a special case of state feedback, that is, when $\mathbf{K}_1 = \mathbf{K}_2\mathbf{M}$, then in actual analysis, state feedback is taken as an example.

In the process of designing the controller, due to the fact that some states or output variables cannot be measured and some have little influence on the damping effect, it is necessary to analyze the contributions of different states (or output) variables to the damping effect of the closed-loop eigenvalues.

3.3.1.1 State Feedback

In the case of full states feedback, the closed-loop state matrix $\mathbf{C} = \mathbf{A} + \mathbf{BK}_1$. Suppose the kth signal is missing and the remaining $n-1$ state variables are used as the feedback signals; then the closed-loop state matrix equation is expressed as

$$\dot{\mathbf{x}}(\mathbf{t}) = (\mathbf{A} + \mathbf{BK}_1 + \mathbf{BK}_1\mathbf{P}_k)\mathbf{x}(\mathbf{t}) \tag{3.2}$$

where $\mathbf{P}_k \in \mathbf{R}^{n\times n}$ is the coefficient matrix whose elements in the kth column are -1 and the rest of the elements are 0, that is,

$$\mathbf{P}_k = \begin{bmatrix} 0 & \cdots & 0 & -1 & 0 & \cdots & 0 \\ \vdots & \ddots & \vdots & \vdots & \vdots & \ddots & \vdots \\ 0 & \cdots & 0 & -1 & 0 & \cdots & 0 \end{bmatrix}$$

When the kth state variable is missing, the variation matrix of the closed-loop state matrix $\mathbf{C}$ is

$$\Delta\mathbf{C} = \mathbf{B}\mathbf{K}_1\mathbf{P}_k \tag{3.3}$$

3.3.1.2 Output Feedback

In the case of full outputs feedback, the closed-loop state matrix $\mathbf{C} = \mathbf{A} + \mathbf{B}\mathbf{K}_1$. Suppose the kth signal is missing and the remaining $n - 1$ output variables are used as the feedback signals; then the closed-loop state matrix equation is expressed as

$$\dot{\mathbf{x}}(\mathbf{t}) = (\mathbf{A} + \mathbf{B}\mathbf{K}_2\mathbf{M} + \mathbf{B}\mathbf{K}_2\mathbf{P}_k\mathbf{M})\mathbf{x}(\mathbf{t}) \tag{3.4}$$

When the kth output variable is missing, the variation matrix of the closed-loop state matrix $\mathbf{C}$ is

$$\Delta\mathbf{C} = \mathbf{B}\mathbf{K}_2\mathbf{P}_k\mathbf{M} \tag{3.5}$$

In (3.3) and (3.5), the variation matrix of a given element $\mathbf{c}$ in matrix $\mathbf{C}$ can be expressed as

$$\mathbf{c}^{\mathbf{a}} = \left(c_1^a, c_2^a \cdots c_m^a\right)^{\mathrm{T}} = \mathbf{c} + \Delta\mathbf{c} \tag{3.6}$$

where $\mathbf{c}^{\mathbf{a}} = (c_1^a, c_2^a \cdots c_m^a)^{\mathrm{T}}$ represents the element set in $\mathbf{C}$ when variation has occurred, $\mathbf{c} = (c_1, c_2 \cdots c_m)^{\mathrm{T}}$ is the element set in $\mathbf{C}$ in the cases of full states or outputs feedback, and $\Delta\mathbf{c} = (\Delta c_1, \Delta c_2 \cdots \Delta c_m)^{\mathrm{T}}$ is the element variant set in $\mathbf{C}$ when the kth state or output variable is missing.

Apply the natural interval extension theory, expand $\mathbf{C}$ at $\mathbf{c}$ using the Taylor series, and then the state matrix after the kth state or output variable is missing is gained:

$$\mathbf{C}\left(\mathbf{c}^{\mathbf{a}}\right) = \mathbf{C}(\mathbf{c}) + \sum_{j=1}^{m}\left(\frac{\partial\mathbf{C}\left(\mathbf{c}^{\mathbf{a}}\right)}{\partial c_j}\right)\left(c_j^a - c_j\right) = \mathbf{C}(\mathbf{c}) + \Delta\mathbf{C}_1^k\left(\mathbf{c}^{\mathbf{a}}\right) \tag{3.7}$$

where $\mathbf{C}(\mathbf{c}^{\mathbf{a}})$ is the closed-loop state matrix made up of the remaining $n - 1$ state or output variables after the kth state or output variable is missing, $\mathbf{C}(\mathbf{c})$ is the closed-loop state or output feedback matrix, and $\Delta\mathbf{C}_1^k\left(\mathbf{c}^{\mathbf{a}}\right) = \sum_{j=1}^{m}\left(\frac{\partial\mathbf{C}\left(\mathbf{c}^{\mathbf{a}}\right)}{\partial c_j}\right)\left(c_j^a - c_j\right)$ is the variation matrix of $\mathbf{C}$ after the kth state or output variable is missing, with c_j^a and c_j representing the jth element in $\mathbf{c}^{\mathbf{a}}$ and $\mathbf{c}$, respectively.

According to the perturbation theory [3], when the closed-loop state matrix $\mathbf{C}$ varies, the corresponding eigenvectors and eigenvalues also vary, as shown in the following:

$$\mathbf{v}_i = \mathbf{v}_{0i} + \Delta\mathbf{v}_{1i} \tag{3.8}$$

$$\lambda_i = \lambda_{0i} + \Delta\lambda_{1i} \tag{3.9}$$

where $\mathbf{v}_i$ and λ_i are the ith eigenvector and eigenvalue of the closed-loop system when the kth state variable is missing, $\mathbf{v}_{0i}$ and λ_{0i} are the ith eigenvector and eigenvalue of the closed-loop system in the case where full states or outputs feedback is applied, and $\Delta\mathbf{v}_{1i}$ and $\Delta\lambda_{1i}$ are the first-order perturbation of the eigenvector and eigenvalue.

Therefore, the relationship between the closed-loop eigenvalues and eigenvectors can be expressed as

$$\left(\mathbf{C}_0+\Delta\mathbf{C}_1^k\right)\left(\mathbf{v}_{0i}+\Delta\mathbf{v}_{1i}\right)=\left(\lambda_{0i}+\Delta\lambda_{1i}\right)\left(\mathbf{v}_{0i}+\Delta\mathbf{v}_{1i}\right) \tag{3.10}$$

Omitting the high-order term $O(\Delta^2)$ in (3.10) and collecting the like items yields

$$\begin{aligned}\Delta^0: \mathbf{C}_0\mathbf{v}_{0i}&=\lambda_{0i}\mathbf{v}_{0i}\\ \Delta^1: \mathbf{C}_0\Delta\mathbf{v}_{1i}+\Delta\mathbf{C}_1^k\mathbf{v}_{0i}&=\lambda_{0i}\Delta\mathbf{v}_{1i}+\Delta\lambda_{1i}\mathbf{v}_{0i}\end{aligned} \tag{3.11}$$

In (3.11), the first-order perturbation vector $\Delta\mathbf{v}_{1i}$ can be expressed as a linear combination of $\mathbf{v}_{0i}$:

$$\Delta\mathbf{v}_{1i}=\sum_{j=0}^{n}p_{1j}\mathbf{v}_{0j} \tag{3.12}$$

where p_{1j} is the linear combination coefficient.

Substituting (3.12) into (3.11) yields

$$\sum_{j=0}^{n}p_{1j}\mathbf{C}_0\mathbf{v}_{0j}+\Delta\mathbf{C}_1^k\mathbf{v}_{0i}=\sum_{j=0}^{n}p_{1j}\mathbf{v}_{0j}\lambda_{0i}+\Delta\lambda_{1i}\mathbf{v}_{0i} \tag{3.13}$$

Multiplying both sides of (3.13) by the left eigenvector of the closed-loop state matrix $\mathbf{r}_{0s}^{\mathrm{T}}$ yields

$$\sum_{j=0}^{n}p_{1j}\mathbf{r}_{0s}^{\mathrm{T}}\mathbf{C}_0\mathbf{v}_{0j}+\mathbf{r}_{0s}^{\mathrm{T}}\Delta\mathbf{C}_1^k\mathbf{v}_{0i}=\sum_{j=0}^{n}p_{1j}\mathbf{r}_{0s}^{\mathrm{T}}\mathbf{v}_{0j}\lambda_{0i}+\Delta\lambda_{1i}\mathbf{r}_{0s}^{\mathrm{T}}\mathbf{v}_{0i} \tag{3.14}$$

Applying the orthogonal condition yields

$$\mathbf{r}_{0s}^{\mathrm{T}}\mathbf{C}_0\mathbf{v}_{0j}=\delta_{sj}\lambda_{0j} \tag{3.15}$$

$$\mathbf{r}_{0s}^{\mathrm{T}}\mathbf{v}_{0j}=\delta_{sj} \tag{3.16}$$

where

$$\delta_{sj}=\begin{cases}1 & (\mathrm{s}=\mathrm{j})\\ 0 & (\mathrm{s}\neq\mathrm{j})\end{cases}$$

Then equation (3.14) is expressed as

$$p_{1j}\lambda_{0i}+\mathbf{r}_{0j}^{\mathrm{T}}\Delta\mathbf{C}_1^k\mathbf{v}_{0i}=p_{1j}\lambda_{0j}+\Delta\lambda_{1i}\delta_{ij} \tag{3.17}$$

When $i = j$, $\lambda_{0i} = \lambda_{0j}$, the first-order perturbation eigenvalue $\Delta\lambda_{1i}$ is obtained:

$$\Delta\lambda_{1i}^{k} = \mathbf{r}_{0i}^{\mathrm{T}}\Delta\mathbf{C}_{1}^{k}\mathbf{v}_{0i} \tag{3.18}$$

Denote $\mathbf{r}_{0i}^{\mathrm{T}}\Delta\mathbf{C}_{1}^{k}\mathbf{v}_{0i}$ in (3.18) as the contribution factor, which indicates how much the ith eigenvalue is affected by the absence of the kth signal. If the contribution factor is positive, this means that when the kth signal is missing, the ith eigenvalue shifts to the right by a half plane. That means that the kth signal increases the damping of the ith eigenvalue. If the contribution factor is zero, this means that when the kth signal is missing, the ith eigenvalue remains unchanged and has nothing to do with the damping of the ith eigenvalue. If the contribution factor is negative, this means that when the kth signal is missing, the ith eigenvalue shifts to the left by a half plane. That means the kth signal can decrease the damping of the ith eigenvalue.

From the above analysis, it can be seen that the contribution factor is utilized to select proper feedback signals as well as to quantify the damping effect of the eigenvalues.

When the system operation mode undergoes minor changes, the contribution factor is expressed as

$$\begin{aligned}\Delta\lambda_{1i}{}' &= \mathbf{r}_{0i}'^{\mathrm{T}}\Delta\mathbf{C}_{1}^{k}\mathbf{v}_{0i}' = \left(\mathbf{r}_{0i} + \Delta\mathbf{r}_{0i}\right)^{\mathrm{T}}\Delta\mathbf{C}_{1}^{k}\left(\mathbf{v}_{0i} + \Delta\mathbf{v}_{0i}\right) \\ &= \mathbf{r}_{0i}^{\mathrm{T}}\Delta\mathbf{C}_{1}^{k}\mathbf{v}_{0i} + \Delta\mathbf{r}_{0i}^{\mathrm{T}}\Delta\mathbf{C}_{1}^{k}\mathbf{v}_{0i} + \mathbf{r}_{0i}^{\mathrm{T}}\Delta\mathbf{C}_{1}^{k}\Delta\mathbf{v}_{0i} + \Delta\mathbf{r}_{0i}^{\mathrm{T}}\Delta\mathbf{C}_{1}^{k}\Delta\mathbf{v}_{0i}\end{aligned} \tag{3.19}$$

Substituting (3.18) into (3.19) yields

$$\Delta\lambda_{1i}{}' = \mathbf{r}_{0i}'^{\mathrm{T}}\Delta\mathbf{C}_{1}^{k}\mathbf{v}_{0i}' = \Delta\lambda_{1i} + \Delta\mathbf{r}_{0i}^{\mathrm{T}}\Delta\mathbf{C}_{1}^{k}\mathbf{v}_{0i} + \mathbf{r}_{0i}^{\mathrm{T}}\Delta\mathbf{C}_{1}^{k}\Delta\mathbf{v}_{0i} + \Delta\mathbf{r}_{0i}^{\mathrm{T}}\Delta\mathbf{C}_{1}^{k}\Delta\mathbf{v}_{0i} \tag{3.20}$$

where $\Delta\lambda_{1i}{}'$ is the contribution factor after a system change, $\mathbf{r}_{0i}'$ and $\mathbf{v}_{0i}'$ are the left and right eigenvectors of the ith eigenvalue, and $\Delta\mathbf{r}_{0i}$ and $\Delta\mathbf{v}_{0i}$ are the variant vectors of the left and right eigenvectors caused by variation of the system state matrix $\Delta\mathbf{A}$.

The system operation mode change being minor, the variation range of the eigenvector is also small. Therefore, the high order item in (3.20) can be omitted:

$$\Delta\lambda_{1ic}{}' = \Delta\lambda_{1ic} + o\left(\Delta^2\right) \approx \Delta\lambda_{1ic} \tag{3.21}$$

It can be seen from (3.21) that minor changes of the system operation mode have little impact on the contribution factor; thus selection methods based on the contribution factor is proved to be robust.

Due to the signal redundancy problem in feedback controller design, optimal selection of feedback signals is necessary in order to achieve an ideal control effect with a relatively small number of feedback signals. The proposed optimal selection method is based on closed-loop control systems with a given control law. Whatever the control law is, with the closed-loop state matrix optimal selection of the feedback signals can be conducted.

Based on the analysis above, the contribution factor characterizes the damping effect of a given signal on a given oscillation mode. A given signal may increase the damping of some modes but may decrease the damping of other modes. Therefore, a comprehensive examination of the contribution factors of a given signal to different modes is necessary for optimal selection. To this end, in this paper the contribution factors of a given signal to different modes is summarized and then this contribution factor sum is used for optimal selection of feedback signals, which characterizes the overall damping effect of a given signal on the whole system. By using the contribution factor to select the optimal signals, the redundancy problem in a state or output feedback controller design can be solved. The control effect of using the selected feedback signals is nearly as good as that of using the whole feedback signals. Meanwhile, the proposed selection method coordinates different types of supplementary controllers and improves the overall damping level of the system.

3.3.2 Simulation Verification

3.3.2.1 Four-Machine System

The IEEE four-machine two-area system shown in Figure 3.1 is taken for simulation analysis. The generator models are detail models, the load model is with 50% constant impedance and 50% constant current hybrid model, and the DC line is between connection buses 7 and 9. The control structure of the HVDC supplementary controller is shown in reference [4], where $\boldsymbol{T_d}$ is the communication delay constant and $\Delta P_{\max}$ and $\Delta P_{\min}$ are the output limits. There are two inner-area oscillation modes and one inter-area oscillation mode with a weak damping ratio. By calculating the geometric indexes such as GMC (geometric controllability) and GMO (geometric observability), the optimal signals of state or output feedback control can also be gained. A comparison between this method and the proposed method based on the contribution factor is made in the revised paper.

Optimal Selection of the State Feedback Control Signal Generator 4 is taken as the reference machine and the HVDC supplementary controller is used for state feedback robust H_2/H_∞ control. The feedback gain $\mathbf{K}$ is obtained and $\left[\Delta\delta_{14}\ \Delta\delta_{24}\ \Delta\delta_{34}\ \Delta\omega_{14}\ \Delta\omega_{24}\ \Delta\omega_{34}\ E'_{q14}\ E'_{q24}\ E'_{q34}\ \cdots\right]$ are the relative feedback states to generator 4. The contribution factors of different state variables to different oscillation modes are calculated using (3.18), as shown in Table 3.1. It can be seen from Table 3.1 that the contribution factor sums of the 4th, 5th, and 6th are the biggest. Therefore, these three signals are selected as the optimal feedback signals. These signals are the relative speed of generators 1, 2, and 3 to generator 4.

The eigenvalues of the open- and closed-loop systems are shown in Figure 3.4. The circles represent eigenvalues of the open-loop system. The dots represent eigenvalues of the closed-loop system using full states as the feedback signals. The triangles represent eigenvalues of the closed-loop system applying the optimal signals as feedback signals.

For nonlinear simulations, a three-phase symmetrical fault at $t = 0.1$ s on line 7–8 next to Bus 7 is applied and the relative angle of generator 1 to generator 4 is shown in Figure 3.5. The relative angular without control is represented by the dashed line, the relative angular with full-state control is represented by the

Table 3.1 Contribution factors.

Mode 1		Mode 2		Mode 3	
No.	Contribution factor	No.	Contribution factor	No.	Contribution factor
4	0.7155	5	15.436	6	8.7763
5	0.5158	23	1.5500	5	3.3328
1	0.1557	4	1.2600	2	0.9253
22	0.0204	9	0.3287	13	0.1762
7	0.0075	3	0.2540	14	0.0774
16	0.0041	15	0.1852	17	0.0658
13	0.0033	14	0.1643	19	0.0190
6	0.0009	1	0.1133	10	0.0123
15	0.0007	24	0.0475	12	0.0040

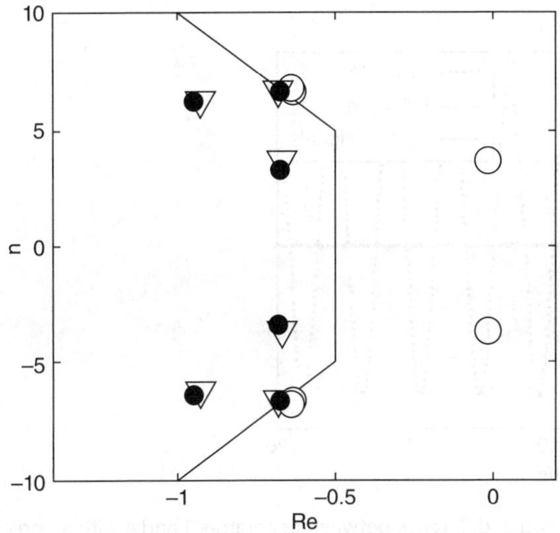

Figure 3.4 Open- and closed-loop eigenvalues using different feedback signals.

dash–dot line, and the angle curve with optimally selected feedback signals for damping control is represented by the solid line. It can be seen that the control effect of applying the optimal signals based on the contribution factor is close to that of using all state variables. Thus, the validity of this optimal selection method is proved.

Shown in Figure 3.6 is the relative angle of generator 1 to generator 4 under the same fault when one of double-circuit lines 8–9 is cut off. It can be seen that, when the system operation conditions change, the optimally selected signals based on the contribution factor can still serve to effectively damp low-frequency

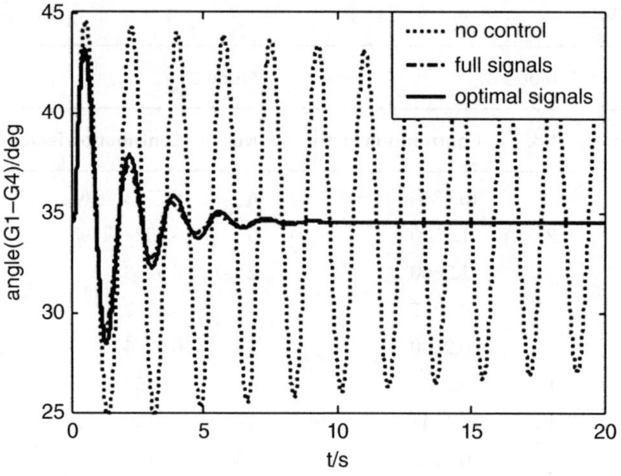

Figure 3.5 Dynamic responses of the angular difference between generators 1 and 4 with a state feedback controller.

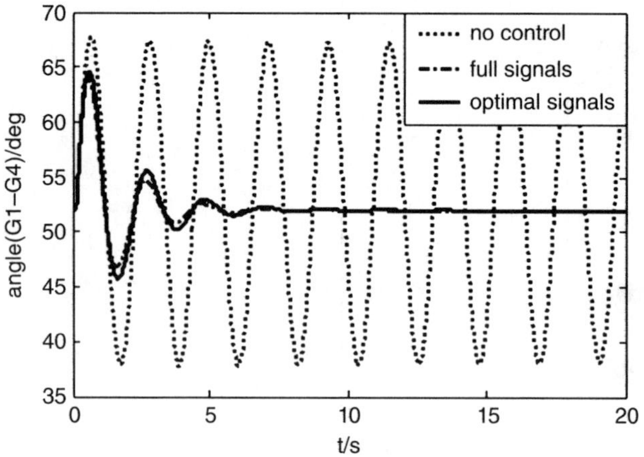

Figure 3.6 Dynamic responses of angular difference between generators 1 and 4 with a single circuit of line 8–9.

oscillation. The results demonstrated that the optimal selection method is robust against changing operating conditions.

Table 3.2 shows the geometric indexes of the generator buses under the same oscillation mode. For the three oscillation modes in the four-machine system, the maximum geometric index falls on the buses of G2, G3, and G1, respectively. This is consistent with the selection result of the method using the contribution factors. However, the method based on the geometric index can only determine the location of the optimal signals. As to which state variable of the corresponding generator is the optimal, it cannot be decided.

Table 3.2 Geometric index.

Mode 1		Mode 2		Mode 3	
No.	**Geometric index**	**No.**	**Geometric index**	**No.**	**Geometric index**
2	0.5446	3	0.9812	1	0.5310
1	0.4540	1	0.0152	2	0.4516
3	0.0014	2	0.0036	3	0.0174

Table 3.3 Contribution factors.

Mode 1		Mode 2		Mode 3	
No.	**Contribution factor**	**No.**	**Contribution factor**	**No.**	**Contribution factor**
3	14.778	3	14.714	3	14.098
2	12.547	2	11.759	2	12.114
4	4.3852	5	0.72753	4	7.0742
6	0.82981	7	0.19609	7	0.54057
7	0.11013	4	0.064475	6	−0.0001
1	−0.72903	6	−0.716	1	−0.077882
5	−1.652	1	−4.4421	5	−3−4443

Optimal Selection of the State Feedback Control Signal When the HVDC supplementary controller is used for output feedback robust H_2/H_∞ damping control, there are many output signals that can be served as feedback signals for effective damping control, for example, the active powers and currents of the tie lines, the generators' speed, etc. Thus $\mathbf{y} = \left[P_{78}, P_{89}, I_{78}, I_{89}, \omega_{14}, \omega_{24}, \omega_{34}\right]^{\mathrm{T}}$, where P_{78} and P_{89} are the active powers of lines 7–8 and 8–9, respectively, and I_{78} and I_{89} are the active currents of lines 7–8 and 8–9, respectively. The contribution factors of different output signals to different oscillation modes are calculated using (3.18) and are shown in Table 3.3. It can be seen from Table 3.3 that the contribution factor sums of output signals 2, 3, and 4 are the biggest. Therefore, these three signals are selected as the optimal feedback signals. These signals are the active power of line 8–9, the active current of line 7–8 and 8–9, respectively.

Suppose the same disturbance occurs in the system and the relative angle of generator 2 to generator 4 is shown in Figure 3.7. The dynamic response without control is represented by the dashed line, and the angle curve with full output damping control is represented by the dash–dot line. The dynamic response with optimally selected feedback signals for damping control is represented by the solid line. It can be seen that the control effect of applying the optimal signals is very close to that of using all output signals. Thus, the validity of this optimal selection method is further proved.

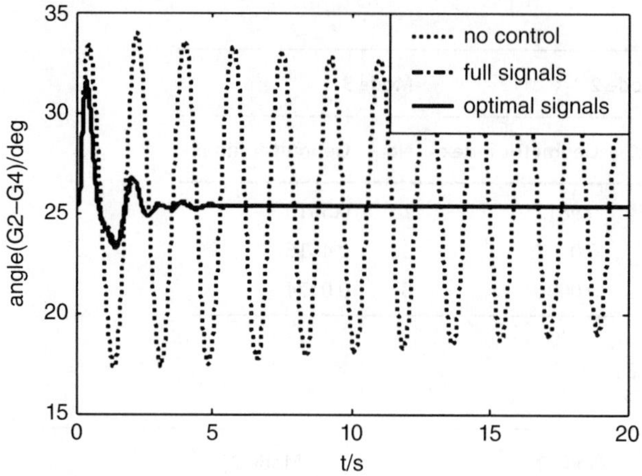

Figure 3.7 Dynamic responses of the angular difference between generators 2 and 4 with the output feedback controller.

Table 3.4 Geometric index.

Mode 1		**Mode 2**		**Mode 3**	
No.	**Geometric index**	**No.**	**Geometric index**	**No.**	**Geometric index**
4	1.1589	4	5.0861	4	1.0074
3	0.9500	2	4.0321	3	0.9156
1	0.1019	3	0.6726	1	0.1736
6	0.0249	7	0.0664	2	0.0576
5	0.0145	6	0.0028	5	0.0222
2	0.0061	5	0.0004	6	0.0159
7	0.0038	1	0.0002	7	0.0026

Table 3.4 shows the geometric indexes of the output signals under the same oscillation mode. For the three oscillation modes in the system, the maximum geometric index corresponds to output signals 2, 3, and 4, respectively, that is, the active power of line 8–9, the active current on line 7–8, and the active current on line 8–9. This is consistent with the selection result of the method using the contribution factors. However, the influence of the same output signal on different oscillation modes is different. The method based on the geometric index can only determine the optimal signal under a certain oscillation mode. As to the overall optimal signal under different oscillation modes, it cannot be decided.

3.3.2.2 Sixteen-Machine System

In order to verify further the validity and generality of the proposed method in wide-area optimal signal selection, the 16-machine 68-bus system is taken for simulation analysis. The control method in reference [5] is used for the damping

Table 3.5 Contribution factors.

Mode 1		Mode 2		Mode 3		Mode 4	
No.	Contribution factor	No.	Contribution factor	No.	Contribution factor	No.	Contribution factor
6	3.8402	14	17.037	6	2.9656	14	9.0644
8	2.4162	6	1.3201	9	1.5913	15	8.4297
9	2.0107	9	1.1384	14	1.4461	1	0.1002
3	1.5512	13	0.9142	5	1.0808	11	0.0495
4	1.4664	8	0.8447	15	1.0389	8	0.0401
5	1.4341	15	0.8050	8	0.9813	6	0.0227
15	1.3952	2	0.6953	7	0.8899	16	0
28	0	30	0	12	−0.252	9	−0.186
29	0	1	−0.126	13	−0.305	12	−0.216
30	0	3	−4.766	11	−0.364	13	−0.531

control design. The system connection diagram is shown in Figure 3.2. An HVDC line is shown between buses 41 and 42, and an SVC at bus 17, with an energy storage device at bus 52. The structure of the SVC supplementary controller is shown in reference [4]. The structure of the energy storage device supplementary controller is shown in reference [6]. The modal analysis of the open-loop system is conducted and the dominant modes of low-frequency oscillation are gained, as well as the corresponding damping ratios of the open loop, shown in Table 3.5. It can be seen that there are four oscillation modes with low damping ratios in the system.

The contribution factors of different wide-area signals to the eigenvalues of the closed-loop system are calculated using (3.18) and are shown in Table 3.5. It can be seen that the contribution factor sums of wide-area signals 6, 14, 9, and 15 are the biggest. Therefore, these four signals are selected as the optimal feedback signals. These signals are the bus frequencies of generators 6, 14, 9, and 15, respectively.

For the 16-machine system, the geometric indexes of the output signals under the same oscillation mode are shown in Table 3.6. For the four oscillation modes in the system, the maximum geometric index corresponds to output signals 6, 14, 15, and 9, respectively, that is the frequency of G6, G14, G15, and G9 buses. Similarly, this is consistent with the selection result of the method using the contribution factors. However, this method can still only determine the optimal signal under a certain oscillation mode. As to the overall optimal signal under different oscillation modes, it cannot be decided.

The damping effect of applying full signals and optimal signals is shown in Table 3.7. It can be seen that when applying optimal signals, the damping ratio of mode 2 increases from 0.2327 to 0.2525, that of mode 3 increases from 0.1689 to 0.2017, that of mode 4 increases from 0.1670 to 0.1848, whereas the damping

Table 3.6 Geometric index.

Mode 1		Mode 2		Mode 3		Mode 4	
No.	**Geometric index**	**No.**	**Geometric index**	**No.**	**Geometric index**	**No.**	**Geometric index**
9	0.0742	14	0.2999	15	0.3237	9	0.3696
6	0.0676	15	0.1582	14	0.1759	5	0.0317
1	0.0494	13	0.0224	6	0.0002	6	0.0266
7	0.0503	9	0.0047	9	0.0002	4	0.0192
3	0.0408	6	0.0041	3	0.0001	3	0.0147
4	0.0408	1	0.0037	4	0.0001	2	0.0117

Table 3.7 Dominant modes of open-loop and closed-loop systems in the test system.

Mode		Open-loop	All signals	Optimal signals
1	Frequency(Hz)	0.4114	0.4618	0.4618
	Damping ratio	0.0040	0.3547	0.3547
2	Frequency(Hz)	0.5210	0.5317	0.5317
	Damping ratio	0.0210	0.2525	0.2525
3	Frequency(Hz)	0.6313	0.5553	0.5553
	Damping ratio	0.0314	0.2017	0.2017
4	Frequency(Hz)	0.7936	0.7073	0.7073
	Damping ratio	0.0314	0.1848	0.1848

ratio of mode 1 decreases from 0.5325 to 0.3547. However, the overall damping effect is satisfactory and good enough to damp inter-area oscillation caused by small disturbances. Therefore, optimal signal selection based on the proposed method can realize damping coordination among multiple dominant oscillation modes.

Suppose the system is in a normal operation condition and at $t = 0.1$ s a three-phase fault occurs on line 8–9. The fault lasts 50 ms and is cleared at $t = 0.15$ s. The dynamic responses of the system are shown in Figure 3.8. The dynamic response without control is represented by the dashed line. The dynamic response with full-signal damping control is represented by the dash–dot line and the dynamic response with optimal-signals damping control is represented by the solid line. Figure 3.8(a), (b), and (c) show the angular between generators 1 and 16, voltage amplitude of the bus with SVC, and voltage amplitude of the bus with an energy storage device, respectively. It can be seen that the optimal signal has a similar effect to that using full signals to damp inter-area oscillation modes. Figure 3.8(d) shows the power flow of the HVDC line. Figure 3.8(e) shows the reactive power compensation of SVC to the system. Figure 3.8(f) shows the active power input of the energy storage device to the system. It can be seen that

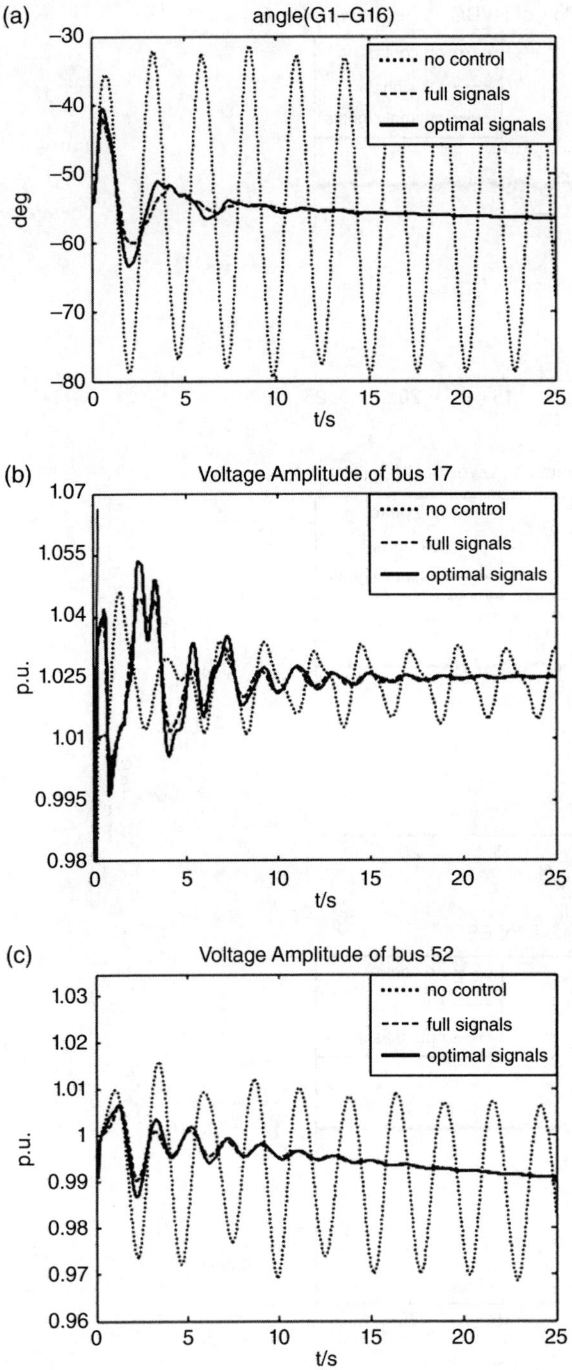

Figure 3.8 Dynamic responses of the 16-machine system.

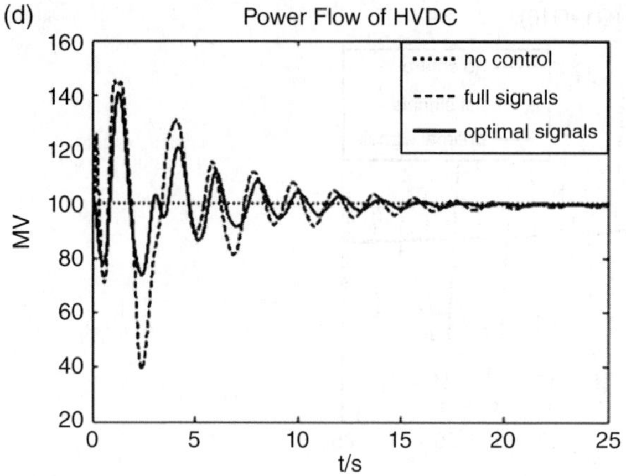

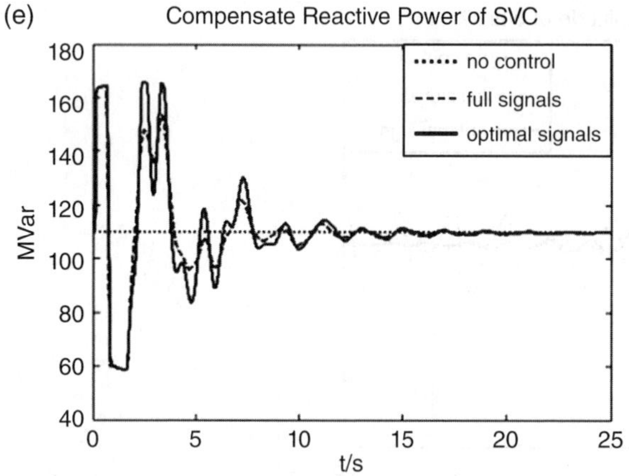

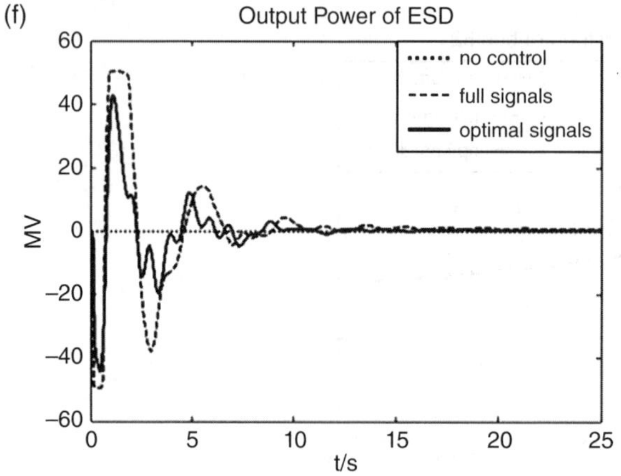

Figure 3.8 (Continued)

damping control based on the optimal signals coordinates the outputs of different supplementary controllers to effectively damp multiple oscillation modes in the system.

The robustness of the optimal signals against changed operating conditions is then investigated. One of the double-circuit lines 1–2 is cut off and a three-phase fault occurs on line 8–9 and lasts for 50 ms, as in the previous case. The dynamic responses of the system are shown in Figure 3.9. It can be seen that the damping control based on the optimal signals can still quickly restore the system to a new stable state in the case of changed operating conditions.

The common wide-area signal selection methods are mainly based on the residue analysis theory of an open-loop system. The wide-area signals are selected before the open-loop system is closed. Therefore the control effect cannot be guaranteed and the types of feedback signals available for selection are limited. On the other hand, based on the proposed method in this paper, the open-loop system is firstly closed and then local signals whose contribution factors are relatively big are selected to realize global feedback. Thus the control effect can be guaranteed. Furthermore, since the transferring ability of different signals to the eigenvalues is considered in the method, different types of signal can be unified.

3.4 Optimal Selection of Wide-Area Controller

The widespread application of FACTS and HVDC has brought new room for suppression of low-frequency oscillation. FACTS devices have a quick adjustment ability and a flexible installation location. The HVDC line has large transmission capacity, flexible operation state, abundant DC modulation method, and multiple signals to be selected for modulation and control. Wide-area PSS (power system stabilizer) is the most economical method for suppressing low-frequency oscillation. Therefore, in the interconnected power system, there is an urgent need to study and solve the optimal selection of the wide-area controller from a global point of view to improve overall performance of damping control.

3.4.1 Mathematical Background

Consider a controlled large-space structure system described by the following linear system model [7]:

$$\ddot{\psi}+\mathbf{M}^{-1}\mathbf{C}\dot{\psi}+\mathbf{M}^{-1}\mathbf{K}\psi=\mathbf{M}^{-1}\mathbf{N}\mathbf{u},\quad \mathbf{u}=-\mathbf{P}\gamma,\ \gamma=\mathbf{N}^{\mathrm{T}}\mathbf{M}^{-\mathrm{T}}\dot{\psi} \tag{3.22}$$

where $\boldsymbol{\psi}$ is the vector of modal coordinates, $\mathbf{u}$ is an m (>1) vector of control input, $\boldsymbol{\gamma}$ is the vector of measurements, $\mathbf{M}$, $\mathbf{C}$, and $\mathbf{K}$ are $n\times n$ mass, damping, and stiffness matrices, respectively, and $\mathbf{N}$ is an $n\times m$ control matrix. The matrix $\Lambda^2=\mathbf{M}^{-1}\mathbf{K}$ is a diagonal matrix with $\Lambda^2=diag\{\omega_1^2\ldots\omega_i^2\ldots\omega_n^2\}$. The damping is proportional to frequency and $\mathbf{D}=\mathbf{M}^{-1}\mathbf{C}=2\boldsymbol{\alpha}\Lambda$, where $\boldsymbol{\alpha}=diag\{\alpha_1\ldots\alpha_i\ldots\alpha_n\}$ with $\alpha_i<1$. $\mathbf{P}$ is a non-negative definite matrix to be determined.

The open-loop eigenvalues of the system are $-\alpha_i\omega_i \pm j\omega_i\sqrt{1-\alpha_i^2}$. The first k modes are the critical low-frequency modes for which additional damping

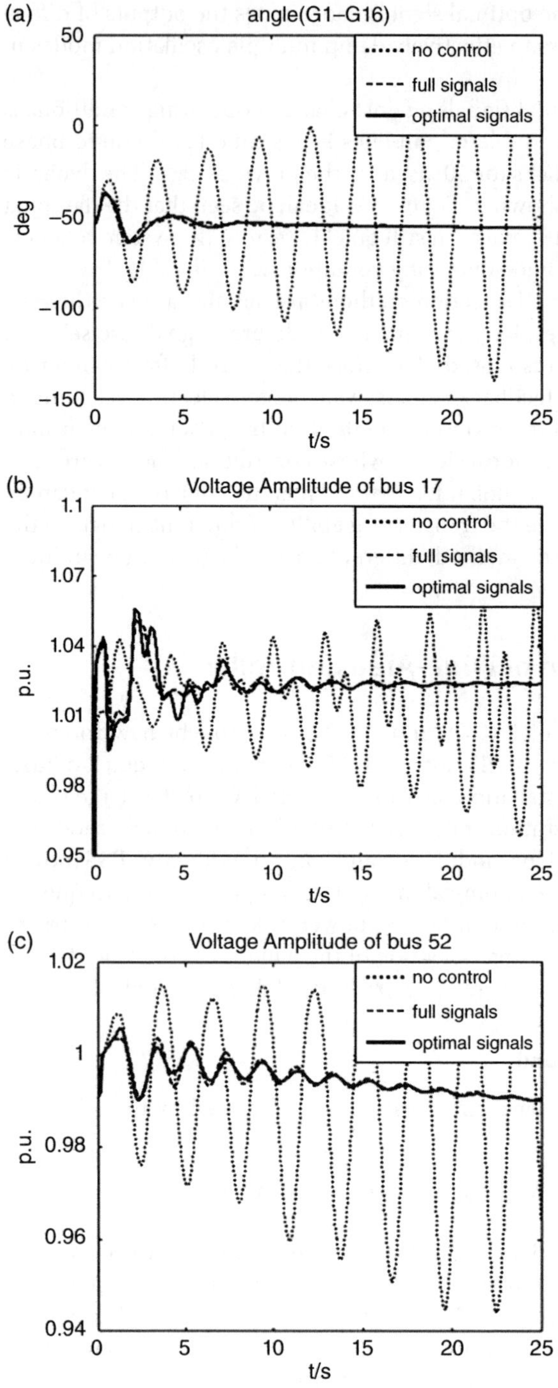

Figure 3.9 Dynamic responses of the 16-machine system with a single circuit of line 1–2.

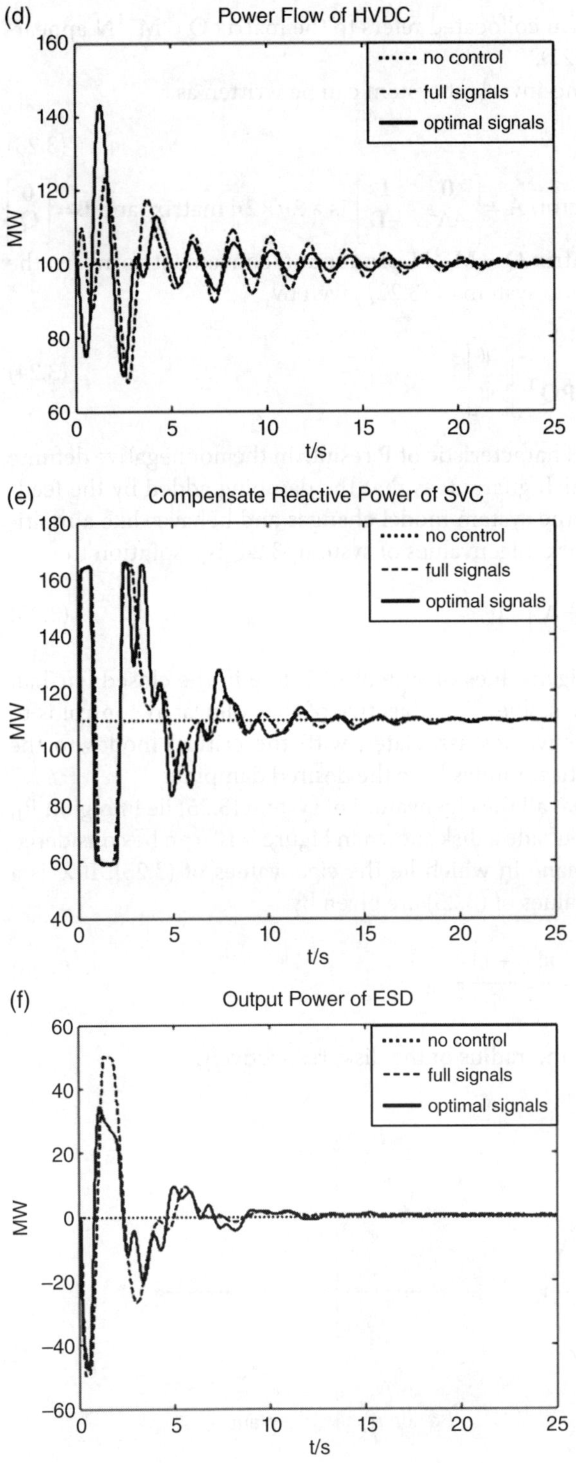

Figure 3.9 (Continued)

should be provided. The term collocated refers to the matrix $\mathbf{Q} = \mathbf{M}^{-1}\mathbf{N}$ appearing with both $\mathbf{u}$ and $\boldsymbol{\gamma}$ in (3.22).

In state space, a linear time-invariant system can be written as

$$\dot{\mathbf{x}} = \mathbf{A}\mathbf{x} + \mathbf{B}\mathbf{u} \tag{3.23}$$

where $\mathbf{x} = [\psi\ \dot{\psi}]^{\mathrm{T}}$ is a $2n$ vector, $\mathbf{A} = \begin{bmatrix} \mathbf{0} & \mathbf{I} \\ -\Lambda^2 & -\mathbf{D} \end{bmatrix}$ is a $2n \times 2n$ matrix, and $\mathbf{B} = \begin{bmatrix} \mathbf{0} \\ \mathbf{Q} \end{bmatrix}$ is a $2n \times m$ matrix. The matrix $\mathbf{Q} = \mathbf{M}^{-1}\mathbf{N}$ produces a convenient form for the eigenvalues of the closed-loop system of (3.22) given by

$$\begin{bmatrix} \dot{\psi} \\ \ddot{\psi} \end{bmatrix} = \begin{bmatrix} \mathbf{0} & \mathbf{I} \\ -\Lambda^2 & -\mathbf{D} - \mathbf{QPQ}^{\mathrm{T}} \end{bmatrix} \begin{bmatrix} \psi \\ \dot{\psi} \end{bmatrix} \tag{3.24}$$

The nonnegative definite characteristic of $\mathbf{P}$ results in the nonnegative definite characteristic of $\mathbf{QPQ}^{\mathrm{T}}$, which guarantees that the damping added by the feedback does no harm even if the system model changes and behaves like a multidimensional resistive network. Eigenvalues of system (3.24) is a solution to

$$\left| \lambda^2 \mathbf{I} + \left(\mathbf{D} + \mathbf{QPQ}^{\mathrm{T}} \right) \lambda + \Lambda^2 \right| = 0 \tag{3.25}$$

It is known that all the eigenvalues of system (3.25) lie in the closed left-half plane (LHP) if the system is stable. The objective of the collocated control is to relocate the open-loop eigenvalues associated with the critical modes to the stable region so that the critical modes have the desired damping.

A feedback gain F such that all the eigenvalues of system (3.25) lie in region R_1, the closed area in the LHP outside a disk shown in Figure 3.10, can be considered by mapping the complex plane in which lie the eigenvalues of (3.25). If $\bar{\lambda}$ is a mapping from λ, the eigenvalues of (3.25) are given by

$$\bar{\lambda} = \frac{r - z_0 + \lambda}{r + z_0 - \lambda} \left(\text{i.e., } \lambda = \frac{r\left(\bar{\lambda} - 1\right)}{\bar{\lambda} + 1} + z_0 \right) \tag{3.26}$$

where z_0 and r is the center and radius of the disk, respectively.

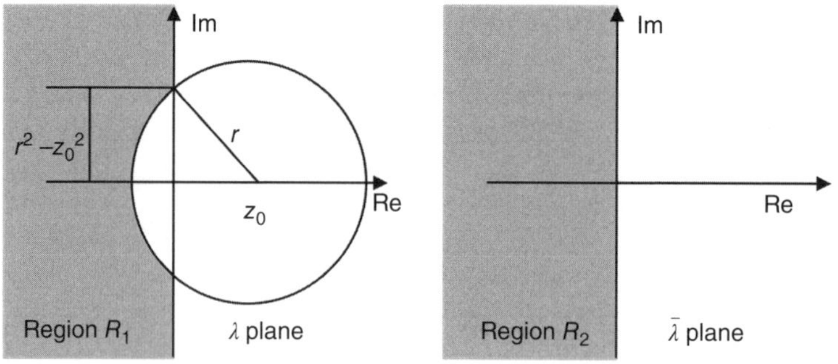

Figure 3.10 The $\bar{\lambda}$ and λ planes.

From this mapping, if the eigenvalues of the $\bar{\lambda}$ system are in the left half of the $\bar{\lambda}$ plane, the eigenvalues of the λ system are in region R_1. Substitute (3.26) into (3.25), which results in

$$\left|\bar{\mathbf{M}}(\mathbf{P},r,z_0)\bar{\lambda}^2+\bar{\mathbf{D}}(\mathbf{P},r,z_0)\bar{\lambda}+\bar{\mathbf{K}}(\mathbf{P},r,z_0)\right|=0 \qquad (3.27)$$

If the solutions of (3.27) lie in region R_2 it is the prerequisite that the solutions of (3.25) lie in region R_1. Due to $r \ge 0$ and $z_0 \ge 0$, $\bar{\mathbf{M}} \ge 0$ for any nonnegative definite matrix $\mathbf{P}$. It can be concluded that the eigenvalues of system (3.25) lie in region R_1 if and only if the nonnegative definite matrix $\mathbf{P}$ satisfies $\bar{\mathbf{D}} \ge 0$ and $\bar{\mathbf{K}} \ge 0$. The interval of $\mathbf{P}$ can be obtained as

$$\frac{1}{z_0}\left(\left(r^2-z_0^2\right)\mathbf{I}-\Lambda^2\right)-\mathbf{D}\le\mathbf{QPQ}^{\mathrm{T}}\le\left(r-z_0\right)\mathbf{I}+\frac{\Lambda^2}{\left(r-z_0\right)}-\mathbf{D} \qquad (3.28)$$

There exist many feedback gains, so other specifications are needed in order to uniquely determine a feedback gain $\mathbf{P}$. Physically, the optimal feedback gain means the least energy consumption occurs when implementing the feedback control. Thus, the least Frobenius norm problem can be formulated as the following constrained optimization problem:

$$\|\mathbf{P}\|_f=\min\left(\sum_i^k\sum_i^k p_{ij}^2\right)^{1/2} \qquad (3.29)$$

where $\|\cdot\|_f$ denotes the Frobenius norm and $\mathbf{P}=\left[p_{ij}\right]\in\Re^{k\times k}$.

Transfer the model form of the power system into the (3.23) form

$$\begin{bmatrix}\dot{\boldsymbol{\delta}}\\ \dot{\boldsymbol{\omega}}\end{bmatrix}=\begin{bmatrix}\mathbf{0} & \omega_0\mathbf{I}\\ \mathbf{A}_{21} & -\mathbf{D}\end{bmatrix}\begin{bmatrix}\boldsymbol{\delta}\\ \boldsymbol{\omega}\end{bmatrix}+\begin{bmatrix}\mathbf{0}\\ \mathbf{Q}\end{bmatrix}\mathbf{u} \qquad (3.30)$$

where $\boldsymbol{\delta}$ and $\boldsymbol{\omega}$ are differences from values at the reference G16. Matrices are formed numerically with perturbations from a load flow equilibrium using Matlab suite PST.

$\mathbf{R}$ is the matrix of the right eigenvectors of $\mathbf{A}_{21}$ such that $\mathbf{R}^{-1}\mathbf{A}_{21}\mathbf{R}=-\Lambda^2$ and $\mathbf{A}_{21}$ is the $n\times n$ matrix where n is the number of machines in the system. Then the transformations of multimodel decomposition are given by

$$\mathbf{x}_1=\mathbf{R}^{-1}\boldsymbol{\delta} \qquad (3.31)$$

$$\mathbf{x}_2=\omega_0\mathbf{R}^{-1}\boldsymbol{\omega} \qquad (3.32)$$

Then $\mathbf{x}_1$ and $\mathbf{x}_2$ are applied to system (3.30), which results in

$$\begin{bmatrix}\dot{\mathbf{x}}_1\\ \dot{\mathbf{x}}_2\end{bmatrix}=\begin{bmatrix}\mathbf{0} & \mathbf{I}\\ -\omega_0\Lambda^2 & -\mathbf{R}^{-1}\mathbf{DR}\end{bmatrix}\begin{bmatrix}\mathbf{x}_1\\ \mathbf{x}_2\end{bmatrix}+\begin{bmatrix}\mathbf{0}\\ \omega_0\mathbf{R}^{-1}\mathbf{Q}\end{bmatrix}\mathbf{u} \qquad (3.33)$$

Using (3.31), (3.32), (3.33), and (3.22), the vector of control input is given by

$$\mathbf{u}=-\omega_0{}^2\mathbf{PQ}^{\mathrm{T}}\mathbf{R}^{-\mathrm{T}}\mathbf{R}^{-1}\boldsymbol{\omega} \qquad (3.34)$$

The matrix $\mathbf{P}$ was chosen using the collocated control algorithm and puts the closed-loop eigenvalues to the left of the disk. Also, the matrix $\mathbf{P}$ can coordinate the wide-area signals and send them to the controllers for damping various inter-area oscillation modes. Due to the nonnegative definite matrix $\mathbf{QPQ}^T$, the collocated control can guarantee that inter-area oscillation modes must be damped. If we measure bus angles $\boldsymbol{\theta}$ and frequencies $\dot{\boldsymbol{\theta}}$ in the system using PMUs then $\boldsymbol{\delta}$ and $\boldsymbol{\omega}$ can be obtained. All the system wide-area rotor speeds of generators are employed as the GPSS input signals.

3.4.2 Example Test System

A 16-machine system with 86 transmission lines and 68 buses study, shown in Figure 3.2, is considered for the damping control design. This is a model of the New England and New York interconnected power systems. The first nine machines (G1–G9) are a simple representation of the New England Test System (NETS) generation. Machines G10–G13 represents the New York Power System (NYPS) generation.

All the generators use a subtransient model, with identical simple exciters, thermal turbines, and governors modeled on all generators. Local power system stabilizers (LPSSs) are installed from generators 1 to 12 just using local rotor speeds. The active loads are 50% constant current and 50% constant impedance; reactive loads all have constant impedance. The tie lines connecting NETS and NYPS carry 294 MW. Group 3 exports 300 MW to NYPS and imports 560 MW from Group 1, while 640 MW flows from Group 2 to NYPS and 400 MW flows from Group 2 to Group 1. The thick lines indicate the tie lines, which caused the low-frequency inter-area oscillations. The detailed description of the test system including network data and dynamic data for the machines, exciters, thermal turbines and governors, buses, branches can be found in reference [1].

The inter-area modes can be obtained from the eigenvalue analysis and are displayed in Table 3.8. There are four inter-area modes in the whole system. The damp ratio of mode 4 is satisfactory for damping out this mode, and is not required to provide additional damping to this mode. However, the first three modes show a low damping ratio which need further control by global power system stabilizers (GPSSs) using remote signals from PMUs.

Table 3.8 Inter-area modes of the study system.

Mode no.	Frequency (Hz)	Damp ratio (ζ)
Mode 1	0.6887	0.1038
Mode 2	0.5910	0.0533
Mode 3	0.4478	0.1034
Mode 4	0.1703	0.4235

3.4.3 GPSS Based on Collocated Controller Design

3.4.3.1 GPSSs Based on Collocated Control Synthesis

The whole damping control system is illustrated in Figure 3.11, including a two-level structure consisting of local power system stabilizers (LPSSs) based on local signals at the primary level and global power system stabilizers (GPSSs) based on the remote signals at the secondary level. The LPSSs are dedicated to damp local oscillations for guaranteeing generator stability. The GPSSs are used to damp the inter-area oscillations. In this level, GPSSs calculate remote signals through dedicated communication links and send them to excitation systems to provide additional torque, as shown in Figure 3.11. One of the advantages of such a hierarchical structure is that the system's basic operation and stability are preserved when the GPSSs fails.

In the sequential design of the PSS, single-input single-output (SISO) is always considered to enhance the damping of just one critical oscillation mode. In that case, the resulting controller is often conservative and the achievable closed-loop performance is limited. To overcome these limitations of a single-objective synthesis technique, the multiple-input multiple-output (MIMO) is naturally considered for GPSSs design, which can easily incorporate various design specifications [8].

3.4.3.2 Selection of Optimum Locations for GPSSs

As the wide-area information is required only for some oscillatory modes and under specific network configurations, no more than the few PSS sites with the highest controllability of these modes need be involved in the supplementary remote-signal-based actions. This controllability analysis [9] method based on modal control theory of a linear time-invariant system is carried out for identifying the optimum locations for GPSSs. Only the generators with the highest controllability of the first three modes are chosen as GPSSs locations. Table 3.9 shows the normalized controllability for the first three modes. It is revealed that generators G13, G14, and G16 (shown in boldface) are the best locations for those inter-area modes.

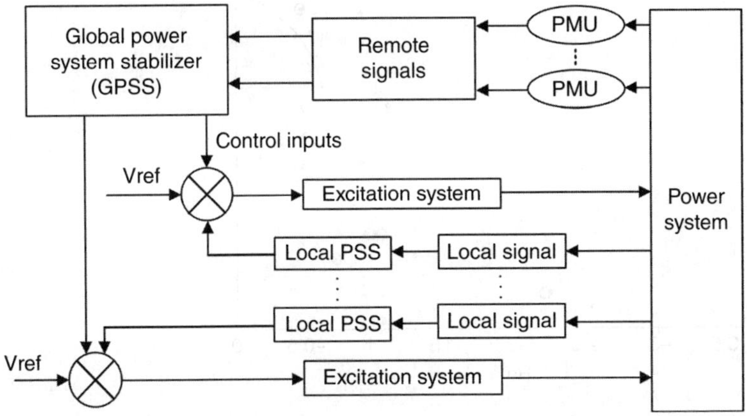

Figure 3.11 GPSSs based on a wide-area controller synthesis configuration.

Table 3.9 Normalized controllability of different generators.

Mode 1		Mode 2		Mode 3	
Generator no.	**Controll-ability**	**Generator no.**	**Controll-ability**	**Generator no.**	**Controll-ability**
13	**1.0000**	**16**	**1.0000**	**14**	**1.0000**
6	0.2315	14	0.5368	15	0.9565
5	0.2077	6	0.0213	13	0.7087
4	0.1786	9	0.0159	16	0.1538
9	0.1692	7	0.0153	6	0.1093
7	0.1563	1	0.0152	12	0.0952

3.4.4 Testing Results and Analysis

3.4.4.1 Frequency Domain Results and Analysis

The frequency domain results and analysis of the study system was carried out in this section. The open- and closed-loop eigenvalues of this large-scale test system are shown in Figure 3.12, which verifies that the collocated control has achieved the goal of moving the critical open-loop eigenvalues outside the disk. The open circle marks represent open-loop eigenvalues, while the dot marks represent closed-loop eigenvalues. Fewer eigenvalues appear in this figure because of the selective modal analysis (SMA) approach [10]. Table 3.10 shows the numerical results with GPSSs based on collocated control. The action of the

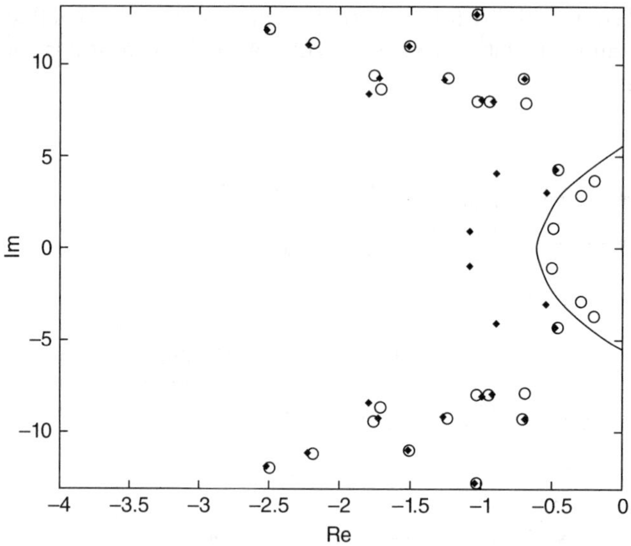

Figure 3.12 Open- and closed-loop eigenvalues for the case of GPSSs: (o) open loop, (·) closed loop.

Table 3.10 Damping ratios and frequencies of open loop and closed loop.

	Open loop		Closed loop	
Mode no.	**Frequency (Hz)**	**Damp ratio**	**Frequency (Hz)**	**Damp ratio**
Mode 1	0.6887	0.1038	0.6883	0.1102
Mode 2	0.5910	0.0533	0.6493	0.2114
Mode 3	0.4478	0.1034	0.4439	0.1763
Mode 4	0.1703	0.4235	0.1517	0.7392

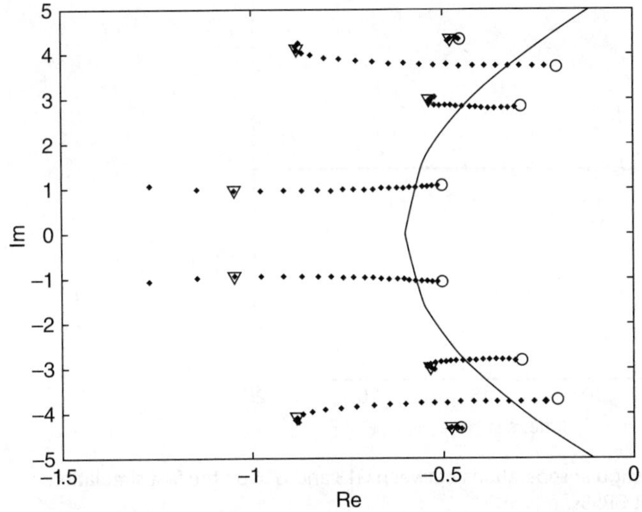

Figure 3.13 The root loci for the case of GPSSs: (o) open loop, (·) root locus.

GPSSs has been found to improve the damping of the first three critical inter-area modes to an adequate level. It is interesting to find that the damp ratio of mode 4 is also improved to a fairly high level.

The root locus with changed feedback gain of eigenvalues near an imaginary axis is shown in Figure 3.13. The open circle marks represent open-loop eigenvalues, while the dot marks represent closed-loop eigenvalues with the increasing feedback gain. Excessive feedback gain should be avoided because it could lead to controller output saturation and a poor large disturbance response of the system. To avoid large feedback gains, the system eigenvalues should not be shifted too far into the left half plane. The triangle marks represent the optimum chosen feedback matrix using collocated control.

3.4.4.2 Time-Domain Results and Analysis

One of the most severe disturbances stimulating poorly damped inter-area oscillations is a three-phase fault in one of the key transmission circuits. To evaluate the performances of the designed GPSSs based on collocated control two nonlinear simulations have been carried out corresponding to some probable fault scenarios.

In the first nonlinear simulation, all the transmission lines are in service and a three-phase fault was applied to the tie line between buses 1 and 2. The fault is cleared by the removal of the faulted circuit after 50 ms (3 cycles). This contingency can excite inter-area oscillations and can be examined for the performance of the damping controllers. Figures 3.14 to 3.17 show the relative angular separations between generators located in separate geographical regions. Inter-area oscillations are mostly manifested in these angular differences and are therefore

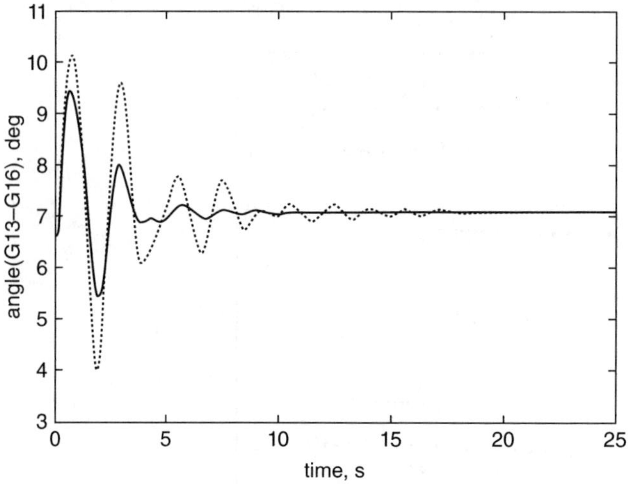

Figure 3.14 The relative angular separations between G13 and G16 on the first simulation: (-) with GPSSs; (--) without GPSSs.

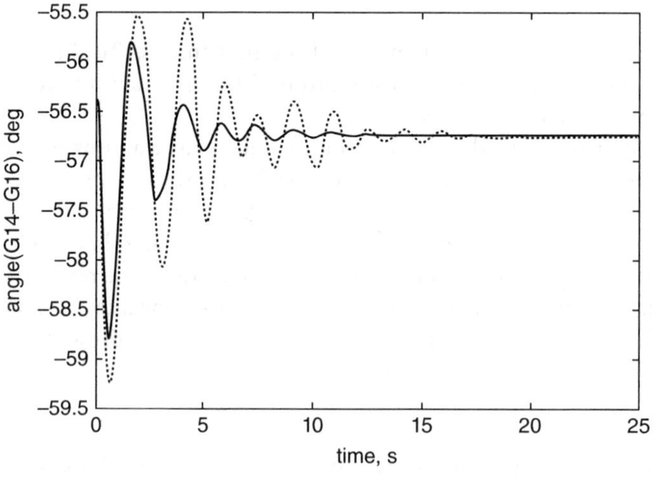

Figure 3.15 The relative angular separations between G14 and G16 on the first simulation: (-) with GPSSs, (--) without GPSSs.

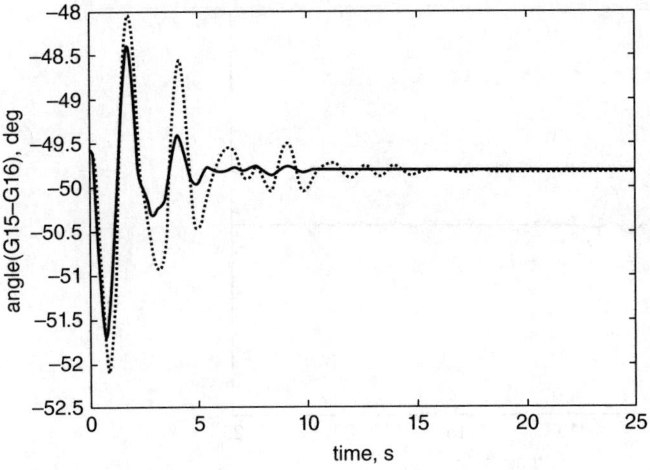

Figure 3.16 The relative angular separations between G15 and G16 on the first simulation: (-) with GPSSs, (--) without GPSSs.

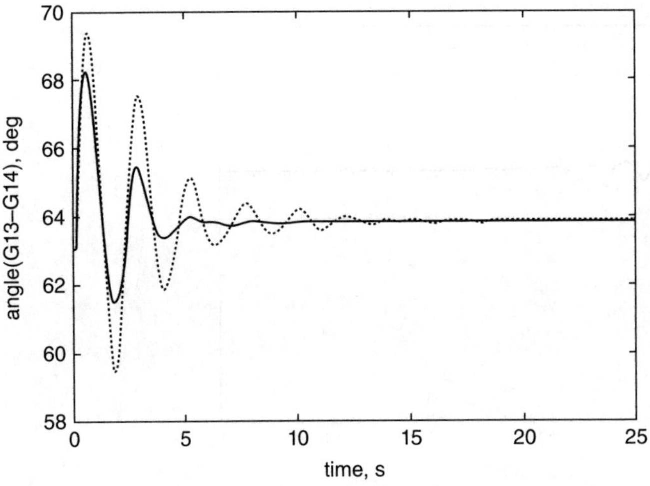

Figure 3.17 The relative angular separations between G13 and G14 on the first simulation: (-) with GPSSs, (--) without GPSSs.

chosen for display. It can be seen that inter-area oscillations settle down within the desired performances specification of 8–10 s for a range of post-fault operating conditions and thus abides by the robustness requirement as well. The simulation results show that GPSSs based on collocated control is able to achieve the desired performance.

In the second nonlinear simulation, tie lines 1–27 are assumed to be out of service and a three-phase fault is simulated for 50 ms in the tie line between buses 1 and 2. The fault was then cleared by taking out the faulted line. Figures 3.18 to 3.21 show the relative angular separations between generators located in separate geographical regions. It can be seen from the responses that

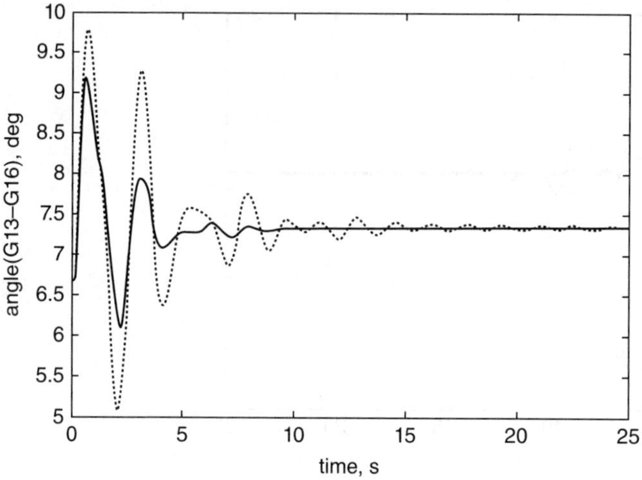

Figure 3.18 The relative angular separations between G13 and G16 on the second simulation: (-) with GPSSs; (--) without GPSSs.

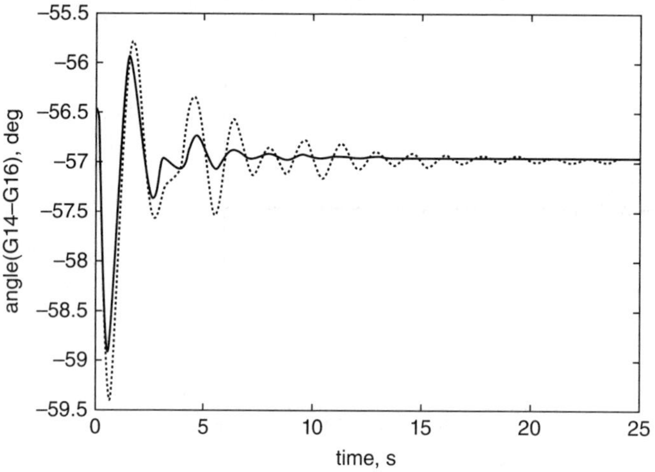

Figure 3.19 The relative angular separations between G14 and G16 on the second simulation: (-) with GPSSs; (--) without GPSSs.

the GPSSs produced additional damping to settle down the inter-area oscillations within 9–11 s. The simulation results illustrate that, even though knowledge of the postdisturbance operating condition is absent, the performance of the GPSSs based on the collocated control scheme is found to be highly acceptable.

In both cases the GPSSs have provided additional damping to settle down the inter-area oscillations within the limited time frame. The control scheme worked satisfactorily following possible disturbances without any prior knowledge about the specific postdisturbance dynamics. The frequency-domain and time-domain results show the potential applicability of collocated control-based GPSSs for large-scale practical power systems.

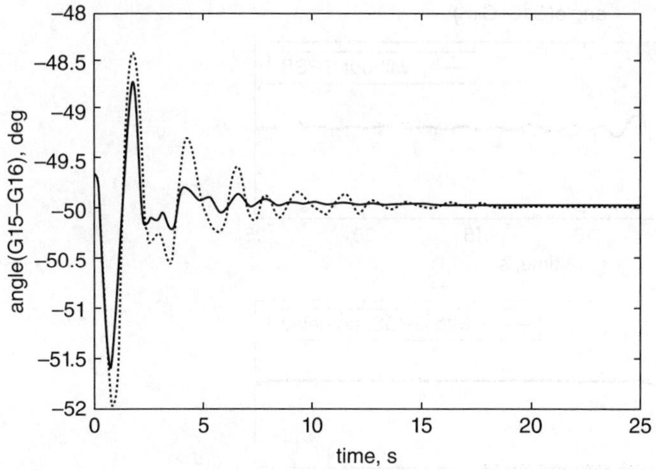

Figure 3.20 The relative angular separations between G15 and G16 on the second simulation: (-) with GPSSs, (--) without GPSSs.

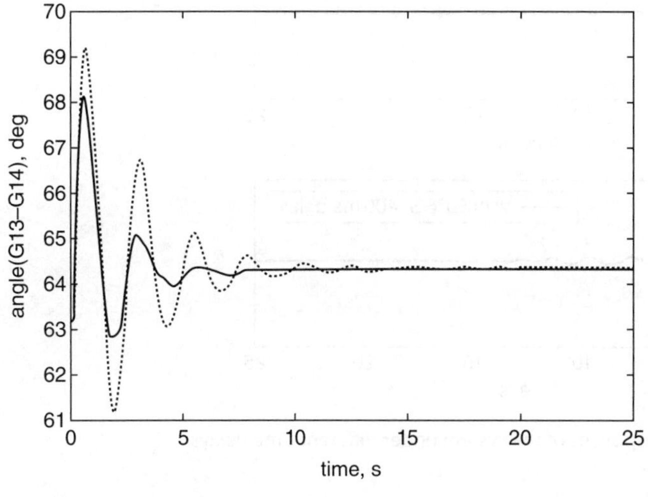

Figure 3.21 The relative angular separations between G13 and G14 on the second simulation: (-) with GPSSs; (--) without GPSSs.

For wide-area control, different time delays of wide-area signals are needed to be considered and tested. Time delays are approximated as a first-order model [11] and nonlinear simulations with 200 ms and 400 ms time delays for the first simulations are carried out to evaluate the control performance with respect to time delays, as shown in Figure 3.22. The simulation results show that relative angular separations between machines G13 and G16 can be settled within the acceptable time frame in cases of different time delays, though the settling time becomes a little longer due to time delays.

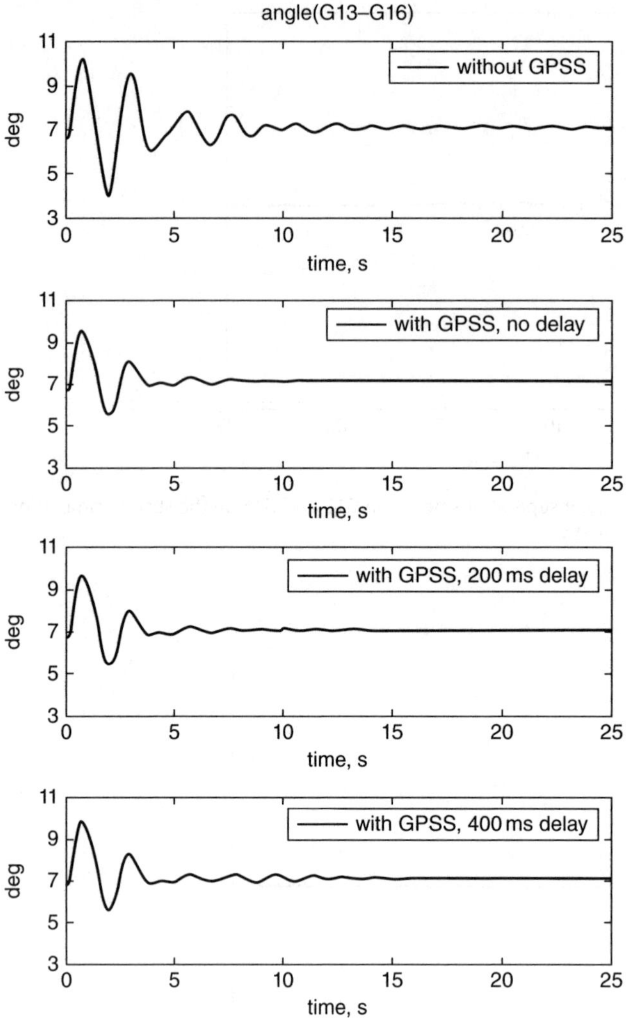

Figure 3.22 Dynamic responses of the system under different time delays.

3.5 Summary

In this chapter, optimal selection of wide-area signals and optimal selection of wide-area controllers are introduced. For the optimal selection of wide-area signals, from the perspective of improving the damping effects of feedback signals, a feedback signal selection method based on contribution factors is introduced. First, the damping effects of different feedback signals on the closed-loop eigenvalue are quantified using the interval extension theory and perturbation theory. On this basis, the concept of contribution factor and contribution factor sum is proposed. Then, according to the contribution factor sum of wide-area signals, the optimal feedback signal is selected. This method could effectively coordinate different types of supplementary controllers and is applicable to different feedback control laws. At the same time, it could improve the damping level of the whole system.

For the optimal selection of wide-area controllers, from the perspective of improving the damping effects of controllers, a coordinating wide-area damping controller optimal selection scheme based on the eigenvalue transfer factor theory is established. This scheme proposes a nonnegative definite eigenvalue transfer factor matrix that could reflect the transfer relationship between multi-dimensional oscillation modes and derives the optimal control input vector that could coordinate the DC supplementary controllers. Then, according to the quantification of the control effects of different controllers by the eigenvalue transfer factor matrix, the optimal wide-area controller is selected. The damping control effect of using the optimal controller is equal to that of using all the controllers. In addition, the proposed scheme is robust when the system operation mode changes.

References

1 Rogers, G. (2000) *Power System Oscillations*, Kluwer Academic Publishers, Dordrecht.
2 Ma, J., Wang, S., Gao, X., and Wang, Z. (2014) Contribution factors based optimal selection of wide-area signals for power system damping control. *Electric Power Components and Systems*, **42**(9), 935–944.
3 Chen, S.H. (1999) *Matrix Perturbation Theory in Structural Dynamic Designs*, Science Press, Beijing.
4 Rao, P.S. and Sen, I. (2000) Robust pole placement stabilizer design using matrix inequalities. *IEEE Transactions on Power Systems*, **15**(1), 313–319.
5 Liu, J., Thorp, J.S., *et al.* (1992) Modal control of large flexible space structures using collocated actuators and sensors. *IEEE Transactions on Automatic Control*, **37**(1), 143–148.
6 Kundur, P. (1994) *Power System Stability and Control*, McGraw-Hill, New York.
7 Ma, J., Wang, T., Wang, Z., *et al.* (2011) Design of global power systems stabilizer to damp interarea oscillations based on wide-area collocated control technique. Power and Energy Society General Meeting.
8 Skogestad, S. and Postlethwaite, I. (2001) *Multivariable Feedback Control*, John Wiley & Sons, Inc., New York.
9 Zhang, Y. and Bose, A. (2008) Design of wide-area damping controllers for interarea oscillations. *IEEE Transactions on Power Systems*, **23**(3), 1136–1143.
10 Verghese, G.C., Perez-Arriage, I.J., and Schweppe, F.C. Selective modal analysis with applications to electric power systems, Part II: The dynamic stability problem. *IEEE Transactions on Power Apparatus and Systems*, **101**(9), 3126–3134.
11 Dotta, D., Silva, E., and Decker, A.S. (2009) Wide-area measurements-based two-level control design considering signal transmission delay. *IEEE Transactions on Power Systems*, **24**(1), 208–216.

4

Stability Analysis of Stochastic System

4.1 Introduction

The traditional power system is a nonlinear dynamic network with various stochastic disturbances such as the stochastic vibration of the prime mover torque, stochastic fluctuation of the load, measurement noise of the control loop, stochastic grid harmonics, and stochastic small-amplitude oscillation of the power angle in a large interconnected power grid, etc. With the continuous development of science and technology, the integration of renewable energy power generation and an electric vehicle has brought more uncertain stochastic disturbances to the power grid. In addition, as the scale of a modern interconnected grid ever increases, the stochastic disturbance caused by load and fault also intensifies and thus the stability of the power system is more easily affected by stochastic disturbances. As traditional stability analysis usually uses deterministic models, it is especially important to study the stability of the power system under stochastic disturbances.

Stochastic disturbances can be classified into three categories: stochastic initial value, stochastic parameter or structure, and stochastic external excitation. The cases of stochastic initial value can be solved using the probability method, so the stability analysis in this section is mainly concerned with the last two categories. The second category of stochastic disturbances can be further classified into stochastic parameter and stochastic structure. Therefore this chapter mainly contains the following three sections:

1) Stability analysis of the system with a stochastic parameter. In low-frequency oscillation modal analysis, an analysis method based on the second-order perturbation theory and an analysis method based on the multiple parameter second-order perturbation sensitivity are introduced. For the stability region, a method designed to calculate the improved small signal stability region (ISSSR) based on the Guardian map is introduced.
2) Stability analysis of the system with a stochastic structure. The stability problems caused by topological structure variations due to an electrical device and component fault are considered. For a single fault, a model-trajectory-based method for analyzing the time-varying oscillation characteristics of fault system is introduced. For multiple faults, a method designed to analyze the

Power System Wide-Area Stability Analysis and Control, First Edition. Jing Ma.

power angle stability of the power system with multiple operating conditions considering cascading failure is introduced.

3) Stability analysis of the system with stochastic excitation. The influence of stochastic fluctuation of a renewable source output power on system stability is discussed and a method designed to analyze the stability of power system with multiple operating conditions considering the stochastic characteristic of wind speed is introduced.

4.2 Stability Analysis of Stochastic Parameter System

4.2.1 Interval Model and Second-Order Perturbation Theory-Based Modal Analysis

Inter-area oscillation modes could be changed by varying the parameters of a large-scale power system. Therefore, complex modal perturbation is proposed in this section in order to make a modal analysis of inter-area oscillations considering uncertainties. First-order perturbation is applicable for the case of small modifications of the parameter. However, if the parameter modification is fairly large, second order perturbation should be used to evaluate the oscillation modes to obtain higher computing accuracy. First, the second-order and first-order perturbations of eigenvalues and eigenvectors are deduced, respectively. Then mode analysis with parameter modification is also investigated using first-order perturbation and second-order perturbation, respectively. Finally, the results demonstrate that second-order perturbation has a much higher accuracy of mode calculation compared with first-order perturbation in the case of large modifications of parameters.

4.2.1.1 Complex Modal Perturbation in Power System

The eigenvalues of an n-machine power system can be obtained by using [1]

$$\mathbf{A}\mathbf{u}_i = \lambda_i \mathbf{u}_i, \qquad \mathbf{w}_i \mathbf{A} = \lambda_i \mathbf{w}_i \tag{4.1}$$

where $\mathbf{A} \in \mathbf{R}^{n \times n}$ is system state matrix, λ_i is the ith eigenvalue, $\mathbf{u}_i$ is the ith right eigenvector with respect to the ith eigenvalue, and $\mathbf{w}_i$ is the ith left eigenvector with respect to the ith eigenvalue.

Assume that a small change of the parameter occurred in the power system, which caused a small change in the system state matrix $\mathbf{A}$, that is,

$$\mathbf{A} = \mathbf{A}_0 + \Delta\mathbf{A} \tag{4.2}$$

where $\mathbf{A}_0$ is the original system state matrix and $\Delta\mathbf{A}$ is the corresponding changes. It is obvious that if the $\Delta\mathbf{A} \to 0$, $\mathbf{A} \to \mathbf{A}_0$. The changes of eigenvalues and eigenvectors of the system is also small while $\Delta\mathbf{A}$ is small. The eigensolutions of (4.2) can be expressed in the form of the power series due to matrix perturbation theory [2,3]:

$$\mathbf{u}_i = \mathbf{u}_{0i} + \mathbf{u}_{1i} + \mathbf{u}_{2i} + \cdots \tag{4.3}$$

$$\lambda_i = \lambda_{0i} + \lambda_{1i} + \lambda_{2i} + \cdots \tag{4.4}$$

where $\mathbf{u}_{0i}$ and λ_{0i} are the eigenvector and eigenvalue of the original system, respectively, λ_{1i} and λ_{2i} are the first- and second-order perturbations of eigenvalues, respectively, and $\mathbf{u}_{1i}$ and $\mathbf{u}_{2i}$ are the first- and second-order perturbations of eigenvectors, respectively.

Substituting (4.2), (4.3), and (4.4) into (4.1) yields

$$(\mathbf{A}_0+\Delta\mathbf{A})(\mathbf{u}_{0i}+\mathbf{u}_{1i}+\mathbf{u}_{2i})=(\lambda_{0i}+\lambda_{1i}+\lambda_{2i})(\mathbf{u}_{0i}+\mathbf{u}_{1i}+\mathbf{u}_{2i}) \tag{4.5}$$

Neglecting the terms of (o^3), then (4.5) can be expressed as follows:

$$o^0:\mathbf{A}_0\mathbf{u}_{0i}=\lambda_{0i}\mathbf{u}_{0i} \tag{4.6}$$

$$o^1:\mathbf{A}_0\mathbf{u}_{1i}+\Delta\mathbf{A}\mathbf{u}_{0i}=\lambda_{0i}\mathbf{u}_{1i}+\lambda_{1i}\mathbf{u}_{0i} \tag{4.7}$$

$$o^2:\mathbf{A}_0\mathbf{u}_{2i}+\Delta\mathbf{A}\mathbf{u}_{1i}=\lambda_{0i}\mathbf{u}_{2i}+\lambda_{1i}\mathbf{u}_{1i}+\lambda_{2i}\mathbf{u}_{0i} \tag{4.8}$$

Based on the expansion theorem, the first-order perturbation $\mathbf{u}_{1i}$ can be expanded by the modal vectors $\mathbf{u}_{0i}$ of the original system:

$$\mathbf{u}_{1i}=\sum_{j=1}^{n}c_{1j}\mathbf{u}_{0j} \tag{4.9}$$

Substituting (4.9) into (4.7) yields

$$\mathbf{A}_0\sum_{j=1}^{n}c_{1j}\mathbf{u}_{0j}+\Delta\mathbf{A}\mathbf{u}_{0i}=\lambda_{0i}\sum_{j=1}^{n}c_{1j}\mathbf{u}_{0j}+\lambda_{1i}\mathbf{u}_{0i} \tag{4.10}$$

The following equation can be obtained by left multiplying the sth left eigenvector $\mathbf{w}_{0s}$ on both sides of (4.10):

$$\sum_{j=1}^{n}c_{1j}\mathbf{w}_{0s}\mathbf{A}_0\mathbf{u}_{0j}+\mathbf{w}_{0s}\Delta\mathbf{A}\mathbf{u}_{0i}=\lambda_{0i}\sum_{j=1}^{n}c_{1j}\mathbf{w}_{0s}\mathbf{u}_{0j}+\lambda_{1i}\mathbf{w}_{0s}\mathbf{u}_{0i} \tag{4.11}$$

By using the orthogonality of left and right eigenvector series conditions the following equations can be obtained:

$$\mathbf{w}_{0s}\mathbf{A}_0\mathbf{u}_{0j}=\delta_{sj}\lambda_{0j} \tag{4.12}$$

$$\mathbf{w}_{0s}\mathbf{u}_{0j}=\delta_{sj} \tag{4.13}$$

$$c_{1s}\lambda_{0s}+\mathbf{w}_{0s}\Delta\mathbf{A}\mathbf{u}_{0i}=c_{1s}\lambda_{0i}+\lambda_{1i}\delta_{is} \tag{4.14}$$

If $i=s$, $\lambda_{0i}=\lambda_{0s}$, (4.15) can be obtained from (4.14):

$$\lambda_{1i}=\mathbf{w}_{0i}\Delta\mathbf{A}\mathbf{u}_{0i} \tag{4.15}$$

If $i\neq s$, $\delta_{is}=0$, (4.16) can be obtained from (4.14):

$$c_{1s}=\frac{1}{\lambda_{0i}-\lambda_{0s}}\mathbf{w}_{0s}\Delta\mathbf{A}\mathbf{u}_{0i} \tag{4.16}$$

If $i = s$ the coefficient c_{1s} can be obtained by the modal normalized condition. From (4.15), if modal vector $\mathbf{u}_{0i}$ of the original system is known, the first-order perturbation of the eigenvalue λ_{1i} can be calculated.

The first-order perturbation is applicable only in the case of a small parameter modification of the system. If the parameter modification is fairly large, the second-order perturbation must be used to obtain higher computing accuracy. Due to the expansion theorem, the second-order perturbation $\mathbf{u}_{2i}$ can be expanded by the modal vector $\mathbf{u}_{0i}$ of the original system as follows:

$$\mathbf{u}_{2i} = \sum_{j=1}^{n} c_{2j}\mathbf{u}_{0j} \tag{4.17}$$

Substituting (4.17) into (4.8) yields

$$\mathbf{A}_0\sum_{j=1}^{n} c_{2j}\mathbf{u}_{0j} + \Delta\mathbf{A}\mathbf{u}_{1i} = \lambda_{0i}\sum_{j=1}^{n} c_{2j}\mathbf{u}_{0j} + \lambda_{1i}\mathbf{u}_{1i} + \lambda_{2i}\mathbf{u}_{0i} \tag{4.18}$$

Equation (4.19) can be obtained by left multiplying the left eigenvector $\mathbf{w}_{0s}$ on both sides of (4.18):

$$\begin{aligned}&\sum_{j=1}^{n} c_{2j}\mathbf{w}_{0s}\mathbf{A}_0\mathbf{u}_{0j} + \mathbf{w}_{0s}\Delta\mathbf{A}\mathbf{u}_{1i}\\ &= \sum_{j=1}^{n} \lambda_{0i}c_{2j}\mathbf{w}_{0s}\mathbf{u}_{0j} + \lambda_{1i}\mathbf{w}_{0s}\mathbf{u}_{1i} + \lambda_{2i}\mathbf{w}_{0s}\mathbf{u}_{0i}\end{aligned} \tag{4.19}$$

Applying the orthogonality conditions of left and right eigenvectors of the state matrix to (4.19) yields

$$c_{2s}\left(\lambda_{0i} - \lambda_{0s}\right) + \lambda_{2i}\delta_{is} = \mathbf{w}_{0s}\Delta\mathbf{A}\mathbf{u}_{1i} - \lambda_{1i}\mathbf{w}_{0s}\mathbf{u}_{1i} \tag{4.20}$$

If $i = s$, $\lambda_{0i} = \lambda_{0s}$, (4.21) can be given from (4.20):

$$\lambda_{2i} = \mathbf{w}_{0i}\Delta\mathbf{A}\mathbf{u}_{1i} - \lambda_{1i}\mathbf{w}_{0i}\mathbf{u}_{1i} \tag{4.21}$$

If $i \neq \mathrm{s}$ and $\delta_{ij} = 1$, (4.22) can be obtained from (4.20):

$$c_{2s} = \frac{1}{\lambda_{0i} - \lambda_{0s}}\left(\mathbf{w}_{0s}\Delta\mathbf{A}\mathbf{u}_{1i} - \lambda_{1i}\mathbf{w}_{0s}\mathbf{u}_{1i}\right) \tag{4.22}$$

For $i = s$, the coefficient c_{2s} can also be obtained by using the normalization condition of the modal vector.

The left eigenvector $\mathbf{w}_i$ and right eigenvectors $\mathbf{u}_i$ should satisfy the normalized condition

$$\mathbf{w}_i\mathbf{u}_i = 1 \tag{4.23}$$

Substituting (4.2) and (4.3) into (4.23) yields

$$\left(\mathbf{w}_{0i} + \mathbf{w}_{1i} + \mathbf{w}_{2i}\right)\left(\mathbf{u}_{0i} + \mathbf{u}_{1i} + \mathbf{u}_{2i}\right) = 1 \tag{4.24}$$

Neglecting the terms (o^3) and equating the same power,

$$o^0 : \mathbf{w}_{0i}\mathbf{u}_{0i} = 1 \tag{4.25}$$

$$o^1 : \mathbf{w}_{0i}\mathbf{u}_{1i} + \mathbf{w}_{1i}\mathbf{u}_{0i} = 0 \tag{4.26}$$

$$o^2 : \mathbf{w}_{0i}\mathbf{u}_{2i} + \mathbf{w}_{1i}\mathbf{u}_{1i} + \mathbf{w}_{2i}\mathbf{u}_{0i} = 0 \tag{4.27}$$

Left multiplying both sides of (4.9) by the left eigenvector $\mathbf{w}_{0s}$ yields

$$\mathbf{w}_{0s}\mathbf{u}_{1i} = \sum_{j=1}^{n} c_{1j}\mathbf{w}_{0s}\mathbf{u}_{0j} \tag{4.28}$$

If $i = s$, (4.29) can be obtained:

$$c_{1i} = \mathbf{w}_{0i}\mathbf{u}_{1i} = \mathbf{w}_{1i}\mathbf{u}_{0i} \tag{4.29}$$

Left multiplying both sides of (4.17) by $\mathbf{w}_{0s}$ yields

$$\mathbf{w}_{0s}\mathbf{u}_{2i} = \sum_{j=1}^{n} c_{2j}\mathbf{w}_{0s}\mathbf{u}_{0j} \tag{4.30}$$

For $i = s$, (4.31) can be obtained from (4.30):

$$c_{2i} = \mathbf{w}_{0i}\mathbf{u}_{2i} = \mathbf{w}_{2i}\mathbf{u}_{0i} \tag{4.31}$$

Using (4.27) and (4.31), the second-order coefficient can be obtained:

$$c_{2i} = -\frac{1}{2}\mathbf{w}_{1i}\mathbf{u}_{1i} \tag{4.32}$$

Therefore, by substituting (4.16) into (4.9), the first-order perturbation of eigenvectors can be obtained:

$$\mathbf{u}_{1i} = \sum_{\substack{s=1 \\ s\neq 1}}^{n} \frac{1}{\lambda_{0i} - \lambda_{0s}} \left(\mathbf{w}_{0s}\Delta\mathbf{A}\mathbf{u}_{0i}\right)\mathbf{u}_{0s} \tag{4.33}$$

By substituting (4.22) and (4.32) into (4.9), the second-order perturbation of eigenvectors can be obtained:

$$\mathbf{u}_{2i} = \sum_{\substack{s=1 \\ s\neq i}}^{n} \frac{1}{\lambda_{0i} - \lambda_{0s}} \left(\mathbf{w}_{0s}\Delta\mathbf{A}\mathbf{u}_{1i} - \lambda_{1i}\mathbf{w}_{0s}\mathbf{u}_{1i}\right)\mathbf{u}_{0s} - \frac{1}{2}\left(\mathbf{u}_{1i}^{\mathrm{T}}\mathbf{u}_{1i}\right)\mathbf{u}_{0s} \tag{4.34}$$

By substitute (4.15) and (4.21) into (4.4), the eigenvalues of the updated system can be calculated, and by substituting (4.33) and (4.34) into (4.3), the eigenvectors of the updated system can be obtained.

It should be noted that the first- and second-order perturbation expressions can not be applied directly to deal with the reanalysis of modes with close eigenvalues. The reasons were that the expressions (4.33) and (4.34) involve division by $\lambda_{0i} - \lambda_{0s}$.

4.2.1.2 Example Test System

A 16-machine system with 86 transmission lines and 68 buses, shown in Figure 3.2, is considered for studying the application of the first- and second-order perturbations. All the generators use the subtransient model. IEEE type DC1 excitation systems are installed in machines 1–9. Machine 16 is the reference and the WECC load model is assumed at all load buses. The real power is 80% constant current, the reactive power is 80% constant admittance, and all loads have the remaining 20% as an induction motor load. The tie lines connecting NETS and NYPS carry 720 MW. Group 3 exports 1163 MW to NYPS and imports 73 MW from Group 1 while 266 MW flows from Group 2 to NYPS and 75.8 MW from Group 2 to Group 1.

The thick lines indicate the tie lines, which caused the low-frequency inter-area oscillations. The inter-area modes can be obtained from the eigenvalue analysis and are displayed in Table 4.1. There are four inter-area modes in the whole system.

A compass plot of the rotor angle state terms of the eigenvector is shown in Figure 4.1. The eigenvector associated with a mode indicates the relative changes in the states that would be observed when that mode of oscillation is excited. It will not only demonstrate the performances of machines under different modes but also can exhibit the relative phase differences between different groups. It can be seen from Figure 4.1 that generators 14, 15, and 16 are oscillating against generators 1 to 13 under mode 1, generator 14 is oscillating against generator 16 under mode 2, generators 12 and 13 are oscillating against generators 1 to 9 under mode 3, and generators 14 and 16 are oscillating against generators 15 under mode 4.

4.2.1.3 Testing Results and Analysis

As an application of the matrix perturbation method, numerical examples are given as follows. Considering the system shown in Figure 3.2, the purpose of this example is to study the application range of the first- and second-order perturbations. In the process of operating the power system, the inertia constants and transient reactance of equivalents will change due to some generators outage or being put into service. Also, there are many researches indicating that the inertia

Table 4.1 Inter-area modes of the study system.

Mode no.	Eigenvalue	Frequency (Hz)	Damp ratio (ζ)
Mode 1	$-0.0251 \pm 2.4779i$	0.3944	0.0101
Mode 2	$-0.0719 \pm 3.2109i$	0.5110	0.0224
Mode 3	$-0.1190 \pm 4.1347i$	0.6581	0.0288
Mode 4	$-0.1677 \pm 4.9721i$	0.7913	0.0337

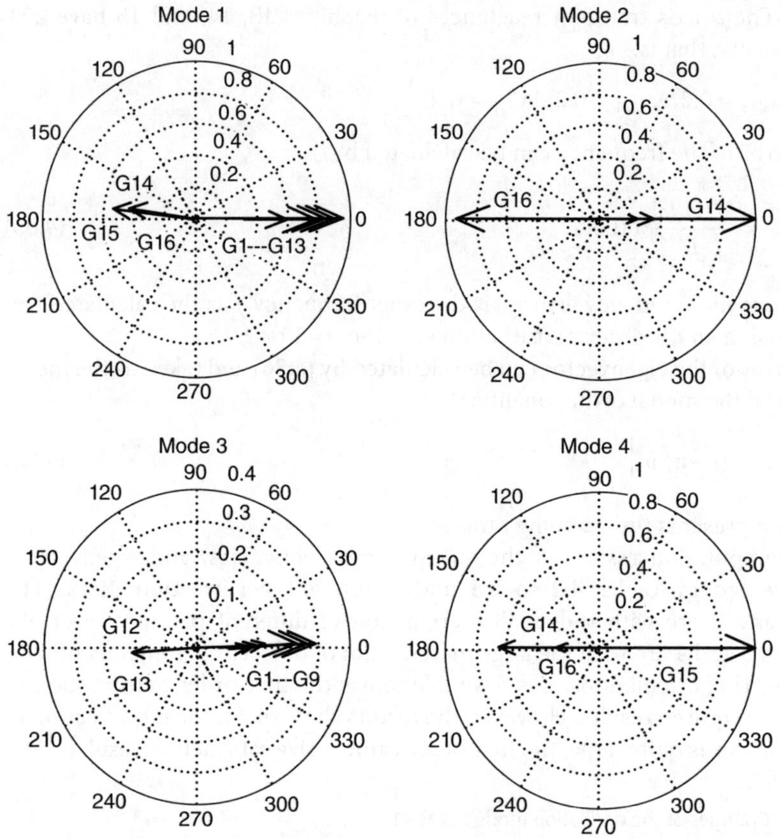

Figure 4.1 Oscillation modal diagram.

constants and transient reactance have a greater impact on the changes of oscillation modes. In order to study the computing accuracy of the first-[t] and second-order perturbations, inertia constants and transient reactance will be chosen as variables to be perturbed with consideration of the reality. Four perturbation cases are as follows:

Case 1: The inertia constants of machines 14, 15, and 16 have 5% increments, that is,

$$\Delta M_{14} = 0.05, \Delta M_{15} = 0.05, \Delta M_{16} = 0.05$$

Case 2: The *d*-axis transient reactances of machines 13, 14, and 15 have 5% increments, that is,

$$\Delta x'_{d13} = 0.05, \Delta x'_{d14} = 0.05, \Delta x'_{d15} = 0.05$$

Case 3: The inertia constants of machines 14, 15, and 16 have 20% increments, that is,

$$\Delta M_{14} = 0.2, \Delta M_{15} = 0.2, \Delta M_{16} = 0.2$$

Case 4: The d-axis transient reactances of machines 13, 14, and 15 have 20% increments, that is,

$$\Delta x'_{d13} = 0.2, \Delta x'_{d14} = 0.2, \Delta x'_{d15} = 0.2$$

The error of the frequency can be calculated by

$$f_{err} = \frac{|f_i - f_0|}{f_0} \times 100\% \tag{4.35}$$

where f_{err} is the frequency error, f_0 is the exact frequency, f_i is the calculated frequency using an ith-order perturbation solution, $i = 1$ or 2.

The error of the eigenvector can be calculated by (4.36) and taken as the measurement of the modal orthogonality:

$$\mathbf{u}_{ierr} = \left|1 - \mathbf{u}_i^T \mathbf{u}_i\right| \times 100\% \tag{4.36}$$

where ξ_i represents the damping ratio.

The computation results of the eigenvalues, frequencies, and eigenvectors are presented in Tables 4.2 to 4.5 under different perturbation cases. The cases 1 and 2 are with the small incremental of different parameters of 5%. The cases 3 and 4 are with the large incremental of different parameters of 20%. In engineering calculations, it is desirable to use the first-order perturbation to obtain the required results. However, the results show that if the modification of the parameter is quite large, the first order cannot give satisfactory results.

Table 4.2 Changes of the oscillation mode in case 1.

Mode no.		1	2	3	4
Exact solution	λ_0	−0.0192± 2.4404i	−0.0732± 3.1359i	−0.1371± 4.2281i	−0.1674± 4.8518i
	f_0	0.3884	0.4991	0.6729	0.7722
	ξ_0	0.0079	0.0233	0.0324	0.0345
First-order solution	λ_1	−0.0193± 2.4405i	−0.0732± 3.1369i	−0.1371± 4.2280i	−0.1674± 4.8533i
	f_1	0.3884	0.4993	0.6729	0.7724
	ξ_1	0.0079	0.0233	0.0324	0.0345
	f_{err1}	0.499e-4	3.024e-4	0.108e-4	3.008e-4
	u_{ierr}	0.294%	0.601%	0.2%	0.59%
Second-order solution	λ_2	−0.0193± 2.4404i	−0.0732± 3.1360i	−0.1371± 4.2281i	−0.1674± 4.8519i
	f_2	0.3884	0.4991	0.6729	0.7722
	ξ_2	0.0079	0.0233	0.0345	0.0345
	f_{err2}	1.825e-6	5.244e-6	0.677e-6	7.071e-6
	u_{ierr}	0.097%	0.045%	0.2%	0.083%

Table 4.3 Changes of the oscillation mode in case 2.

Mode no.		1	2	3	4
Exact solution	λ_0	−0.0211± 2.4571i	−0.0748± 3.2023i	−0.1372± 4.2285i	−0.1687± 4.9682i
	f_0	0.3911	0.5097	0.6730	0.7907
	ξ_0	0.8605%	2.3352%	3.2433%	3.3928%
First-order solution	λ_1	−0.0211± 2.4573i	−0.0748± 3.2029i	−0.1372± 4.2285i	−0.1686± 4.9684i
	f_1	0.3911	0.5097	0.6730	0.7907
	ξ_1	0.8606%	2.3342%	3.2433%	3.3920%
	f_{err1}	0.937e-4	1.758e-4	0.042e-4	0.442e-4
	u_{ierr}	0.81%	0.95%	0.088%	0.16%
Second-order solution	λ_2	−0.0212± 2.4571i	−0.0748± 3.2023i	−0.1372 ± 4.2285i	−0.1686± 4.9682i
	f_2	0.3911	0.5097	0.6730	0.7907
	ξ_2	0.8609%	2.3340%	3.2433%	3.3914%
	f_{err2}	1.489e-6	0.008e-6	0.063e-6	0.238e-6
	u_{ierr}	0.058%	0.09%	0.008%	0.003%

Table 4.4 Changes of the oscillation mode in case 3.

Mode no.		1	2	3	4
Exact solution	λ_0	−0.0176± 2.4218i	−0.0731± 3.0645i	−0.1370± 4.2274i	−0.1671± 4.7398i
	f_0	0.3854	0.4877	0.6728	0.7544
	ξ_0	0.7286%	2.3831%	3.2395%	3.5233%
First-order solution	λ_1	−0.0178± 2.4223i	−0.0730± 3.0680i	−0.1370± 4.2273i	−0.1671± 4.7453i
	f_1	0.3855	0.4883	0.6728	0.7552
	ξ_1	0.7328%	2.3797%	3.2397%	3.5198%
	f_{err1}	0.002	0.0011	0	0.0011
	u_{ierr}	5.67%	2.28%	0.39%	5.01%
Second-order solution	λ_2	−0.0178± 2.4218i	−0.0730± 3.0646i	−0.1370± 4.2274i	−0.1671± 4.7401i
	f_2	0.3854	0.4877	0.6728	0.7544
	ξ_2	0.7370%	2.3816%	3.2396%	3.5241%
	f_{err2}	1.387e-5	3.841e-5	0.448e-5	5.182e-5
	u_{ierr}	0.81%	0.426%	0.39%	0.549%

Table 4.5 Changes of the oscillation mode in case 4.

Mode no.		1	2	3	4
Exact solution	λ_0	−0.0219± 2.4457i	−0.0791± 3.1708i	−0.1374± 4.2273i	−0.1713± 4.9562i
	f_0	0.3893	0.5046	0.6728	0.7888
	ξ_0	0.8937%	2.4945%	3.2488%	3.4552%
First-order solution	λ_1	−0.0223± 2.4487i	−0.0796± 3.1777i	−0.1374± 4.2275i	−0.1712± 4.9589i
	f_1	0.3897	0.5057	0.6728	0.7892
	ξ_1	0.9102%	2.5057%	3.2489%	3.4507%
	f_{err1}	0.001	0.0022	0.0000	0.0005
	u_{ierr}	2.94%	3.25%	0.32%	6.0%
Second-order solution	λ_2	−0.0227± 2.4459i	−0.0801± 3.1708i	−0.1374± 4.2273i	−0.1711± 4.9561i
	f_2	0.3893	0.5046	0.6728	0.7888
	ξ_2	0.9287%	2.5265%	3.2493%	3.4497%
	f_{err2}	7.539e-5	0.048e-5	0.288e-5	1.033e-5
	u_{ierr}	0.352%	0.088%	0.24%	0.26%

When the changes of the parameters are 5% as shown in cases 1 and 2, the first-order perturbation solution has errors of around 10^{-4} for frequencies and errors of around 10^{-3} for eigenvectors. The second-order perturbation solution has average errors of around 10^{-6} for frequencies and errors of around 10^{-4} for eigenvectors. Under this situation, the first-order perturbation is effective, whereas the second-order perturbation gives more excellent results, as shown in Tables 4.2 and 4.3. The estimate of the error caused by uncertainty verifies the accuracy of second-order perturbation applied in power system oscillation modes evaluation.

Furthermore, if the changes to system parameters reach 20%, as shown in cases 3 and 4, as can be seen from the results shown in Tables 4.4 and 4.5, the first-order perturbation solution has errors around 10^{-3} for frequencies and errors around 10^{-2} for eigenvectors. The second-order perturbation solution has errors around 10^{-5} for frequencies and errors around 10^{-3} for eigenvectors. Therefore, if the changes of system parameters are fairly large, the first-order perturbation cannot be used, but the second-order perturbation is still effective.

4.2.2 Power System Small-Signal Stability Region Calculation Method Based on the Guardian Map Theory

In view of the potential problems in the system, such as sustained oscillation and a low damping ratio, which are not considered in traditional SSSR, an improved small-signal stability region (ISSSR) calculation method based on the guardian map is proposed. First, according to the map theory the system is mapped from the ISSSR to the negative half-plane. Then by a direct sum operation the guardian map is constituted. The guardian map approach, which is able to solve the exact stability region of the Hurwitz matrix, is used for fast and accurate calculation of the boundary of ISSSR. In addition, the influence of the generator excitation system parameters on the ISSSR is also analyzed, which is highly valuable in system dispatching instructions and small disturbance instability prevention.

4.2.2.1 Improved Small-Signal Stability Region (ISSSR)

The linearized system model at the equilibrium point can be expressed as differential equations [4]:

$$\begin{cases} \Delta\dot{\boldsymbol{x}} = \boldsymbol{A}(\boldsymbol{p})\Delta\boldsymbol{x} + \boldsymbol{B}(\boldsymbol{p})\Delta\boldsymbol{y} \\ \boldsymbol{0} = \boldsymbol{C}(\boldsymbol{p})\Delta\boldsymbol{x} + \boldsymbol{D}(\boldsymbol{p})\Delta\boldsymbol{y} \end{cases} \tag{4.37}$$

where $\boldsymbol{x}$ is the system state variable, $\boldsymbol{y}$ is the system algebra variable, and $\boldsymbol{p}$ is the parameter variable. By eliminating $\boldsymbol{\Delta y}$, the reduced-order linear differential equations can be gained:

$$\Delta\dot{\boldsymbol{x}} = \tilde{\boldsymbol{A}}(\boldsymbol{p})\Delta\boldsymbol{x} \tag{4.38}$$

$$\tilde{\boldsymbol{A}}(\boldsymbol{p}) = \boldsymbol{A}(\boldsymbol{p}) - \boldsymbol{B}(\boldsymbol{p})\boldsymbol{D}^{-1}(\boldsymbol{p})\boldsymbol{C}(\boldsymbol{p}) \tag{4.39}$$

According to Lyapunov stability theory [5], when the system operates in a stable fashion, the real parts of the eigenvalues of $\boldsymbol{A(p)}$ should all be negative. When $\boldsymbol{p}$ changes, if there is a real eigenvalue passing through the imaginary axis from the origin, then the system is affected by saddle node bifurcation (SNB); if there is a pair of complex conjugate eigenvalues passing through the imaginary axis, then the system is affected by Hopf bifurcation (HB). If the algebra Jacobian matrix $\boldsymbol{D}$ is irreversible, then the system is affected by singularity induced bifurcation (SIB). Thus, the small-signal stability region (SSSR) of the power system can be defined as

$$\begin{aligned} &\Omega\text{SSSR} = \{\boldsymbol{p} \mid \textbf{\textit{the real parts of eigenvalues of }} \tilde{\boldsymbol{A}}(\boldsymbol{p}) \\ &\textbf{\textit{are negative and }} \boldsymbol{D}(\boldsymbol{p}) \textbf{\textit{ is reversable}}\} \end{aligned} \tag{4.40}$$

The boundary of the SSSR is the union sets of the three kinds of bifurcation points above:

$$\partial\Omega SSSR = \left\{\overline{\text{SNB} \cap \text{SIB} \cap HB}\right\} \tag{4.41}$$

Traditional SSSR is good at judging whether the system is unstable. However, for operating points with a low damping ratio and a small oscillation attenuation coefficient, though being theoretically stable, they are so close to the unstable status that, when a disturbance occurs, they are liable to oscillate due to small-oscillation attenuation. In real practice the system should avoid operating under such conditions. Therefore, the concept of an improved small-signal stability region (ISSSR) is put forward. Suppose the system eigenvalue vector gained from $\tilde{\boldsymbol{A}}(\boldsymbol{p})$ in (4.39) is

$$\lambda = [\lambda 1, \lambda 2, \ldots, \lambda i, \ldots, \lambda n,] \tag{4.42}$$

$$\lambda i = \alpha i \pm j\Omega i \tag{4.43}$$

where α_i is the attenuation coefficient and Ω_i is the actual oscillation frequency. Define the system damping ratio as ξ_i:

$$\xi i = \frac{-\alpha}{\sqrt{\alpha^2 + \Omega^2}} \tag{4.44}$$

The ISSSR is defined as the set of system operating points that meet the following requirements:

1) The real parts of the system eigenvalues, that is, the attenuation coefficient $\alpha_i \leq -\varepsilon_1$, where ε_1 is a positive real number.
2) The system damping ratio $\xi_i \geq \varepsilon_2$, where ε_2 is a positive real number.

The boundary of the ISSSR on the complex plane is determined by the above requirements, shown in Figure 4.2, where $\theta = \arcsin \varepsilon_2$.

The boundary determined by requirement (4.37) ensures that the real parts of the system eigenvalues are smaller than $-\varepsilon_1$, so that SNB and SIB will not occur, and has a relatively fast oscillation attenuation speed. The boundary determined by requirement (4.38) ensures that the damping ratio of every operating point in the stability region is bigger than ε_2.

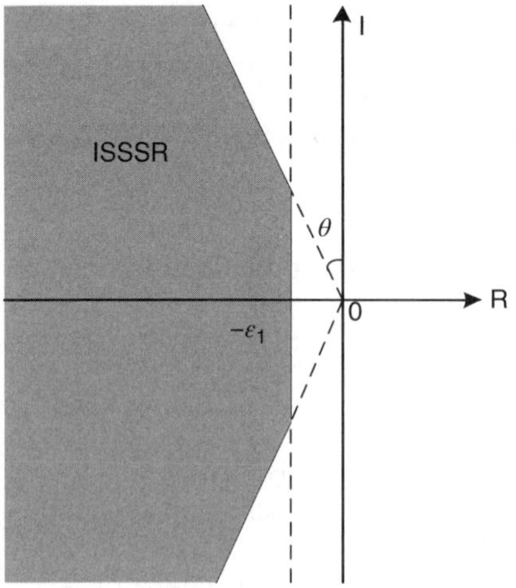

Figure 4.2 Improved small-signal stability region (ISSSR) and its complex plane boundary.

4.2.2.2 The Stability Region Calculation Method Based on Guardian Map Theory

Theoretical Foundation In order to calculate the boundary of ISSSR, the guardian map theory is introduced, which transforms the stability problem of a matrix into the nonsingular problem of its correlation matrix [6].

Definition 4.1 (guardian map). Suppose $\mathbf{S} \subseteq \mathbf{R}^{n\times n}$ is an open interval. If there is a map ν: $\mathbf{R}^{n\times n} \to \mathbf{R}$ that enables $\nu\ (\boldsymbol{A}) \neq 0$ for all $\boldsymbol{A} \in \mathbf{S}$ and $\nu\ (\boldsymbol{A}) = 0$ for all $\boldsymbol{A} \in \partial\mathbf{S}$ ($\partial\mathbf{S}$ is the boundary of $\mathbf{S}$), then ν is called the guardian map of $\mathbf{S}$ [7].

It can be seen from Definition 4.1 that there are different ways to constitute a guardian map. Since the direct sum operation is fast and easy, it is used in this section to constitute the guardian map. To this end, the following lemma is introduced.

Lemma 4.1 If $\boldsymbol{A} \in \mathbf{R}^{n\times n}$ and $\boldsymbol{B} \in \mathbf{R}^{m\times m}$, then $mspec(\boldsymbol{A} \oplus \boldsymbol{B}) = \{\lambda_i + \mu_j\text{: } \lambda_i \in mspec(\boldsymbol{A}), \mu_j \in mspec(\boldsymbol{B}), i = 1,2,\ldots,n; j = 1,2,\ldots,m\}$, where *mspec* means the set of eigenvalues of the matrix [8].

According to Lemma 4.1, for a given matrix $\boldsymbol{A} \in \mathbf{R}^{n\times n}$, define $\bar{\boldsymbol{A}} = \boldsymbol{A} \oplus \boldsymbol{A}$ if $\boldsymbol{A}$ is a Hurwitz matrix. Then the following conclusions can be derived:

1) $\bar{\boldsymbol{A}}$ is also a Hurwitz matrix.
2) $det\bar{\boldsymbol{A}} \neq 0$.

Based on the above theories, for a Hurwitz matrix set $\mathbf{S}$, if $\boldsymbol{A} \in \mathbf{S}$, then $\bar{\boldsymbol{A}} \in \mathbf{S}$, i.e., $\nu\ (\boldsymbol{A}) \neq 0$; if $\boldsymbol{A}$ is on the boundary of $\mathbf{S}$, then $\boldsymbol{A} = 0$ and $\bar{\boldsymbol{A}} = 0$, i.e., for all $\boldsymbol{A} \in \partial\mathbf{S}$, $\nu\ (\boldsymbol{A}) = 0$. Therefore, the direct sum operation is able to constitute the guardian map of the Hurwitz matrix $\boldsymbol{A}$:ν_1:$\mathbf{R}^{n\times n} \to \mathbf{R}$, as shown by

$$\nu_1(\boldsymbol{A}) = det(\boldsymbol{A} \oplus \boldsymbol{A}) \tag{4.45}$$

Stability Region Calculation Method of Single Parameter System The single parameter linearized system at the equilibrium point can be expressed as differential equations:

$$\begin{cases} \dot{\boldsymbol{x}} = \boldsymbol{A}(\rho)\boldsymbol{x} \\ \boldsymbol{A}(\rho) = \boldsymbol{A}_0 + \boldsymbol{A}g \end{cases} \tag{4.46}$$

where $\boldsymbol{x}$ is the system state variable, ρ is the disturbance parameter, $\boldsymbol{A}_0$ is the system state matrix before the disturbance, and $\boldsymbol{A}_g$ is the variation of the system state matrix caused by the disturbance.

To transform the stability problem of the original matrix $\boldsymbol{A}_0$ in (4.46) into the nonsingular problem of its correlation matrix, the order set $\mathcal{B}(\boldsymbol{M})$ is defined.

Definition 4.2 For a given matrix $\boldsymbol{M} \subseteq \mathbf{R}^{n\times n}$, suppose $\tilde{\lambda}_i(\boldsymbol{M})$, $i = 1,2,\ldots,p$, represent the real parts of the nonzero eigenvalues of $\boldsymbol{M}$, and $r_0 = -\infty$, $r_{p+1} = +\infty$, $r_i = -1/\tilde{\lambda}_i(\boldsymbol{M})$, and $i = 1,2,\ldots,p$. Then the order set $\mathcal{B}(\boldsymbol{M}) = \{r_0, r_1, r_2, \ldots, r_p, r_{p+1}\}$, where $r_i < r_{i+1}$.

Through the guardian map, the correlation matrix of A_0 can be found and then the stability range of the disturbance parameter ρ can be obtained by applying B(M) (see Theorem 4.1 and Proof for details).

Theorem 4.1 Suppose A_0, $A_g \in \mathbf{R}^{n\times n}$ and $det(A_0 \oplus A_0) \neq 0$. Define $\overline{A_0} := A_0 \oplus A_0$ and $\overline{A_g} := A_g \oplus A_g$. If there is a real number $\rho_i \in (r_i, r_{i+1})$(where r_i and r_{i+1} are successive elements in $\mathcal{B}\left(\overline{A_0}^{-1}\overline{A_g}\right)$) that enables $A_0 + \rho_i A_g$ to be a Hurwitz matrix, then for all $\rho \in (r_i, r_{i+1})$, $A_0 + \rho A_g$ is a Hurwitz matrix [7].

Proof. The map $\nu_1 : \mathbf{R}^{n\times n} \to \mathbf{R}$ in (4.9) is a guardian map of the $n \times n$ Hurwitz matrix A. Let $A(\rho) = A_0 + \rho A_g$, according to the definition of $\mathcal{B}\left(\overline{A_0}^{-1}\overline{A_g}\right)$, if r_i, r_{i+1} are the noninfinite successive elements of $\mathcal{B}\left(\overline{A_0}^{-1}\overline{A_g}\right)$. Then $\nu_1(A(r_i)) = 0$, $\nu_1(A(r_{i+1})) = 0$, and for all $r_i < \rho < r_{i+1}$, $\nu_1(A(\rho)) \neq 0$. Meanwhile, since ν_1 is a guardian map, then for all $r_i < \rho < r_{i+1}$, $A(\rho)$ is a Hurwitz matrix.

For any disturbance parameter ρ, the stability criterion of the system is given in Theorem 4.1. As for the stability region of the system concerning ρ, it can be gained through Theorem 4.2.

Theorem 4.2 For given matrices A_0, $A_g \in \mathbf{R}^{n\times n}$, supppose $det(A_0 \oplus A_0) \neq 0$. Define $\overline{A_0} := A_0 \oplus A_0$, $\overline{A_g} := A_g \oplus A_g$, index set $\Gamma = \{i$: when $\rho_i \in (r_i, r_{i+1})$, $A_0 + \rho_i A_g$ is Hurwitz matrix$\}$ and the open interval set:

$$\Omega_{\in} = \bigcup_{\Gamma} (r_i, r_{i+1}) \tag{4.47}$$

where r_i and r_{i+1} are successive elements in $\mathcal{B}\left(\overline{A_0}^{-1}\overline{A_g}\right)$ and $i \in \Gamma$; then $A_0 + \rho A_g$ is a Hurwitz matrix when and only when $\rho \in \mathbf{\Omega}_{\in}$.

Proof. To prove sufficiency, choose $\rho \in \mathbf{\Omega}_{\in}$ and let $\rho \in (r_i, r_{i+1})$ for some $i \in \Gamma$. From Theorem 4.1 and the fact that $A_0 + \rho_i A_g$ is Hurwitz for $\rho_i \in (r_i, r_{i+1})$, it follows that $A_0 + \rho A_g$ is Hurwitz. To prove necessity, assume that $A_0 + \rho A_g$ is Hurwitz. It follows that $\rho \notin \mathcal{B}\left(\overline{A_0}^{-1}\overline{A_g}\right)$. Therefore, there exists $i \in \{0, 1, \ldots, p-1, p\}$ such that $r_i < \rho < r_{i+1}$. Since $A_0 + \rho A_g$ is Hurwitz, it follows that $i \in \Gamma$. Hence, $\rho \in (r_i, r_{i+1}) \subseteq \mathbf{\Omega}_{\in}$.

The range $\mathbf{\Omega}_{\in}$ gained in Theorem 4.2 is the stability region of the single parameter system.

Stability Region Calculation Method of Multiparameter System The multiparameter system linearized at the equilibrium point can be expressed by

$$\begin{cases} \dot{\boldsymbol{x}} = A(\rho_1, \rho_2, \ldots, \rho_m)\boldsymbol{x} \\ A(\rho_1, \rho_2, \ldots, \rho_m) = A_0 + \sum_{i=1}^{m} A_{g,i} \end{cases} \tag{4.48}$$

Similar to the single parameter system, $\boldsymbol{x}$ is the system state variable, $\rho_1, \rho_2, \ldots, \rho_m$ are the disturbance parameters, A_0 is the system state matrix before the disturbances, and $A_{g,i}$ is the variation of the system state matrix caused by disturbance ρ_i.

To obtain the stability region of the multiparameter system in (4.48), the polar coordinate conversion method in Lemma 4.2 is introduced.

Lemma 4.2 For a given vector $(\rho_1,\rho_2,\ldots,\rho_k)^T \in \mathbf{R}^k$, $k \geq 2$, there is always a real number r and $k-1$ scalar quantities $\theta_i \in [0, \pi)$ that enable

$$(\rho_1,\rho_2,\ldots,\rho_k)^T = r\omega(\theta) \tag{4.49}$$

where $\theta = (\theta_2,\ldots,\theta_k)^T \in \left[0, \pi\right)^{k-1}$ and

$$\begin{aligned}\omega(\theta) = (&\cos\theta_2, \sin\theta_2\cos\theta_3, \sin\theta_2\sin\theta_3\cos\theta_4,\\ &\ldots, \sin\theta_2\sin\theta_3\ldots\sin\theta_{k-1}\cos\theta_k,\\ &\sin\theta_2\sin\theta_3\ldots\sin\theta_k)^T \in [-1,1]^k\end{aligned} \tag{4.50}$$

After the polar coordinate conversion in Lemma 4.2, the stability region calculation method for the multiparameter system based on the guardian map can be gained by theintroduction of Theorem 4.3.

Theorem 4.3 Suppose there is a given A_0 and the variation $A_{g,i} \in \mathbf{R}^{n \times n}$, $i \in \{1,\ldots, k-1, k\}$ caused by a certain parameter ρ_i, $det(A_0 \oplus A_0) \neq 0$.

Define $\overline{A_0} := A_0 \oplus A_0$, $(\rho_1,\rho_2,\ldots,\rho_k)^T = r\omega(\theta)$ and $Ag(\theta) = \sum_{i=1}^{k} A_{g,i}\omega_i(\theta)$, where $\omega_i(\theta)$ is the element in the vector $\omega(\theta)$. $\overline{A_g(\theta)} = A_g(\theta) \oplus A_g(\theta)$. Define the following set:

$$\Omega_{\in}(\theta) = \bigcup_{i \in \Gamma(\theta)} (r_i, r_{i+1}) \tag{4.51}$$

where r_i and r_{i+1} are successive elements in $\mathcal{B}\left(\overline{A_0}^{-1}\overline{A_g(\theta)}\right)$, $i \in \{0, 1,\ldots,p-1, p\}$. $\Gamma(\theta)$ is an index set: $\Gamma(\theta) = \{i$: For $r_i' \in (r_i, r_{i+1})$, $A_0 + r_i' A_g(\theta)$ is a Hurwitz matrix$\}$. $\Omega_{\in}(\theta)$ is the system stability region with θ dimensions. The stability region of the multiparameter system is further defined as

$$\Omega_{\in}' = \left\{ \begin{array}{l} \bigcup_{\theta \in [0,\pi)^{k-1}} \{y(\theta) \in \mathbf{R}^k : y(\theta) = r\omega(\theta), \\ r \in \Omega_{\in}(\theta)\} \end{array} \right\} \tag{4.52}$$

so that only when $(\rho_1,\rho_2,\ldots,\rho_k)^T \in \Omega_{\in}'$ is $A_0 + \sum_{i=1}^{k} \rho_i A_{g,i}$ a Hurwitz matrix. Therefore $\Omega_{\in}(\theta)$ in Theorem 4.3 is the stability region of the multiparameter system.

In conclusion, for a given system with the initial operating state, the ISSSR of the system can be gained using the following steps:

Step 1. Form A_0 according to the current operating state of the system and decide the disturbance parameters, thus finding the matrix $A_g(\theta)$.

Step 2. Calculate $\overline{A_0}$, $\overline{A_g(\theta)}$, and the eigenvalues of the matrix $\overline{A_0}^{-1}\overline{A_g(\theta)}$.

Step 3. According to Definition 4.2, choose the unrepeated eigenvalues of $\overline{A_0}^{-1}\overline{A_g(\theta)}$ that have nonzero real parts to constitute $\mathcal{B}\left(\overline{A_0}^{-1}\overline{A_g(\theta)}\right)$.

Step 4. For $r_i' \in (r_i, r_{i+1})$, determine if $A_0 + r_i' A_g(\theta)$ is a Hurwitz matrix. If yes, then in the open interval (r_i, r_{i+1}) the system is stable on the negative half-plane.

Step 5. Calculate the union of all the open intervals where the system is stable. The result is the stability region of the multiparameter system.

4.2.2.3 Map from Negative Half-Plane to ISSSR

The stability region calculation method based on the guardian map theory is only applicable to the system whose system matrix is a Hurwitz matrix, that is, the system eigenvalues are all on the negative half-plane. Therefore, the negative half-plane needs to be mapped to the ISSSR. Being irregular in shape, the ISSSR cannot be directly mapped. Thus, the ISSSR is first decomposed to the intersection of three regions, shown in Figure 4.3.

By mapping the system from the negative half-plane to the three regions in Figure 4.3, respectively, the corresponding stability regions can be gained using the guardian map based on the stability region calculation method. The operating points in the stability region corresponding to Region 1 can meet requirement (4.37) in the ISSSR definition and the operating points in the stability regions corresponding to Region 2 and Region 3 can meet requirement (4.38) in the ISSSR definition. Since the system eigenvalues are symmetrical across the real axis, the stability regions corresponding to Region 2 and Region 3 are the same.

Mapping the system from the negative half-plane to Region 1 can be achieved by a translation map:

$$\upsilon_1 : A_{region1} = A_0 + \varepsilon_1 E \tag{4.53}$$

where A_0 is the system matrix of the original system, $A_{region1}$ is the system matrix of the mapped system, and E is the unit matrix.

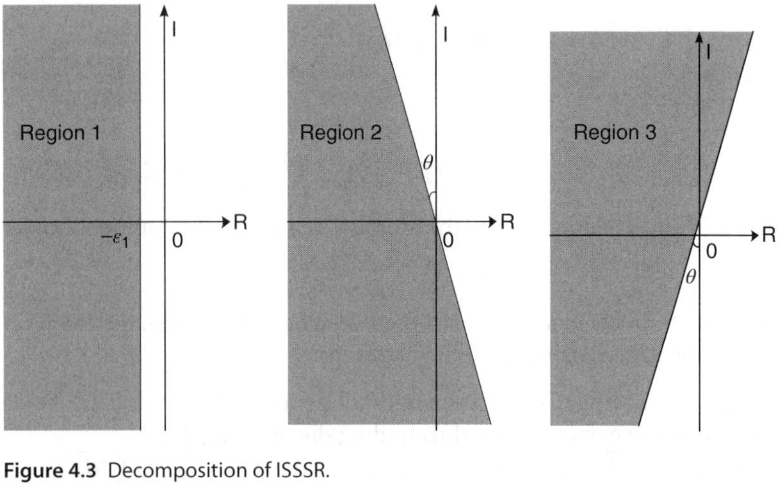

Figure 4.3 Decomposition of ISSSR.

Mapping the system from the negative half-plane to Region 2 can be achieved by a rotation map:

$$\upsilon_2 : \boldsymbol{A}_{region2} = \boldsymbol{A}_0 \times e^{i\theta} \tag{4.54}$$

where $\boldsymbol{A}_0$ is the system matrix of the original system, and $\boldsymbol{A}_{region2}$ is the system matrix of the mapped system.

With $\boldsymbol{A}_{region1}$ and $\boldsymbol{A}_{region2}$ obtained, the corresponding stability regions can be calculated using the guardian map based stability region calculation method. The intersection of the corresponding stability regions is the ISSSR of the system.

4.2.2.4 Simulation Verification Applying a Four-Machine 11-Node System

Simulation System The IEEE four-machine 11-node system is taken for a simulation analysis, shown in Figure 3.2. The generator applies the sixth-order transient model. The system order is reduced using the selective mode analysis (SMA), and during the order reduction the state variables that are related to low-frequency oscillation are preserved. The excitation system applies fast excitation. The load under the baseline model applies 50% constant impedance and 50% constant current model. The major parameters are listed in Table 4.6.

Test of a Single-Parameter System The active load of bus 14 is taken as the parameter variable to calculate the single-parameter stable region of the test system. The detailed calculation steps are as follows:

Step 1. Take G4 as the reference node and form A_0 according to the current operating state of the system:

$$A0 = \begin{bmatrix} 0 & 376.9911 & 0 & 0 & 0 & 0 \\ -0.0731 & -0.7390 & 0.0650 & 0.2729 & 0.0047 & 0.0768 \\ 0 & 0 & 0 & 376.9911 & 0 & 0 \\ 0.0588 & 1.1603 & -0.0872 & -0.3437 & 0.0091 & -0.1340 \\ 0 & 0 & 0 & 0 & 0 & 376.9911 \\ 0.0082 & -0.0208 & 0.0110 & 0.0472 & -0.0824 & -0.5546 \end{bmatrix}$$

Step 2. Make the disturbance parameters P_{14} vary by 5%, so that $\boldsymbol{A}_g$ is formed:

$$Ag = \begin{bmatrix} 0 & 0 & 0 & 0 & 0 & 0 \\ 0.0310 & 4.3047 & 0.0334 & -3.4122 & -0.0218 & -0.5780 \\ 0 & 0 & 0 & 0 & 0 & 0 \\ 0.0341 & -0.3701 & 0.0275 & 2.6194 & -0.0266 & -0.4590 \\ 0 & 0 & 0 & 0 & 0 & 0 \\ -0.0042 & 0.8639 & -0.0140 & 1.8658 & 0.0076 & -0.6408 \end{bmatrix}$$

Table 4.6 Major parameters of a four-generator 13-node system.

Generator	Major parameters											
	BASE MVA	X_d/pu	X'_d/pu	X''_d/pu	T''_{d0}/s	T''_{d0}/s	X_q/pu	X'_q/pu	X''_q/pu	T'_{q0}/s	T''_{q0}/s	H/s
1	900	1.8	0.3	0.25	8	0.03	1.7	0.55	0.25	0.4	0.05	6.5
2	900	1.8	0.3	0.25	8	0.03	1.7	0.55	0.25	0.4	0.05	6.5
3	900	1.8	0.3	0.25	8	0.03	1.7	0.55	0.25	0.4	0.05	6.5
4	900	1.8	0.3	0.25	8	0.03	1.7	0.55	0.25	0.4	0.05	6.5

Step 3. Calculate $\overline{A0}$, $\overline{A_g}$, and the eigenvalues of the matrix $\overline{A0}^{-1}\overline{A_g}$.

Step 4. According to Definition 4.2, choose the unrepeated eigenvalues of $\overline{A0}^{-1}\overline{A_g}$ that have nonzero real parts to constitute $\mathcal{B}\left(\overline{A0}^{-1}\overline{A_g}\right)$.

Step 5. For all intervals in $\mathcal{B}\left(\overline{A0}^{-1}\overline{A_g}\right)$, if any $r_i' \in (r_i, r_{i+1})$ enables $A_0 + r_i' A_g$ to be a Hurwitz matrix, then in the open interval (r_i, r_{i+1}) the system is stable. Keep a record of all such open intervals in $\mathcal{B}\left(\overline{A0}^{-1}\overline{A_g}\right)$, and the stable region of single parameter P14 is thus calculated:

$$\Omega_{\in} = \left(-0.454, -0.0718\right)$$

The result means that when P14 changes from –45.4% to –7.18%, system will remain stable.

Calculation of the Boundary of ISSSR The active loads of bus 4 and bus 14 – P_4 and P_{14} – are taken as the parameter variables to calculate the boundary of ISSSR. Set $\varepsilon_1 = 0.05$ and $\theta = 2°$ (see Figure 4.4).

The ISSSR in Figure 4.4 is then discomposed according to Figure 4.3, and the corresponding stability regions of Region 1 and Region 2 are calculated using the method in Section 2.3. Taking Region 1 as an example, the detailed calculation steps are as follows:

Step 1. Take G4 as the reference node and form A_0 according to the current operating state of the system. By the translation map according to (4.52), $A_{region1}$ is gained. The disturbance parameters P_4 and P_{14} both vary by 5%, so that $A_{g,1}$ and $A_{g,2}$ are formed. Since the parameter space is two-dimensional, θ only needs to be one-dimensional and varies from 0° to 180°. Then the polar coordinate conversion method is applied to form $A_g(\theta)$: $A_g(\theta) = \cos(\theta)A_{g,1} + \sin(\theta)A_{g,2}$:

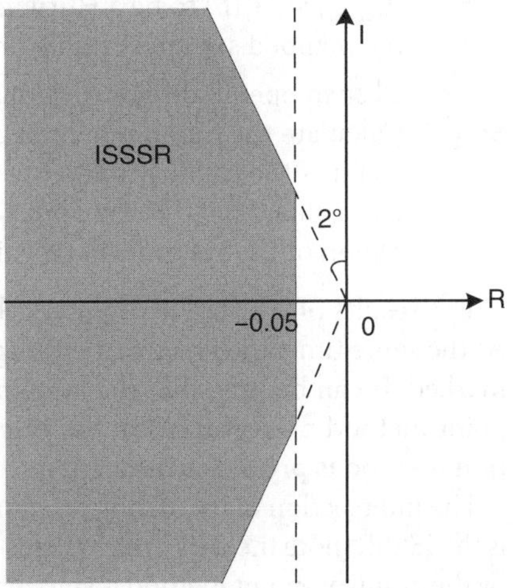

Figure 4.4 The ISSSR used in the simulation case.

$$A_{region1} = \begin{bmatrix} 0.0500 & 376.9911 & 0 & 0 & 0 & 0 \\ -0.0731 & -0.6890 & 0.0650 & 0.2729 & 0.0047 & 0.0768 \\ 0 & 0 & 0.0500 & 376.9911 & 0 & 0 \\ 0.0588 & 1.1603 & -0.0872 & -0.2937 & 0.0091 & -0.1340 \\ 0 & 0 & 0 & 0 & 0.0500 & 376.9911 \\ 0.0082 & -0.0208 & 0.0110 & 0.0472 & -0.0824 & -0.5046 \end{bmatrix}$$

$$A_{g,1} = \begin{bmatrix} 0 & 0 & 0 & 0 & 0 & 0 \\ 0.0310 & 4.3047 & 0.0334 & -3.4122 & -0.0218 & -0.5780 \\ 0 & 0 & 0 & 0 & 0 & 0 \\ 0.0341 & -0.3701 & 0.0275 & 2.6194 & -0.0266 & -0.4590 \\ 0 & 0 & 0 & 0 & 0 & 0 \\ -0.0042 & 0.8639 & -0.0140 & 1.8658 & 0.0076 & -0.6408 \end{bmatrix}$$

$$A_{g,2} = \begin{bmatrix} 0 & 0 & 0 & 0 & 0 & 0 \\ -0.0021 & 2.7900 & 0 & -1.2272 & -0.0015 & -0.3514 \\ 0 & 0 & 0 & 0 & 0 & 0 \\ 0.0123 & -1.1209 & 0.0075 & 0.1515 & -0.0049 & 0.1260 \\ 0 & 0 & 0 & 0 & 0 & 0 \\ 0.0024 & 0.1898 & -0.0026 & 0.2547 & 0 & -0.0180 \end{bmatrix}$$

Step 2. Calculate $\overline{A_{region1}}$, $\overline{A_g(\theta)}$ and the eigenvalues of the matrix $\overline{A_{region1}}^{-1}\overline{A_g(\theta)}$.

Step 3. According to Definition 4.2, choose the unrepeated eigenvalues of $\overline{A_{region1}}^{-1}\overline{A_g(\theta)}$ that have nonzero real parts to constitute $\mathcal{B}\left(\overline{A_{region1}}^{-1}\overline{A_g(\theta)}\right)$.

Step 4. For all intervals in $\mathcal{B}\left(\overline{A_{region1}}^{-1}\overline{A_g(\theta)}\right)$, if any $r_i' \in (r_i, r_{i+1})$ enables $A_{region1}+r_i'A_g(\boldsymbol{\theta})$ to be a Hurwitz matrix, then in the open interval (r_i, r_{i+1}) the mapped system is stable on the negative half-plane. Keep a record of all such open intervals in $\mathcal{B}\left(\overline{A_{region1}}^{-1}\overline{A_g(\theta)}\right)$.

Step 5. Calculate the union of all the open intervals recorded in Step 4, and the result is the stability region of the mapped system when θ applies a certain value. When θ has applied all the values in $[0, \pi)$, the final stability region of the system will be gained.

The results of the stability region calculation are shown in Figures 4.5 and 4.6. At the same time, the system operating points in Region 1 and Region 2 are star-marked. It can be seen that the system operating points obtained by the tracing point method nearly all fall in the calculated stability regions. Thus the calculation method is proved to be accurate.

The intersection of the stability regions corresponding to Region 1 and Region 2 is the ISSSR (note the dark area in Figure 4.7). At the same time, the ISSSR obtained by the point-by-point method is star-marked. It can be seen that the stability region

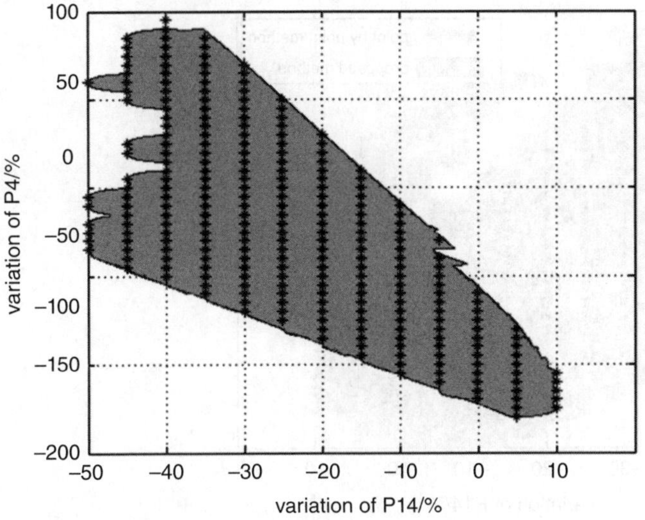

Figure 4.5 Stability region corresponding to Region 1.

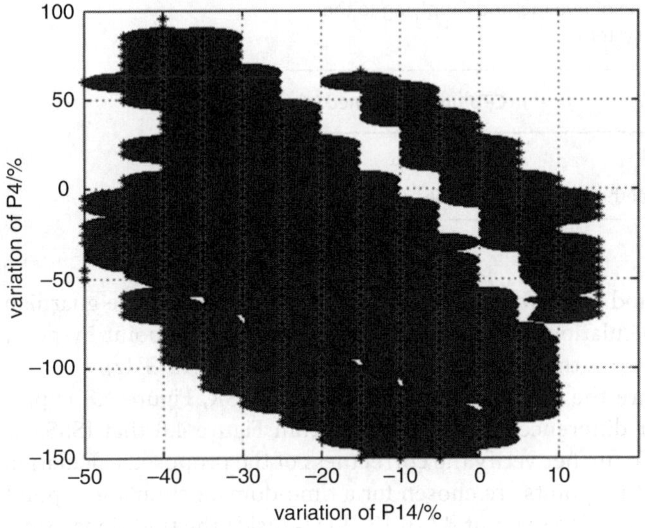

Figure 4.6 Stability region corresponding to Region 2.

formed by the point-by-point method closely fits the calculated ISSSR, with only a few operating points falling outside the ISSSR. However, these operating points are close to the instability region, so they should be avoided in real practice. Therefore, the proposed method to calculate the ISSSR is proved accurate.

To prove the rapidity of the proposed method, a comparison of the calculation speed between the guardian map-based ISSSR calculation method and the

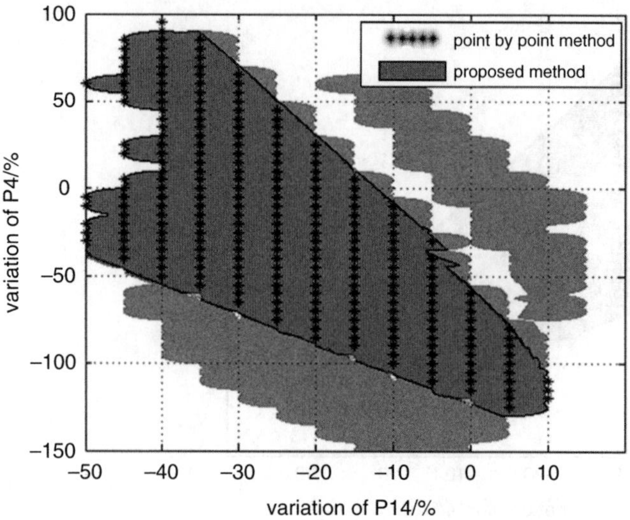

Figure 4.7 ISSSR of the four-generator 13-node system.

Table 4.7 Calculation time of two methods applying to the four-generator 13-node system.

Method	CPU time consumed (s)
Point-by-point method	3540.561875
Guardian map-based method	186.515613

point-by-point method is shown in Table 4.7. It can be seen that the guardian map-based ISSSR calculation method is more efficient than the point-by-point method. The CPU used is the Intel core i3 processor 330 M (2.13 GHz).

In order to compare the ISSSR with the traditional SSSR, Figure 4.8 is presented to show their differences. It can be seen from Figure 4.8 that ISSSR is smaller than SSSR. To further verify the correctness of the proposed calculation method, three operating points are chosen for a time-domain simulation: point A (−0.13, 0) is inside the ISSSR; point B (−0.12, 0.1) is inside the traditional SSSR but outside the ISSSR; and point C (−0.11, 0.2) is outside the SSSR (see Figure 4.9).

Suppose a three-phase fault occurs at the transmission line between bus 3 and bus 101 at $t = 0.1$ s and is cleared at $t = 0.2$ s. Simulation results of points A, B, and C are shown later in Figures 4.10 to 4.12. The recorded variable is the angular difference between G1 and G3.

It can be seen from Figure 4.10 that when the system operates at point A, the angular difference between G1 and G3 gradually becomes stable after the disturbance. However, when the system operates at point B, the angular difference between G1 and G3 experiences sustained constant-amplitude oscillation after the disturbance, as can be seen in Figure 4.11.

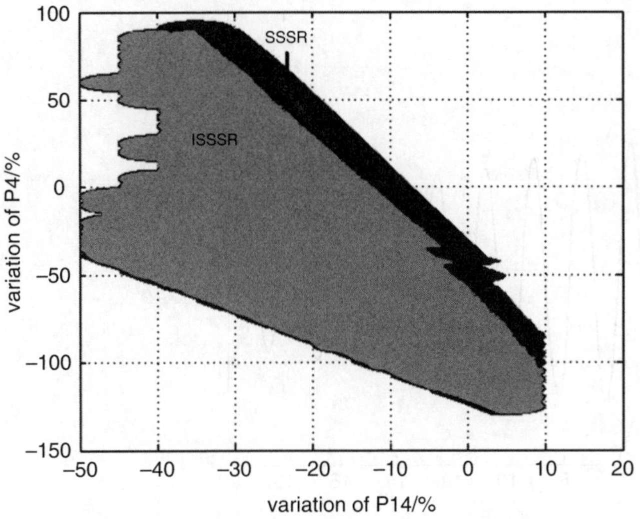

Figure 4.8 ISSSR compared with SSSR.

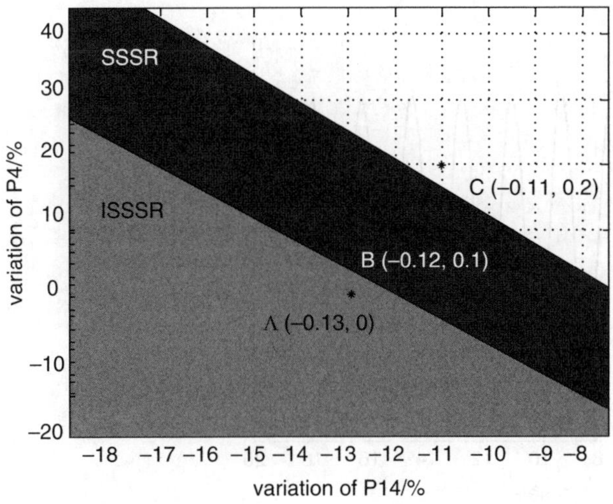

Figure 4.9 Points inside and outside the ISSSR.

Though the system may not go unstable soon, the damping ratio being obviously inadequate, serious problems may arise if the oscillation continues. When the system operates at point C, the system damping is negative. Therefore, when the disturbance occurs, the angular difference between G1 and G3 experiences increased-amplitude oscillation and the system gradually becomes unstable (see Figure 4.12).

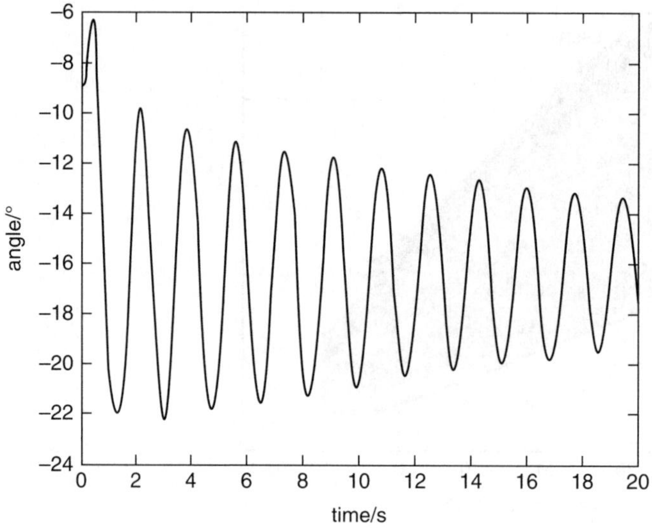

Figure 4.10 Time-domain simulation of point A.

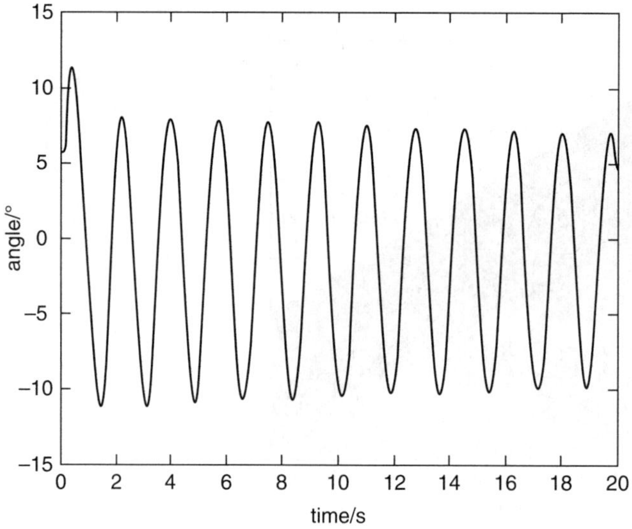

Figure 4.11 Time-domain simulation of point B.

Influence of Excitation System Parameters on ISSSR The calculation of the ISSSR is mainly discussed above, but the influence of excitation system parameters on ISSSR is not mentioned. In order to apply the ISSSR to system operation, the influence of excitation system parameters on ISSSR is studied. Figure 4.13 shows the ISSSRs of systems with an excitation system applying different magnification factors. The excitation magnification factors corresponding to different boundaries in Figure 4.13 are shown in Table 4.8.

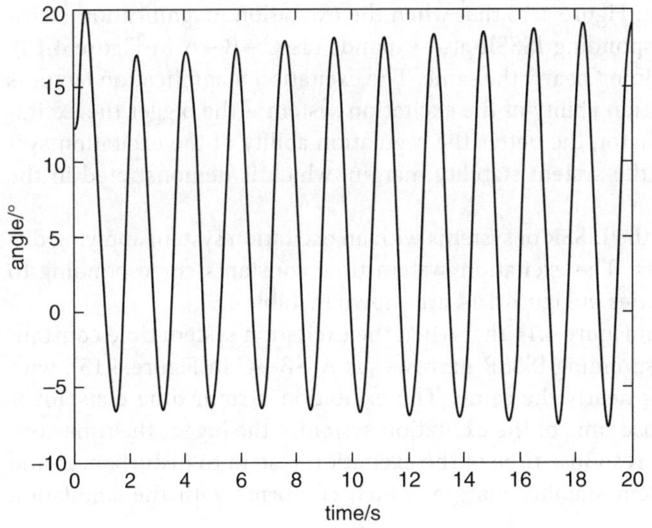

Figure 4.12 Time-domain simulation of point C

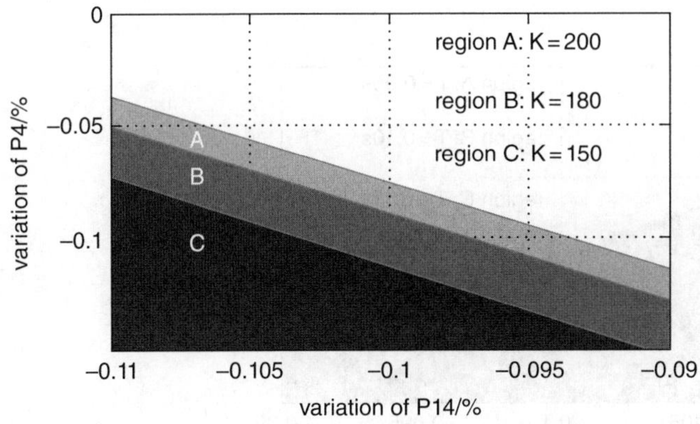

Figure 4.13 ISSSR under different excitation magnification factors.

Table 4.8 Corresponding excitation magnification factors of different boundaries in Figure 4.13.

Excitation magnification factor K	Generator number in four-machine system			
	1	2	3	4
KA	200	200	200	200
KB	180	180	180	180
KC	150	150	150	150

It can be seen from Figure 4.13 that when the excitation magnification factor increases, the corresponding ISSSR also expands (as C→B→A in Figure 4.13), with the shape remaining nearly the same. The excitation magnification factor is related to the regulation ability of the excitation system – the bigger the excitation magnification factor, the better the regulation ability of the excitation system and the bigger the system stability margin, which is demonstrated in the simulation results.

Figure 4.14 shows the ISSSR of systems with an excitation system applying different time constants. The excitation system time constants corresponding to the different boundaries in Figure 4.14 are shown in Table 4.9.

It can be seen from Figure 4.14 that when the excitation system time constant increases, the corresponding ISSSR narrows (as A→B→C in Figure 4.13), with the shape remaining nearly the same. The excitation system time constant is related to the response time of the excitation system – the bigger the time constant, the longer the response time of the excitation system to disturbances and the smaller the system stability margin, which conforms with the simulation results [5].

When the load cannot be used for regulation, the excitation system magnification factor and time constant can be changed to increase the system stability margin and ensure the safe and stable operation of the system.

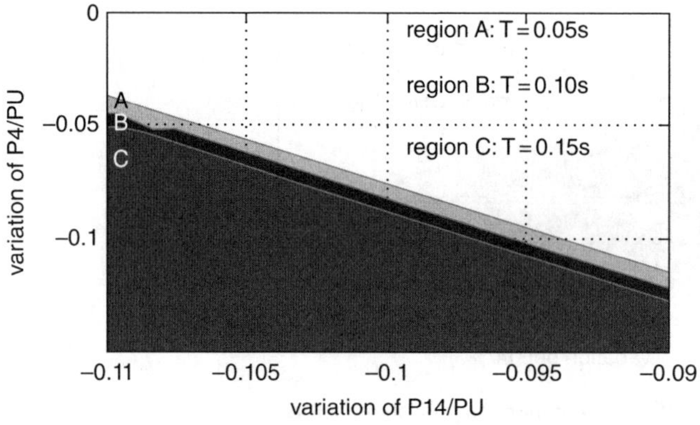

Figure 4.14 ISSSR under different excitation system time constants.

Table 4.9 Corresponding excitation system time constants of different boundaries in Figure 4.14.

Excitation system time constant T	Generator number in 4-machine system			
	1	2	3	4
TA	0.05	0.05	0.05	0.05
TB	0.10	0.10	0.10	0.10
TC	0.15	0.15	0.15	0.15

4.2.2.5 Simulation Verification Applying 16-Machine 68-Node System

A 16-Machine 68-Node System A 16-machine 68-node system, which is a reduced-order model of the New England/New York interconnected system, is employed to further test the performance of the proposed ISSSR calculation method. The system order is reduced using the selective mode analysis (SMA), and during the order reduction the state variables, which are related to low-frequency oscillation, are preserved.

Calculation of the Boundary of ISSSR The active load of bus 1 and bus 27 – P1 and P27 – are taken as the parameter variables to calculate the boundary of ISSSR. Set $\varepsilon_1 = 0.01$ and $\theta = 2°$ and follow the same steps stated in Section 4.2.2.4. The results of the ISSSR calculation are shown in Figure 4.15. At the same time, the ISSSR obtained by the point-by-point method is star-marked. It can be seen that the system operating points obtained by the point-by-point method nearly all fall in the calculated stability regions and thus the calculation method is proved to be accurate.

To prove the rapidity of the proposed method, a comparison of the calculation speed between the guardian map-based ISSSR calculation method and the point-by-point method is shown in Table 4.10. It can be seen that the guardian map-based ISSSR calculation method is more efficient than the point-by-point method. The CPU used is the Intel core i3 processor 330 M (2.13 GHz).

In order to compare the ISSSR with the traditional SSSR, Figure 4.16 is presented to show the differences. It can be seen from Figure 4.16 that ISSSR is

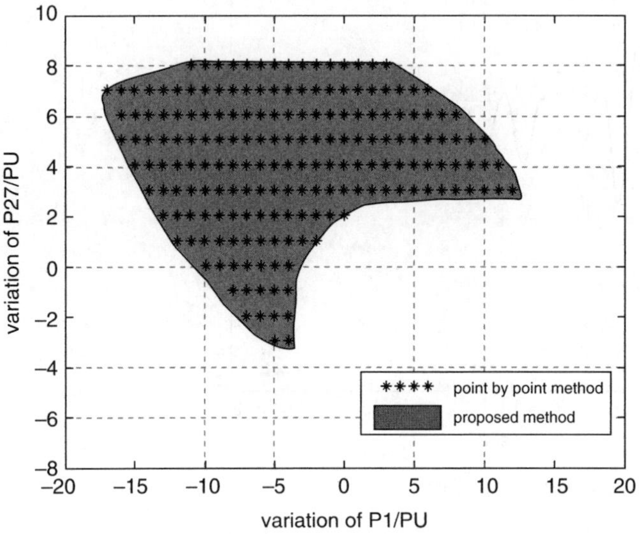

Figure 4.15 ISSSR of the 16-machine 68-node system.

Table 4.10 Calculation times of the two methods applied to the 16-machine system.

Method	CPU time consumed (s)
Point-by-point method	15453.452137
Guardian map-based method	1254.896421

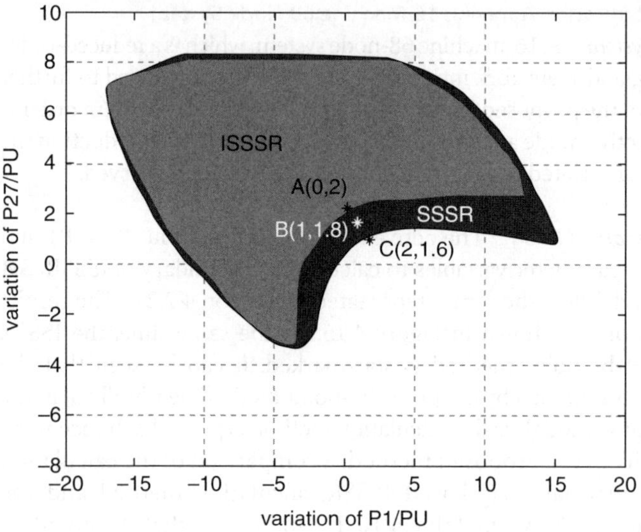

Figure 4.16 ISSSR compared with SSSR.

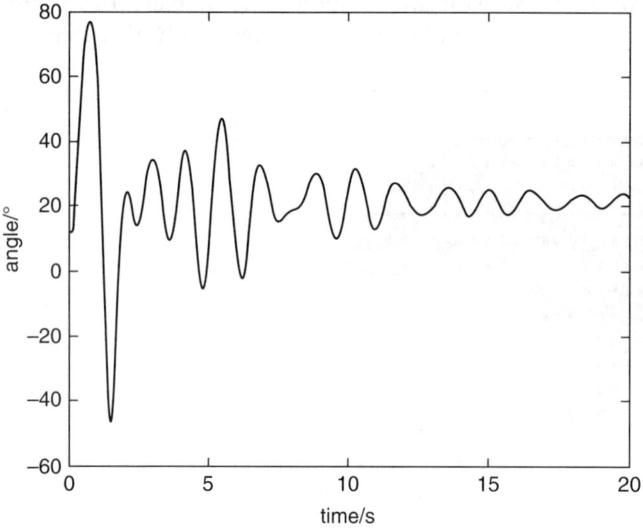

Figure 4.17 Time-domain simulation of point A.

smaller than SSSR. To further verify the correctness of the proposed calculation method, three operating points are chosen for time-domain simulation: point A (0, 2) is inside the ISSSR, point B (1, 1.8) is inside the traditional SSSR but outside the ISSSR, and point C (2, 1.6) is outside the SSSR.

Suppose a three-phase fault occurs at the transmission line between bus 28 and bus 29 at $t = 0.1$ s and is cleared at $t = 0.2$ s. Simulation results of points A, B, and C are shown in Figures 4.17 to 4.19. The recorded variable is the angular difference between G9 and G12.

It can be seen from Figure 4.17 that when the system operates at point A, the angular difference between G9 and G12 gradually becomes stable after the

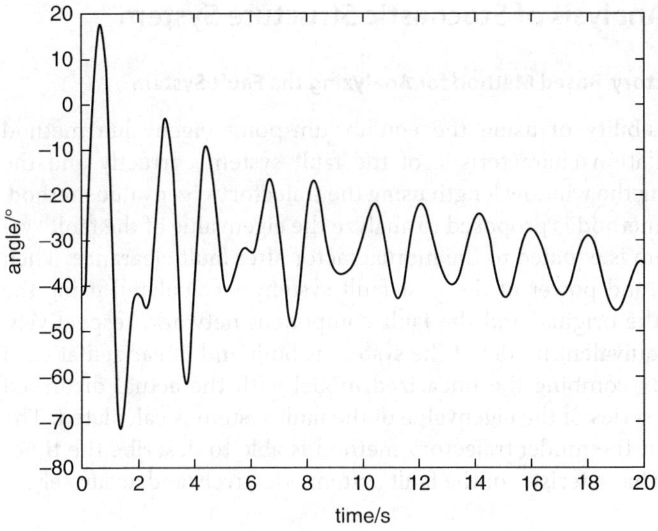

Figure 4.18 Time-domain simulation of point B.

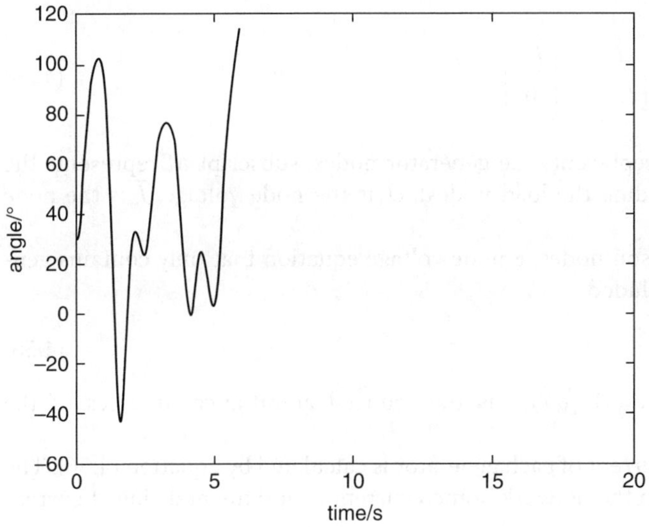

Figure 4.19 Time-domain simulation of point C.

disturbance. However, when the system operates at point B, the angular difference between G9 and G12 experiences sustained oscillation after the disturbance, as can be seen in Figure 4.18. Though the system is still stable, obvious oscillations still exist even 20 seconds after the fault is cleared. The damping ratio is obviously inadequate and serious problems may arise if the oscillation continues. When the system operates at point C, the system damping is negative. Therefore, when the disturbance occurs, the angular difference between G9 and G12 experiences increased-amplitude oscillation and lost angular stability at $t = 6\,s$ (see Figure 4.19). Simulation results verify the correctness of ISSSR calculated using the proposed method and show the advantages of ISSSR over traditional SSSR.

4.3 Stability Analysis of Stochastic Structure System

4.3.1 Model-Trajectory-Based Method for Analyzing the Fault System

Due to the impossibility of using the equilibrium-point eigenvalue method to reflect the oscillation characteristic of the fault system correctly and the difficulty in choosing the window length using the trajectory eigenvalue method, a model trajectory method is proposed to analyze the eigenvalue of the fault system. First, the system is equated to the mutual factor after fault clearance. Then the branch current and power of the post-fault system is calculated using the branch current of the original and the fault component network, respectively. Furthermore, the equivalent model of the system is built and linearized at each measuring point. By combing the linearized model with the actual disturbed trajectory, the time series of the eigenvalue of the fault system is calculated. The results illustrate that the model trajectory method is able to describe the time-varying oscillation characteristic of the fault system effectively and accurately.

4.3.1.1 Stability Analysis of the Fault System

Equivalence of the Fault Model The node voltage equation of the system is given by [9]

$$\begin{bmatrix} Y_{NN} & Y_{NM} \\ Y_{MN} & Y_{MM} \end{bmatrix} \begin{bmatrix} \vec{U}_N \\ \vec{U}_M \end{bmatrix} = \begin{bmatrix} \vec{I}_N \\ 0 \end{bmatrix} \tag{4.55}$$

where subscript N represents the generator nodes, subscript M represents the liaison nodes (including the load nodes), $\vec{U}$ is the node voltage, $\vec{I}$ is the node inject current.

Removing the liaison nodes, a node voltage equation that only contains generator nodes is concluded:

$$Y'_{NN}\vec{U}_N = \vec{I}_N \tag{4.56}$$

where $Y'_{NN} = Y_{NN} - Y_{NM}Y_{MM}^{-1}Y_{MN}$ is the reduced admittance matrices of the system.

The node inject current of each generator is calculated by equation (4.56). The relationship between the network branch current $\vec{I}_b$ and the node inject current $\vec{I}_N$ is as follows:

$$\vec{I}_b = YP^T Y_n^{-1}\vec{I}_N = C\vec{I}_N \tag{4.57}$$

where $\mathbf{Y}$ is the branch admittance matrix, $\mathbf{P}$ is the node incidence matrix, and $\mathbf{Y}_\mathrm{n}$ is the node admittance matrix of the whole network. $\mathbf{C} = \mathbf{YP}^T\mathbf{Y}_\mathrm{n}^{-1}$ represents the common factor matrix of the normal network.

The equivalent circuit diagram of a two-generator system is shown in Figure 4.20, where the number of each node and branch is labeled accordingly. Generators 1 and 2 are connected by a double line and loads L1 and L2 are at buses 1 and 2, respectively.

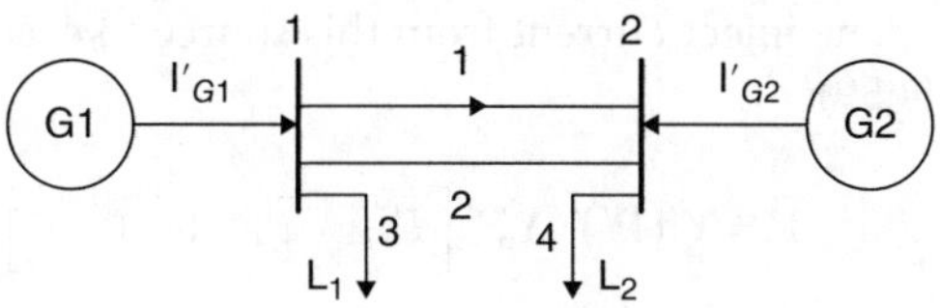

Figure 4.20 Two-machine system.

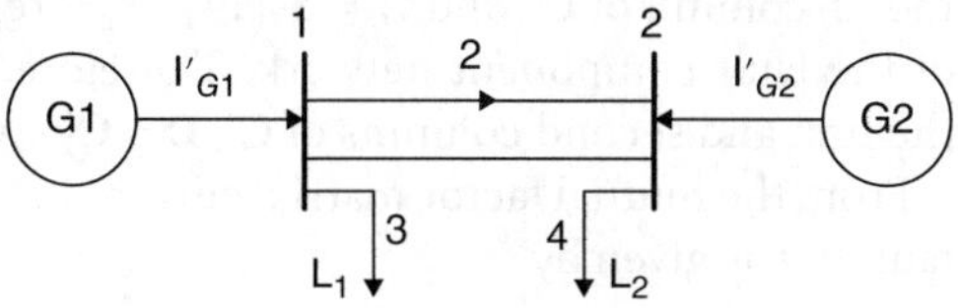

Figure 4.21 Two-machine system after the fault line is removed.

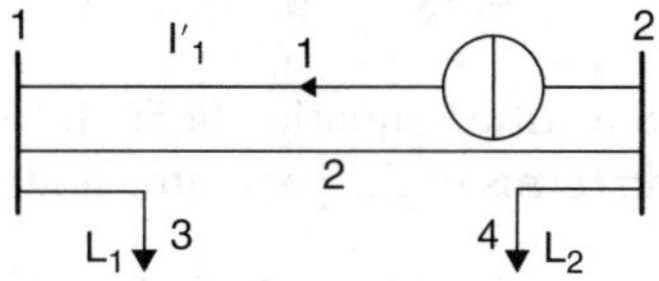

Figure 4.22 The system only contains the fault component.

At t_0, a three-phase short-circuit fault occurs on system tie line 1, where the proportion of the distance between the short circuit point and node 1 to the length of the whole line is α. The duration of the short circuit is t_c and after that the fault line is removed. Then the network becomes as shown in Figure 4.21.

The fault occurs at $t = t_c$, at the fault clearance instant, $\mathrm{I}(t_{c-}) = \mathrm{I}(t_{c+})$. Thus the inject current of the network before the fault line clearance equals the inject current after the fault line clearance. Suppose the inject current of the network before the fault line clearance is $\vec{I}_{N2}$. The power flow distribution of the network after the fault line clearance is regarded as a combination of the branch power flow before the clearance and the fault component caused by the clearance due to the superposition principle. This means that Figure 4.21 is the superposition equivalent of Figure 4.20 and Figure 4.22.

From equation (4.57), the current in branch i when the node inject current is $\vec{I}_{N2}$ is as follows:

$$\vec{I'}_i = \mathbf{C}_i \vec{\mathbf{I}}_{\mathbf{N2}} \tag{4.58}$$

where $\vec{I'}_i$ is the current in branch I and C_i is the ith row of the matrix.

When the fault line i is cleared from the system, there is just one source left in the fault component network, which is $\vec{I'}_i$. Therefore the power flow in other lines is only related to this source. Thus the node voltage equation of the network is

$$\mathbf{Y_n'}\vec{\mathbf{U}}_{\mathbf{n}} = \begin{bmatrix} \vec{I'}_i & -\vec{I'}_i & 0 & 0 & 0 \end{bmatrix}^T \tag{4.59}$$

where $\mathbf{Y_n'}$ is the node admittance matrix of the fault component network.

The inject current from this source to each branch is obtained from equation (4.59):

$$\vec{\mathbf{I}''} = \mathbf{Y}'(\mathbf{P}')^T \mathbf{Y}_{\mathbf{n}}'^{-1}\begin{bmatrix} \vec{I'}_i & -\vec{I'}_i & 0 & 0 & 0 \end{bmatrix}^T = (\mathbf{C}_1' - \mathbf{C}_2')\vec{I'}_i = \mathbf{D}\vec{I'}_i \tag{4.60}$$

where $\vec{\mathbf{I}''}$ is the current in each branch of the fault component network, $\mathbf{Y}'$ is the branch admittance matrix of this network, $\mathbf{P}'$ is the node incidence matrix, $\mathbf{C}_i'$ is the ith column of $\mathbf{C}'$, and $\mathbf{C}' = \mathbf{Y}'(\mathbf{P}')^{\mathbf{T}} \mathbf{Y}_{\mathbf{n}}'^{-1}$ represents the common factor matrix of the fault component network. Therefore, $\mathbf{C}_1'$ and $\mathbf{C}_2'$, respectively, represent the first and second columns of $\mathbf{C}'$; $\mathbf{D} = \mathbf{C}_1' - \mathbf{C}_2'$.

From the mutual factor matrix, current in the kth line after the clearance of the fault line is given by

$$\begin{aligned} \vec{I}_k &= \vec{I'}_k + \vec{I''}_k \\ \mathbf{C}_k\vec{\mathbf{I}}_{N2} + \mathbf{D}_k\mathbf{C}_i\vec{\mathbf{I}}_{N2} &= (\mathbf{C}_k + \mathbf{D}_k\mathbf{C}_i)\vec{\mathbf{I}}_{N2} \end{aligned} \tag{4.61}$$

Substitute equation (4.56) into equation (4.61) and the branch current expressed in terms of the generator node voltage is then

$$\begin{aligned} \vec{I}_k &= (\mathbf{C}_k + \mathbf{D}_k\mathbf{C}_i)\vec{\mathbf{I}}_{N2} = (\mathbf{C}_k + \mathbf{D}_k\mathbf{C}_i)\mathbf{Y}_{\mathbf{NN}}'\vec{\mathbf{U}}_{\mathbf{N}} \\ &= \mathbf{E}\begin{bmatrix} U_{G1}\angle\delta_1 & U_{G2}\angle\delta_2 & \cdots & U_{Gn}\angle\delta_n \end{bmatrix}^T \\ &= U_{G1}|\mathbf{E}_1|\angle(\delta_1 + \varphi_1) + \cdots + U_{Gn}|\mathbf{E}_n|\angle(\delta_n + \varphi_n) \end{aligned} \tag{4.62}$$

where $\mathbf{E} = (\mathbf{C}_k + \mathbf{D}_k\mathbf{C}_i)\mathbf{Y}_{\mathrm{NN}}'$, $|\mathbf{E}_i|$, ϕ_i (i = 1, 2,..., n) is the amplitude and phase angle of the ith element in $\mathbf{E}$.

Equations (4.59) to (4.62) aim to equate the injected current during the fault and analyze the line flow after the fault clearance. Since the instantaneous introduction of the injected current will cause the system power flow to change in the whole time period afterwards, by using time-varying electrical variables, the dynamic characteristic of the system could be reflected.

For the two-generator system shown in Figure 4.20, when a fault clearance occurs on branch 1, current in branch 2 changes. From the deduction made above, the current in branch 2 is

$$\vec{I}_2 = U_{G1}|\mathbf{E}_1|\angle(\delta_1 + \varphi_1) + U_{G2}|\mathbf{E}_2|\angle(\delta_2 + \varphi_2) \tag{4.63}$$

Then the power transmitting between the two generators can be expressed by

$$\begin{aligned} P_{12} &= real\left(\vec{U}_{G1}\vec{I}_2^*\right) \\ &= real\left(U_{G1}^2|\mathbf{E}_1|\angle -\varphi_1 + U_{G1}U_{G2}|\mathbf{E}_2|\angle(\delta_1 - \delta_2 - \varphi_2)\right) \\ &= U_{G1}^2|\mathbf{E}_1|\cos(\varphi_1) + U_{G1}U_{G2}|\mathbf{E}_2|\cos(\delta_1 - \delta_2 - \varphi_2) \\ P_{21} &= U_{G2}^2|\mathbf{E}_2|\cos(\varphi_2) + U_{G1}U_{G2}|\mathbf{E}_1|\cos(\delta_2 - \delta_1 - \varphi_1) \end{aligned} \tag{4.64}$$

An equivalence of the network after the fault clearance is built. Then the current and power of each branch after fault clearance is calculated using the node voltage of the original network and the branch current of the fault component network. Furthermore, the eigenvalue of the system is able to be analyzed.

Time-Varying Oscillation Characteristic Analysis of the Fault System A two-generator system is presented as an example of analyzing the eigenvalue problem of the system after a fault clearance. The structure of the system is as shown in Figure 4.20. Introducing the classical rotor motion equation of the two-generator system and neglecting the mechanical power change, the linearized equation of the system under small perturbation is expressed as follows:

$$\begin{cases} \Delta\dot{\delta}_i = \omega_0 \Delta\omega_i \qquad (i=1,2) \\ M_1 \Delta\dot{\omega}_1 = -\Delta P_{e1} - K_{D1}\Delta\omega_1 \\ M_2 \Delta\dot{\omega}_2 = -\Delta P_{e2} - K_{D2}\Delta\omega_2 \end{cases} \tag{4.65}$$

The rotor motion equation is linearized near the operating point and (4.65) is obtained, where the electromagnetic powers P_{e1} and P_{e2} are the sum of the transmission power between generators and the load power. The transmission power between generators after the fault clearance is expressed in (4.64) according to the superposition principle. Due to the power balance, the electromagnetic power output of the generator is expressed as

$$\begin{cases} P_{e1} = P_{L1} + P_{12} = P_{L1} + \cdots \\ \quad U_{G1}^2 |\mathbf{E}_1| \cos(\phi_1) + U_{G1}U_{G2}|\mathbf{E}_2|\cos(\delta_1 - \delta_2 - \phi_2) \\ P_{e2} = P_{L2} + P_{21} = P_{L2} + \cdots \\ \quad U_{G2}^2 |\mathbf{E}_2| \cos(\phi_2) + U_{G1}U_{G2}|\mathbf{E}_1|\cos(\delta_2 - \delta_1 - \phi_1) \end{cases} \tag{4.66}$$

The relationship between the fault current and dynamic behavior of the power system is reflected in equations (4.62) to (4.66). The influence of disturbance on the dynamic behavior of a power system is through the electromagnetic power p_e. According to (4.62) to (4.66), the electromagnetic power output of the generator p_e is the sum of p_{ij} (the transmission power between generators) and p_L (the load power); p_{ij} is affected by the branch current $\vec{I}_k$. After the fault is cleared, $\vec{I}_k = (\mathbf{C}_k + \mathbf{D}_k\mathbf{C}_i)\vec{\mathbf{I}}_{N2}$, where $\vec{\mathbf{I}}_{N2}$ is the node injection current at any time after branch i is cut. It can be seen from (4.62) that the node injection current could be expressed by the node voltages, which are time-variant. Using the measured values of the node voltages, the dynamic and time-variant oscillatory characteristics of the power system after the fault clearance can be analyzed.

The linearized expression of the electromagnetic power output of the generator under small perturbation is given by

$$\begin{cases} \Delta P_{e1} = U_{G1}U_{G2}|\mathbf{E}_2|[-\sin(\delta_1 - \delta_2 - \phi_2)\Delta\delta_1 + \cdots \\ \sin(\delta_1 - \delta_2 - \phi_2)\Delta\delta_2] = k_{11}\Delta\delta_1 + k_{12}\Delta\delta_2 \\ \Delta P_{e2} = U_{G1}U_{G2}|\mathbf{E}_1|[\sin(\delta_2 - \delta_1 - \phi_1)\Delta\delta_1 - \cdots \\ \sin(\delta_2 - \delta_1 - \phi_1)\Delta\delta_2] = k_{21}\Delta\delta_1 + k_{22}\Delta\delta_2 \end{cases} \tag{4.67}$$

where

$$\begin{aligned} k_{11} &= -k_{12} = -U_{G1}U_{G2}\left|\mathbf{E}_2\right|\sin\left(\delta_1 - \delta_2 - \phi_2\right) \\ k_{21} &= -k_{22} = U_{G1}U_{G2}\left|\mathbf{E}_1\right|\sin\left(\delta_2 - \delta_1 - \phi_1\right) \end{aligned} \tag{4.68}$$

Substitute equation (4.67) in equation (4.65) and then the state equation of the system after fault clearance is obtained:

$$\begin{cases} \Delta\dot{\delta}_i = \omega_0 \Delta\omega_i & (i = 1,2) \\ M_1 \Delta\dot{\omega}_1 = -\left(k_{11}\Delta\delta_1 + k_{12}\Delta\delta_2\right) - K_{D1}\Delta\omega_1 \\ M_2 \Delta\dot{\omega}_2 = -\left(k_{21}\Delta\delta_1 + k_{22}\Delta\delta_2\right) - K_{D2}\Delta\omega_2 \end{cases} \tag{4.69}$$

The state equation **A** of the system which is deduced from equation (4.69) is

$$\mathbf{A}_t = \begin{bmatrix} \mathbf{0} & \omega_0 & \mathbf{I} \\ \dfrac{-k_{11t}}{M_1} & \dfrac{-k_{12t}}{M_1} & \dfrac{-K_{D1t}}{M_1} \\ \dfrac{-k_{21t}}{M_2} & \dfrac{-k_{22t}}{M_2} & \dfrac{-K_{D2t}}{M_2} \end{bmatrix} \tag{4.70}$$

where **0** is a 2 × 2 zero matrix and **I** is a 2 × 2 unit matrix.

The oscillation eigenvalue of the system in real time is calculated by substituting the actual measured value of the system at the corresponding time in equation (4.70). The measured value here refers to U_{G1}, U_{G2}, ϕ_1, and ϕ_2 in (4.66). Thus the whole time series of the eigenvalue is obtained.

It needs to be noted that when a perturbation occurs in the system, the amplitude and phase angle of the generator node are no longer constants. Some elements in the system state equation **A** change as well, and according to equation (4.70), the eigenvalue of the system changes simultaneously. Also, as the broadness of the change range of the voltage amplitude and phase angle increases, the change range of the eigenvalue broadens as well.

Similar to the two-generator system, the transmission power between generators in the multigenerator system can also be calculated by adding the line flow before line cutting and the fault component caused by line cutting, and then the post-fault real-time state matrix can be obtained for analysis.

4.3.1.2 Testing Results and Analysis

Simulation of a Two-Generator Equivalence System The three-phase short circuit is the most severe fault in the system, which is why it is taken as an example for analysis. The analysis method for other kinds of fault such as a single-phase short circuit or a phase-to-phase short circuit is similar to that of the three-phase fault, that is, by adding the line flow before line cutting and the fault component caused by line cutting to calculate the transmission power between generators, and then getting the post-fault real-time state matrix for analysis. Thus, other kinds of faults are not analyzed separately in this section.

As shown in Figure 4.20, the classic second-order model is used for the two-generator system, where the load applies the constant power model. At the time of 0.1 s, a three-phase short-circuit fault occurs at a certain point on line 1. The proportion of the distance between the fault point and node 1 to the length of the whole line is α. The duration of the fault is t_c and then the fault line is cleared. The time-varying oscillation characteristic of the fault system is analyzed while considering the following fault types as shown in Table 4.11.

The disturbed trajectories of the system are obtained from simulation and then the real-time eigenvalue change of the system is calculated by substituting the corresponding actual measured value in equation (4.70), as shown in Figures 4.23 to 4.29. The full line represents the results of the equilibrium method, while the dashed line represents the accurate eigenvalue of the system and the dotted line represents results of the method proposed in this section.

It can be seen from Figures 4.23 and 4.24 that when the fault duration is 0.1 s, the influence of fault location variation on the eigenvalues is obvious. When α = 0.5, the real part of the accurate eigenvalue varies between –0.1402 and –0.1374 and the virtual part between 3.6458 and 6.8467. With the proposed method, the calculated real part varies between –0.1422 and –0.1365 and the virtual part between 3.4253 and 6.7919. When α = 0.9, the real part of the accurate eigenvalue varies between –0.1404 and –0.1372 and the virtual part between 3.3187 and 6.8653. With the proposed method, the calculated real part varies between –0.1425 and –0.1362 and the virtual part between 2.7051 and 6.8423.

It can be seen from Figures 4.25 and 4.26 that when the fault duration is 0.175 s, the system falls into a marginally stable state after the fault clearance and the influence of fault location variation on the eigenvalues is small. The eigenvalues both vary alternately between a real number and a complex number. The real part of the accurate eigenvalue varies between –0.1417 and 3.8980 and the virtual part between 0 and 6.8726. With the proposed method, the calculated real part varies between –0.1453 and 3.9951 and the virtual part between 0 and 6.8471. Though the variation ranges of the accurate value and the calculated value are close, their values at the same instant are largely different.

It can be seen from Figures 4.23 to 4.26 that the pre-fault equilibrium point eigenvalue of the two-machine system is a constant: –0.1394 + 7.0484i. After the fault clearance, the eigenvalue calculated using the proposed method will fluctuate around the equilibrium point, with a relatively large fluctuation quantity for both the real part and the imaginary part. Meanwhile, the average values of the

Table 4.11 Fault type of a two-machine system.

Fault type	Fault position α	Duration t_c
Case 1	0.5	0.1
Case 2	0.9	0.1
Case 3	0.5	0.175
Case 4	0.9	0.175

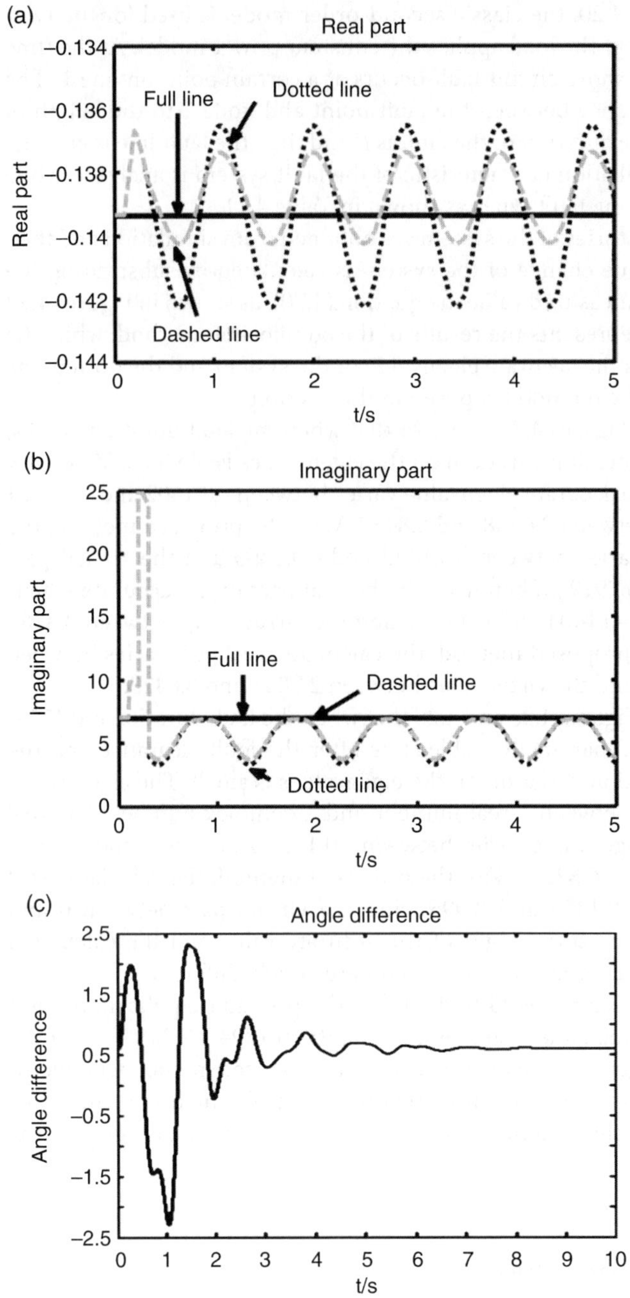

Figure 4.23 The simulation results when α is 0.5 and t_c is 0.1 s.

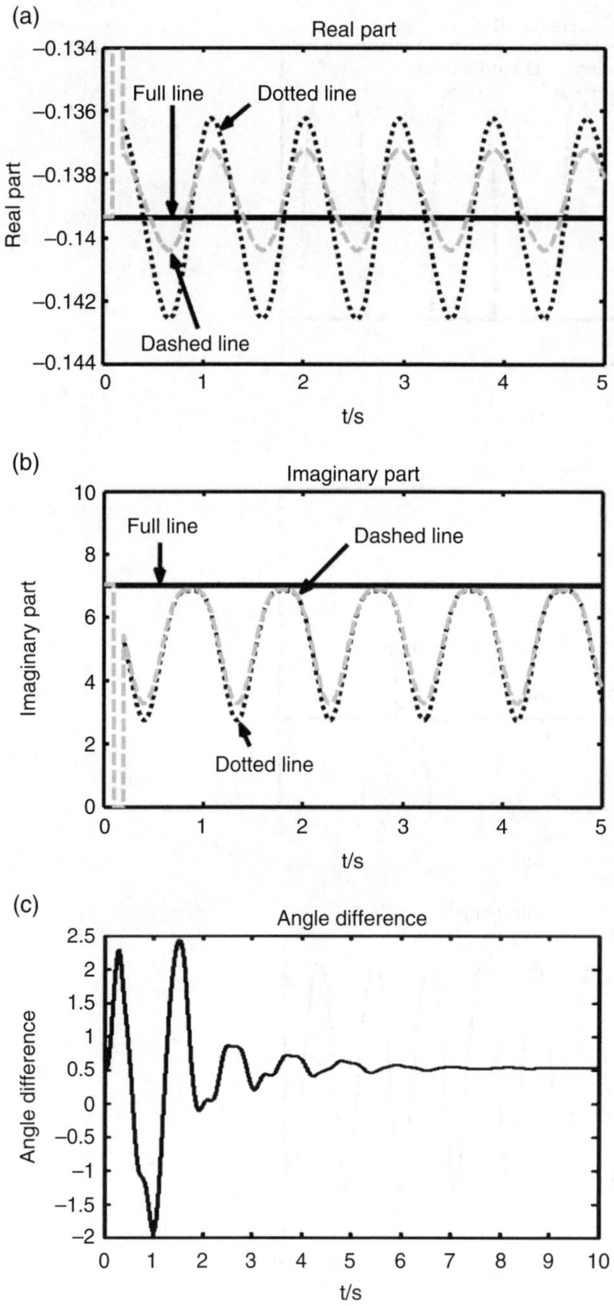

Figure 4.24 The simulation results when α is 0.9 and t_c is 0.1 s.

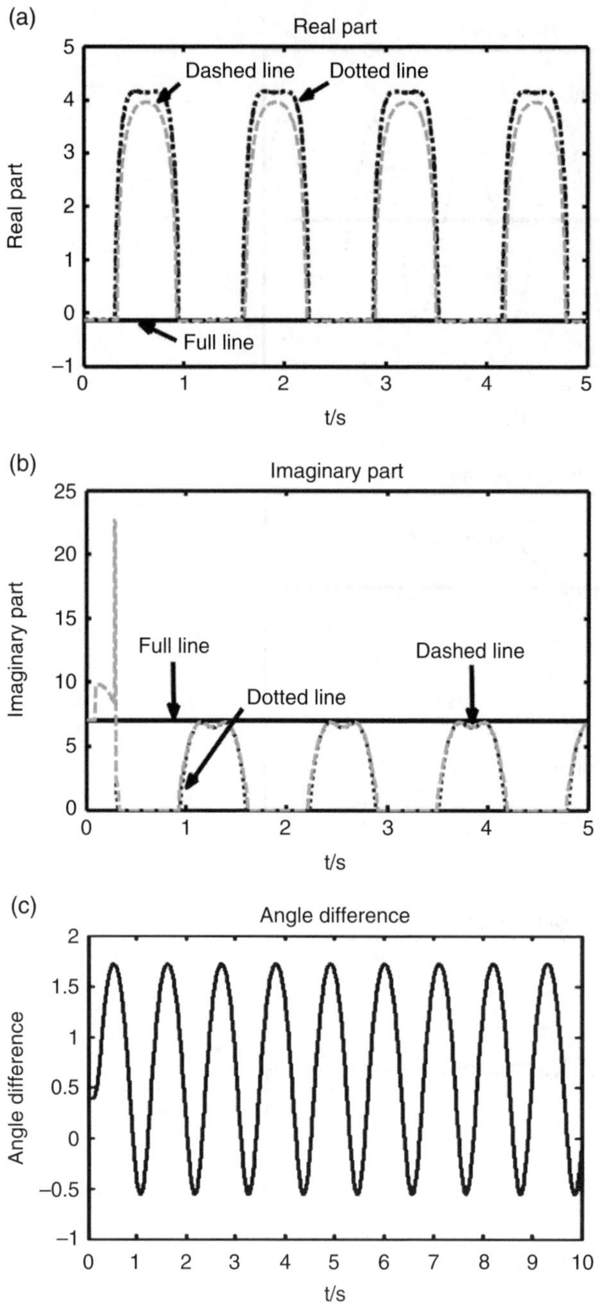

Figure 4.25 The simulation results when α is 0.5 and t_c is 0.175 s.

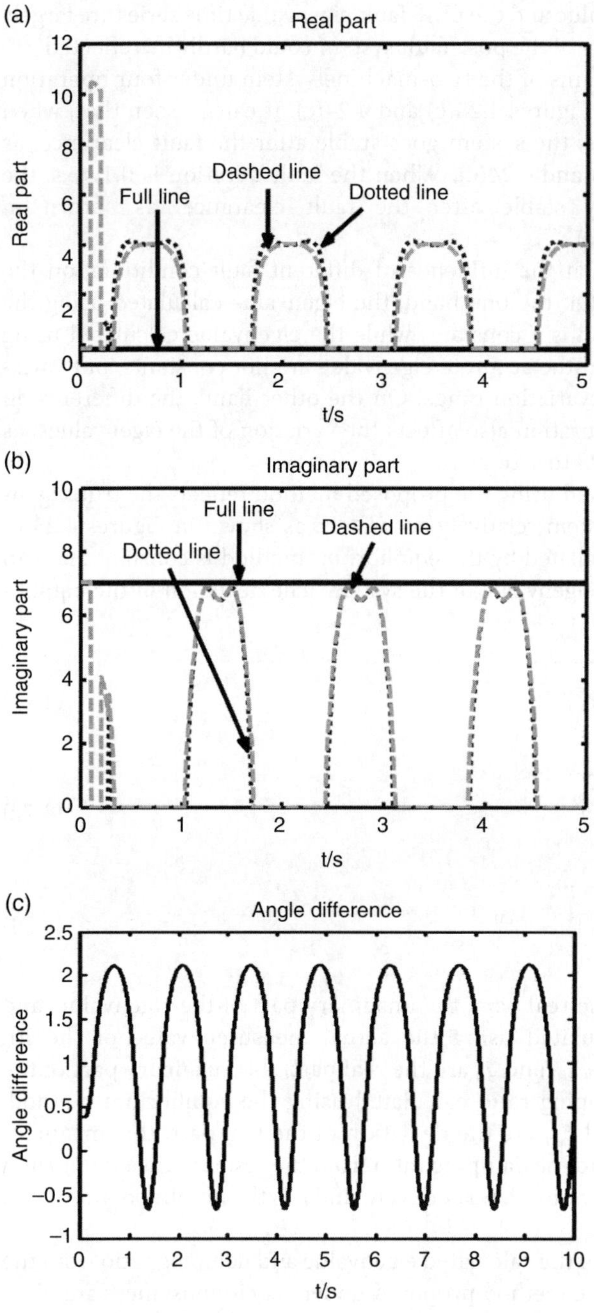

Figure 4.26 The simulation results when α is 0.9 and t_c is 0.175 s.

equilibrium point eigenvalue and the post-fault eigenvalue time series are largely different; thus the stability of the post-fault system could hardly be reflected.

The power angle diagrams of the two-machine system under four operation conditions are shown in Figures 4.23(c) and 4.24(c). It can be seen that, when the fault duration is 0.1 s, the system goes stable after the fault clearance, as shown in Figures 4.23(c) and 4.24(c). When the fault duration is 0.17 5 s, the system goes marginally stable after the fault clearance, as shown in Figures 4.25(c) and 4.26(c).

It can be concluded that the influence of different fault conditions on the eigenvalues is different. On the one hand, the eigenvalue calculated using the equilibrium point method is a constant, while the eigenvalue calculated using the proposed method and the accurate eigenvalue are not constants, but variables with a considerable variation range. On the other hand, the difference in fault location and fault duration also affects the variation of the eigenvalues, as can be seen in Figures 4.23 to 4.26.

The eigenvalue calculated using the proposed method reflects the true eigenvalue variation of the system relatively accurately, as shown in Figures 4.23 to 4.26. The eigenvalue calculated by the equilibrium method is constant and can not reflect the real-time eigenvalue of the system. The deviation of the equilibrium method is given by

$$R_{err} = \sqrt{\frac{1}{n}\sum_{i=1}^{n}\left(R_i - R_0\right)^2} \tag{4.71}$$

$$I_{err} = \sqrt{\frac{1}{n}\sum_{i=1}^{n}\left(I_i - I_0\right)^2} \tag{4.72}$$

$$\xi_{err} = \sqrt{\frac{1}{n}\sum_{i=1}^{n}\left(\xi_i - \xi_0\right)^2} \tag{4.73}$$

where R_i, I_i, and ξ_i are the real part, the imaginary part of the eigenvalue, and the damping ratio calculated using the actual measured value of the ith moment, respectively. R_0, I_0, and ξ_0 are the real part, the imaginary part of the eigenvalue, and the damping ratio calculated using the equilibrium method, respectively. R_{err}, I_{err}, and ξ_{err} are the deviation of the real part, the imaginary part of the eigenvalue, and the damping ratio from the result of the equilibrium method to the method proposed, respectively, and n is the number of the actual measured point.

The relative deviation of the calculated eigenvalue and damping ratio from the equilibrium method to the method proposed under six circumstances are illustrated in Table 4.12.

It can be seen that as the duration of the short-circuit fault and the fault position increase, the deviation of the eigenvalue and damping ratio calculated by the two methods increases as well. Table 4.12 also illustrates that the equilibrium method cannot reflect the real-time eigenvalue change of the fault system since it does not consider the uncertain fault factor.

Table 4.12 Mean square error of the equilibrium point method.

Relative deviation	Real part	Imaginary part	Damping ratio (%)
Case 1	0.0001	0.1541	0.0913
Case 2	0.0001	0.1566	0.0935
Case 3	0.0002	0.0083	0.5309
Case 4	0.1680	0.2904	4.2019

Simulation Analysis of Sixteen-Generator System The IEEE 16-generator 68-bus New England to New York interconnected system is introduced to verify the correctness and effectiveness of the proposed method. It can be divided into five large areas and the generators are adopted as a six-order detailed model; the excitation type is IEEE-DC1. A three-phase short-circuit fault occurs at a certain point on line 28–29 at 0.1 s. The proportion of the distance between the fault point and node 28 to the whole length of the line is α. The change of eigenvalue of the system after fault clearance is analyzed using four cases is shown in Table 4.13.

Substitute the trajectory obtained from the system simulation into equation (4.70) and the time series of the eigenvalue is calculated. The time series of the inter-area oscillation mode of area 4 and area 5 are shown in figures from Figures 4.27 to 4.30. In these figures, the full line represents the result of the equilibrium method while the dashed line represents the accurate eigenvalue of the system and the dotted line represents the result of the proposed method.

It can be seen from Figures 4.27 and 4.28 that, when the fault duration is 0.05 s, the influence of fault location variation on the eigenvalues is small. The real part of the accurate eigenvalue varies between –0.0258 and –0.0047 and the virtual part between 0.15 and 2.4590. With the proposed method, the calculated real part varies between –0.04 and 0 and the virtual part between 0 and 2.4590.

It can be seen from Figures 4.28 and 4.30 that, when the fault duration is 0.085 s, the system falls into a marginally stable state after the fault clearance and the influence of the fault location variation on the eigenvalues is obvious. Though the variation ranges of the accurate value and the calculated value are close, their values at the same instant are largely different.

It can be seen from Figures 4.27 to 4.30 that the pre-fault equilibrium point eigenvalue of the 16-machine system is a constant: –0.0053 + 2.4690i. After the

Table 4.13 Fault type of a 16-machine system.

Fault type	Fault position α	Duration t_c
Case 1	0.5	0.05
Case 2	0.9	0.05
Case 3	0.5	0.085
Case 4	0.9	0.085

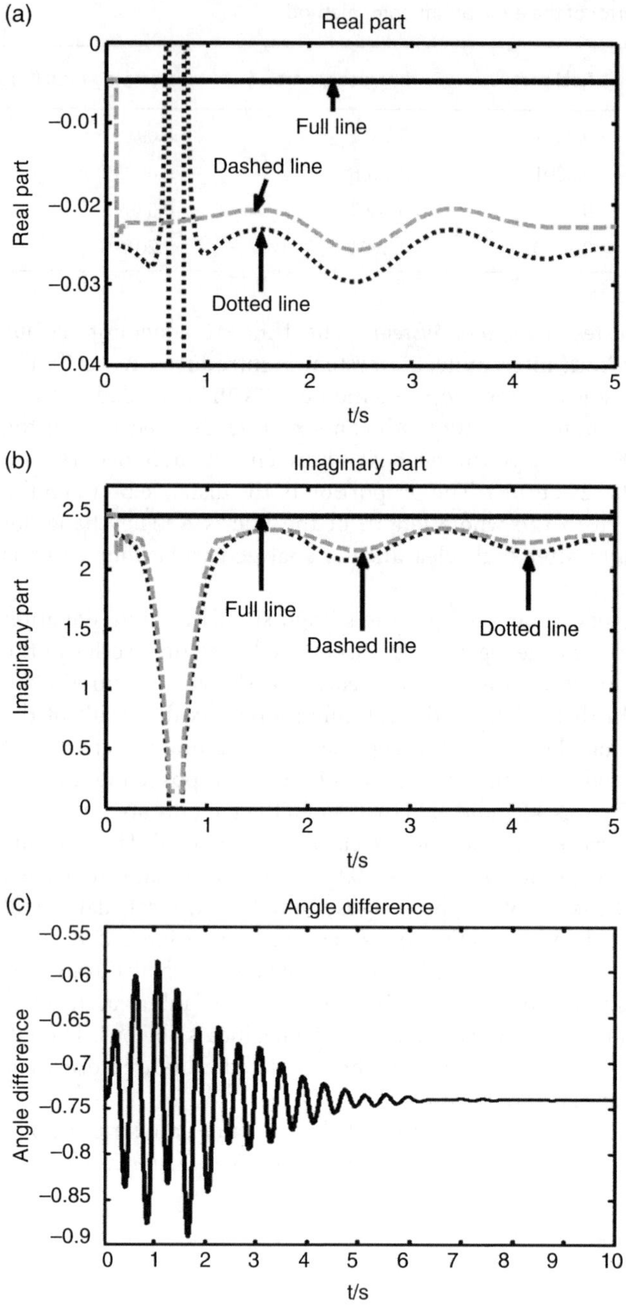

Figure 4.27 The simulation results when α is 0.5 and t_c is 0.05 s.

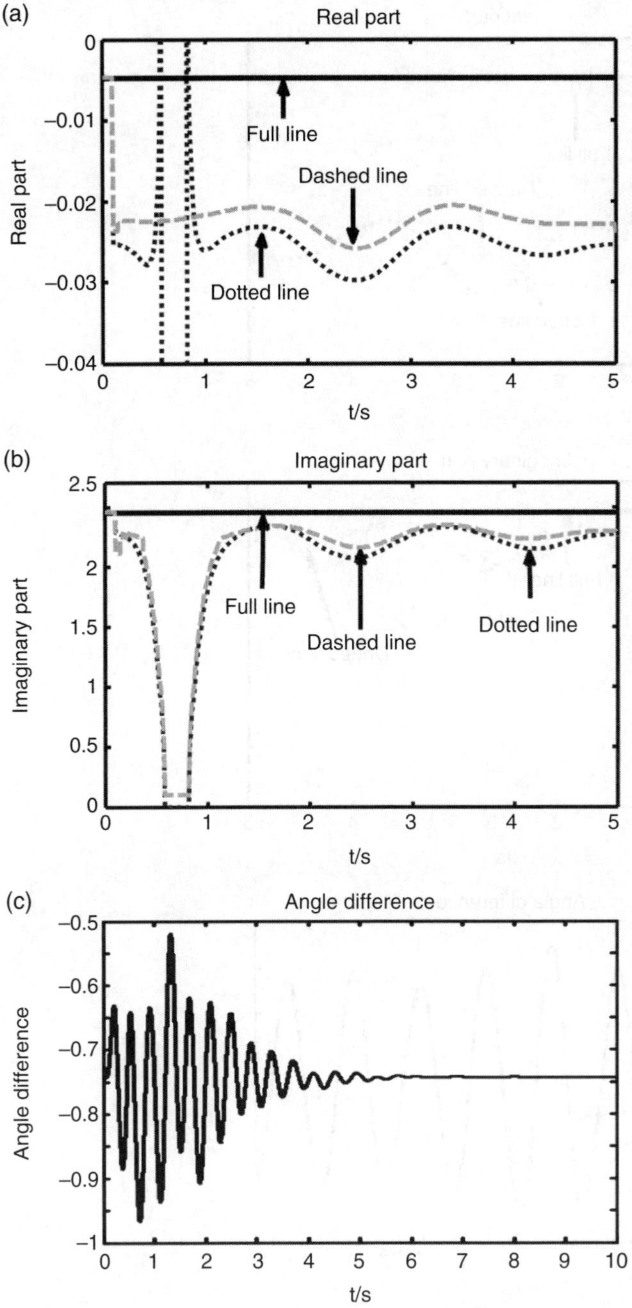

Figure 4.28 The simulation results when α is 0.9 and t_c is 0.05 s.

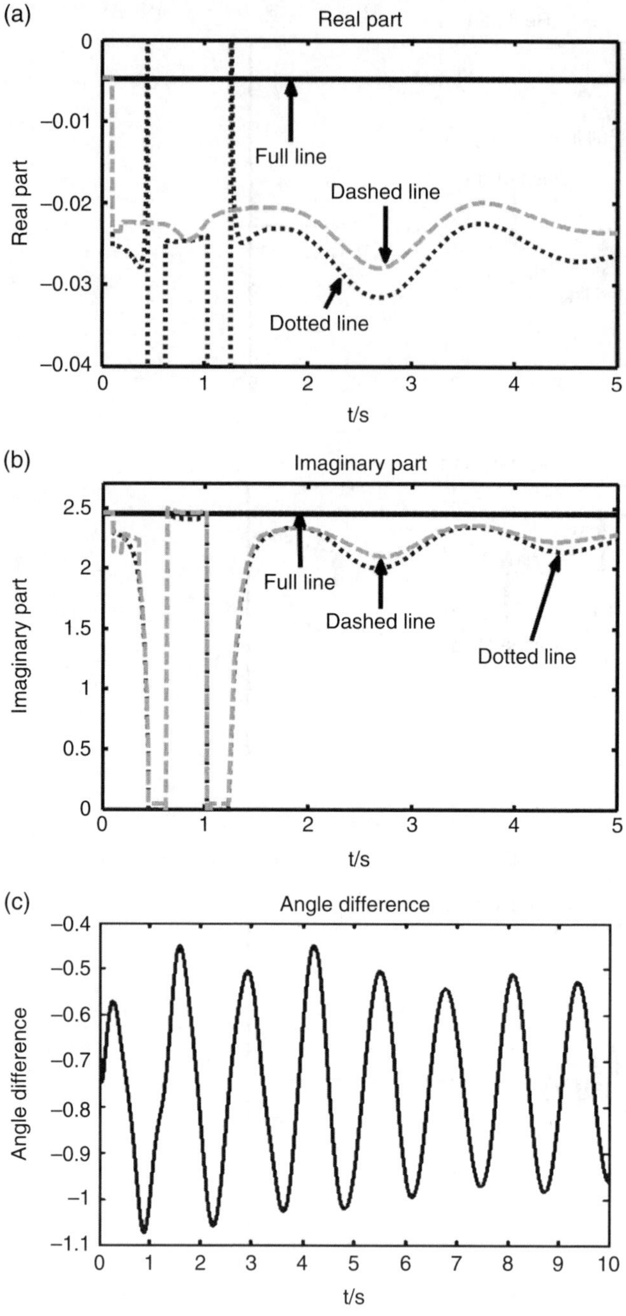

Figure 4.29 The simulation results when α is 0.5 and t_c is 0.085 s.

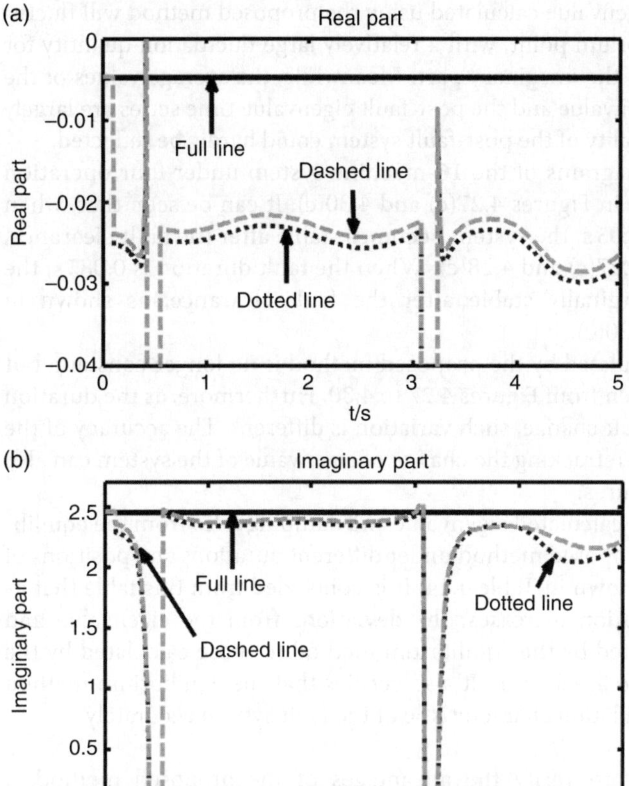

Figure 4.30 The simulation results when α is 0.9 and t_c is 0.085 s.

fault clearance, the eigenvalue calculated using the proposed method will fluctuate around the equilibrium point, with a relatively large fluctuation quantity for both the real part and the imaginary part. Meanwhile, the average values of the equilibrium point eigenvalue and the post-fault eigenvalue time series are largely different; thus the stability of the post-fault system could hardly be reflected.

The power angle diagrams of the 16-machine system under four operation conditions are shown in Figures 4.27(c) and 4.30(c). It can be seen that, when the fault duration is 0.05 s, the system becomes stable after the fault clearance, as shown in Figures 4.27(c) and 4.28(c). When the fault duration is 0.085 s, the system becomes marginally stable after the fault clearance, as shown in Figures 4.29(c) and 4.30(c).

The eigenvalue calculated by the proposed method is no longer constant, but varies with time, as seen from Figures 4.27 to 4.30. Furthermore, as the duration and position of the fault change, such variation is different. The accuracy of the proposed method when tracking the change of eigenvalue of the system can also be seen from these figures.

The deviation of the calculated eigenvalue and damping ratio from the equilibrium method to the proposed method under different durations and positions of short circuit fault is shown in Table 4.14. It is concluded from this table that as the short-circuit duration increases, the deviations from the eigenvalue and damping ratio calculated by the equilibrium method to those calculated by the proposed method increase as well. It also verifies that the equilibrium method cannot reflect the oscillation characteristic of the fault system accurately.

Comparison In order to verify the advantages of the proposed method, a comparison is made between the proposed method and the trajectory eigenvalues method in reference [10].

The trajectory eigenvalues method in reference [10] is based on piecewise linearization of the disturbed trajectory; that is, along the real disturbed trajectory, the mathematical model of the nonlinear system is linearized at the beginning of each integration step and then the time series of the eigenvalues of each piecewise linear system is calculated. By mapping the above piecewise linearization dynamic equations to a series of a time-variant single machine system and using the extended equal area criterion (EEAC) to recognize the oscillation modes in each integration step, the eigenvalue time series of EEAC map trajectory section can be gained.

The time-variant eigenvalue series obtained with this method is able to reflect the time-variation trend of the oscillatory characteristics on the whole. However, in solving the EEAC, the time-response curves of the multiple-machine system need to be converged to a leading cluster and the rest cluster according to the complementary cluster center of inertia (CCCOI) transform. This equivalence will lower the accuracy of the result. For the 16-machine system, for example, the mean square errors (MSE) of the two methods taken from the accurate value are shown in Table 4.15. It can be seen that the accuracy of the proposed method is obviously higher than that of the trajectory eigenvalue method, especially in the real part.

Table 4.14 Mean square error of the equilibrium point method.

Relative deviation	Real part	Imaginary part	Damping ratio (%)
Case 1	0.0011	0.0101	0.0504
Case 2	0.0011	0.0101	0.0504
Case 3	0.0012	0.0122	0.0536
Case 4	0.0013	0.0151	0.0586

Table 4.15 Comparison of MSE (mean square error) between the trajectory method and the proposed method.

Methods	Fault position	Fault Duration(s)	Real part	Imaginary part
Trajectory eigenvalues	0.5	0.05	0.0418	4.54×10^{-4}
	0.9	0.085	0.0409	4.40×10^{-4}
	0.5	0.05	0.0396	4.47×10^{-4}
	0.9	0.085	0.0388	3.90×10^{-4}
Proposed method	0.5	0.05	0.0029	1.87×10^{-4}
	0.9	0.085	0.0031	1.69×10^{-4}
	0.5	0.05	0.0027	1.64×10^{-4}
	0.9	0.085	0.0017	1.06×10^{-4}

In addition, the trajectory eigenvalue method needs to rank the stability degrees of the mappings according to the stability margin and find the dominant mapping in the system. The bigger the system, the longer this will take. According to the simulation results, when the simulation time step is 0.005 s (on a Lenovo M490 PC, which has a 64-bit operating system and 4G of memory), it takes the proposed method 0.242 s to calculate the eigenvalue time series of the 16-machine system, while it takes the trajectory eigenvalue method much longer – 15.453 s.

In conclusion, compared with current methods, the proposed method is faster and more accurate.

4.3.2 Angle Stability Analysis of Power System Considering Cascading Failure

In recent years, as power system becomes ever more complicated, the stability problem of a power system with multiple operating conditions caused by topology variation due to a power device and component failure has become increasingly prominent, greatly endangering the safety and stability of a power system. Thus a method to analyze the angle stability of a power system with multiple operating conditions considering cascading failure is proposed in this section. The proposed method could identify the stability of a power system with multiple operating conditions effectively, with no need to obtain the operating trajectory of the system. Compared with a traditional time-domain simulation method, the computation amount is reduced. In addition, it is simple and easy to implement.

4.3.2.1 Model of Power System with Multiple Operating Conditions Considering Cascading Failure

Theoretical Analysis of Cascading Failure Existing research [11,12] demonstrates that maloperation of line protection is the main cause of cascading failure. Currently, protection configuration on transmission lines mainly consists of of pilot protection and distance protection. Pilot protection serves as the primary protection, which clears the fault immediately after a fault occurs on the protected line. Distance protection serves as the backup protection, which ensures the safety of this line and the neighboring line [11]. When the system is in normal operation, the lines are moderately loaded. Once a fault occurs in the system, relay protection will quickly operate to cut the corresponding fault line and the load on this line will be transferred to the parallel section or other lines. After the load transfer, if the power flow on certain lines exceeds the setting value of distance protection, these lines will be tripped from service due to overload and power flow transfer will take place in the system again, leading to another stage of cascading fault.

Suppose that the line state set r_j only contains two elements "0" and "1", which represents the line in service and the line outage, respectively. Introduce α_{ij}, β_{ij}, and γ_{ij} as the evaluation indexes for the power flow variation rate, overload severity degree, and the coupling relationship between flow variation amount and the original flow of line j after line i is cut off at the mth stage of the cascading failure [13]:

$$\alpha_{ij} = \left| \frac{S_{j(t_b)} - S_{j(t_a)}}{S_{j(t_a)}} \right| \tag{4.74}$$

$$\beta_{ij} = \left| \frac{S_{j(t_b)}}{S_{j\max}} \right| \tag{4.75}$$

$$\gamma_{ij} = \left| \frac{S_{j(t_b)} - S_{j(t_a)}}{S_{i(t_a)}} \right| \tag{4.76}$$

where $S_{i(t_a)}$ represents the power flow on line i before the fault, $S_{j(t_a)}$ represents the power flow on line j before the fault, $S_{j(t_b)}$ represents the power flow on line j after the fault occurs on line i, and $S_{j\max}$ is the maximum power flow allowed on line j.

Introduce the weighting factor q_j, which represents the proportion of the power flow on line j in the whole power flow:

$$q_j = \frac{\left|S_{j(t_b)}\right|}{\sum_{i=1}^{l}\left|S_{i(t_b)}\right|} \tag{4.77}$$

At the mth stage of the cascading failure, the state transfer probability considering the incorrect operation of relay protection can be expressed as

$$p_{ij} = p_{ij\_ori}\left[(1-p_r)(1-p_c)+p_{err}\right] \tag{4.78}$$

$$p_{ij\_ori} = \frac{D_{ij}}{\sum_{j=1}^{l} D_{ij} - D_{ii}} \tag{4.79}$$

$$D_{ij} = q_j D_i \alpha_{ij} \beta_{ij} \gamma_{ij} \tag{4.80}$$

$$D_i = \left|\frac{S_i}{S_{i\max}}\right| \tag{4.81}$$

where p_r is the probability of line overload protection refusing to operate, p_c is the probability that the circuit breaker corresponding to line overload protection refuses to operate, p_{err} is the probability of a line breaking off when relay protection devices operate incorrectly, S_i is the practical power flow on line i, $S_{i\max}$ is the maximum power flow on line i, D_i is the load rate on the line i, D_{ij} is the correlation degree between the former and the later stage of cascading failure, and p_{ij_ori} is the state transfer probability neglecting the incorrect operation of relay protection devices.

In a cascading failure, after each stage of fault occurs, the line flow transfer amount is only related to the structure parameters of the former stage fault network. Thus, the calculation of line state transfer probability is also related to the former stage of the fault alone. Therefore, the discrete Markov system model could be used for modeling a power system with multiple operating conditions when considering cascading failure.

Model of Power System with Multiple Operating Conditions Based on the Discrete Markov Process The model of a power system with multiple operating conditions can be expressed with the discrete Markov jump system as

$$\begin{cases} x_{k+1} = \left(A(s_k)+\Delta A_k(s_k)\right)x_k + B(s_k)u_k + G(s_k)\omega_k \\ z_k = C(s_k)x(t) + D(s_k)u_k + L(s_k)\omega_k \end{cases} \tag{4.82}$$

where $x_k \in R^n$ is the state vector, $u_k \in R^p$ is the control input vector, $z_k \in R^r$ is the control output vector, ω_k is a process noise sequence, the mean value of which is zero, and $\{s(t), t \geq 0\}$ is a Markov chain defined in the finite space $S = \{1,2,\cdots,l\}$, which corresponds to possible system operating conditions in the cascading failure and describes the variation of system operating condition with the development of the fault. The state probability of the Markov process p_{ij} is

$$\mathbf{Pr}\{s_{k+1} = j | s_k = i\} = p_{ij} \qquad \sum_{j=1} p_{ij} = 1(p_{ij} \geq 0) \tag{4.83}$$

For each $s(t) = i \in S$, denote $A(s_k)$, $B(s_k)$, $G(s_k)$, $C(s_k)$, $D(s_k)$, and $L(s_k)$ as A_i, B_i, G_i, C_i, D_i, and L_i, respectively. The uncertain parameters satisfy the following matching conditions:

$$\Delta A_k(s_k) = \Delta A_k(i) = H_i F_k(i) M_i \tag{4.84}$$

where H_i and M_i are stationary real matrices which reflect the structure information of system uncertain parameters. $\boldsymbol{F}_k(i)$ is unknown time-variant matrix which reflects the degree of uncertainty. H_i, M_i and $\boldsymbol{F}_k(i)$ can be calculated according to the matrix decomposition theory. Matrix $\boldsymbol{F}_k(i)$ satisfies:

$$F_k^T(i) F_k(i) \leq I \tag{4.85}$$

In (4.82) to (4.85), the model of a power system with multiple operating conditions considering cascading failure is established. Based on this model, a stability criterion is derived in this section to analyze the stability of a power system varying between multiple operating conditions.

4.3.2.2 Angle Stability Analysis of Power System with Multiple Operating Conditions Considering Cascading Failure

Definition 4.3 [13]. For the system in (4.82), when $\boldsymbol{u}_k = 0$ and $\omega_k = 0$, if there exists a finite positive constant $\tilde{\boldsymbol{M}}(\boldsymbol{x}_0, s_0)$ that satisfies the inequality in (4.86) for any initial condition $(\boldsymbol{x}_0, s_0)$, then the discrete Markov stochastic system in (4.82) is robust stochastic stable:

$$\lim_{N\to\infty} E\left\{\sum_{k=0}^{N} |\boldsymbol{x}_k|^2 \middle| \boldsymbol{x}_0, s_0\right\} \leq \tilde{\boldsymbol{M}}(\boldsymbol{x}_0, s_0) \tag{4.86}$$

where $E\{...\}$ represents the expectation of $\{...\}$.

Theorem 4.4 For a given positive constant $\gamma > 0$, if there exists a set of symmetric positive definite matrices $Q_i > 0$ $(i \in S)$ that satisfy the following matrix inequalities:

$$\begin{bmatrix} -Q_i & 0 & Q_i A_i^T \Omega_i & Q_i C_i^T \\ * & -\gamma^2 I & G_i^T \Omega_i & L_i^T \\ * & * & -\psi & 0 \\ * & * & * & -I \end{bmatrix} < 0 \tag{4.87}$$

$$\Omega_i = \left[\sqrt{p_{i1}}I \quad \sqrt{p_{i2}}I \quad \cdots \quad \sqrt{p_{il}}I\right] \tag{4.88}$$

$$\psi = diag\left(Q_1, Q_2, \ldots, Q_l\right) \tag{4.89}$$

then when $u(t) \equiv 0$, the system in (4.82) is robust stochastic stable and satisfies the disturbance attenuation degree γ, that is,

$$\sum_{k=0}^{N} E\left[\|z_k\|^2\right] \leq \gamma^2 \sum_{k=0}^{N} E\left[\|\omega_k\|^2\right] \tag{4.90}$$

Proof. According to (4.87), for $i \in S$, the following LMI is true:

$$\begin{bmatrix} -Q_i & Q_i A_i^T \Omega_i \\ * & -\psi \end{bmatrix} < 0 \tag{4.91}$$

Therefore, if it could be proved that the system in (4.82) is stochastic stable when (4.91) is satisfied, it could then be proved that the stochastic stability criterion in Theorem 4.4 is true.

Premultiplying the left side and postmultiplying the right side of (4.91) by matrix $diag(P_1, I)$, then

$$\begin{bmatrix} -P_i & A_i^T \Omega_i \\ * & -\Sigma \end{bmatrix} < 0 \tag{4.92}$$

where

$$\Sigma = diag\left(P_1^{-1}, P_2^{-1}, \ldots, P_l^{-1}\right) \tag{4.93}$$

According to the Schur complement theorem and (4.92), there exists a scalar quantity $\delta > 0$ that satisfies the following inequality:

$$A_i^T \Omega_i \sum^{-1} \Omega_i^T A_i - P_i + \delta I < 0 \tag{4.94}$$

Define the state-related matrix $M_k(s_k)$ as

$$M_k\left(s_k\right) = E\left[x_k\left(s_k\right) x_k^T\left(s_k\right)\right] \tag{4.95}$$

and construct the following Lyapunov functional:

$$V_k\left(M_k\left(s_k\right), s_k\right) = tr\left[M_k\left(s_k\right) P\left(s_k\right)\right]\left(P\left(s_k\right) > 0\right) \tag{4.96}$$

Thus for $i \in S$,

$$\begin{aligned} V_{k+1}\left(M_{k+1}(i), i\right) &= tr\left[A_i M_k(i) A_i^T P_i\right] \\ &= tr\left[M_k(i) A_i \Omega_i \sum^{-1} \Omega_i^T A_i^T\right] \\ &< tr\left[M_k(i) P_i\right] = V_k\left(M_k(i), i\right) \end{aligned} \tag{4.97}$$

According to (4.97),

$$\begin{aligned}
&\frac{E\left\{V_{k+1}\left(M_{k+1}\left(s_{k+1}\right),s_{k+1}\right)\middle|x_k,s_k\right\}-V_k\left(M_k\left(s_k\right),s_k\right)}{V_k\left(M_k\left(s_k\right),s_k\right)}\\
&\qquad=\frac{E\left\{V_{k+1}\left(M_{k+1}\left(s_{k+1}\right),s_{k+1}\right)\middle|x_k,s_k\right\}}{V_k\left(M_k\left(s_k\right),s_k\right)}-1\\
&\qquad=\alpha-1\\
&\qquad<-\min_{s_k\in S}\left(\frac{\lambda_{\min}(\Sigma)}{\lambda_{\min}\left(P\left(s_k\right)\right)}\right)
\end{aligned} \tag{4.98}$$

where

$$\alpha=\frac{E\left\{V_{k+1}\left(M_{k+1}\left(s_{k+1}\right),s_{k+1}\right)\middle|x_k,s_k\right\}}{V_k\left(M_k\left(s_k\right),s_k\right)} \tag{4.99}$$

Consider that

$$\begin{aligned}
&\frac{E\left\{V_{k+1}\left(M_{k+1}\left(s_{k+1}\right),s_{k+1}\right)\middle|x_k,s_k\right\}}{V_k\left(M_k\left(s_k\right),s_k\right)}>0\\
&\min_{s_k\in S}\left(\frac{\lambda_{\min}(\Sigma)}{\lambda_{\min}\left(P\left(s_k\right)\right)}\right)>0
\end{aligned} \tag{4.100}$$

Thus, α satisfies $1>\alpha>0$ and

$$E\left\{V_{k+1}\left(M_{k+1}\left(s_{k+1}\right),s_{k+1}\right)\middle|x_k,s_k\right\}<\alpha V_k\left(M_k\left(s_k\right),s_k\right) \tag{4.101}$$

Conduct iteration on (4.101) from 0 to k, so that

$$E\left\{V_{k+1}\left(M_{k+1}\left(s_{k+1}\right),s_{k+1}\right)\middle|x_0,s_0\right\}<\alpha^k V_0\left(M_0\left(s_0\right),s_0\right) \tag{4.102}$$

According to (4.102),

$$\begin{aligned}
&E\left\{\sum_{k=0}^{N}V_k\left(M_k\left(s_k\right),s_k\right)\middle|x_0,s_0\right\}\\
&\qquad<\left(1+\alpha+\cdots+\alpha^N\right)V_0\left(M_0\left(s_0\right),s_0\right)\\
&\qquad<\frac{1}{1-\alpha}V_0\left(M_0\left(s_0\right),s_0\right)
\end{aligned} \tag{4.103}$$

Thus

$$E\left\{\sum_{k=0}^{N}\left\|x_k\right\|^2\middle|x_0,s_0\right\}<\frac{\min_{s_k\in S}\lambda_{\min}\left(P\left(s_k\right)\right)}{1-\alpha}V_0\left(M_0\left(s_0\right),s_0\right) \tag{4.104}$$

Denote

$$\tilde{N}(x_0, s_0) = \frac{\min\limits_{s_k \in S} \lambda_{\min}(P(s_k))}{1-\alpha} V_0(M_0(s_0), s_0) \tag{4.105}$$

According to (4.104) and (4.105),

$$\lim_{N\to\infty} E\left\{\sum_{k=0}^{N} \|x_k\|^2 |x_0, s_0\right\} < \tilde{N}(x_0, s_0) \tag{4.106}$$

that is, the system in (4.82) is stochastic stable. In addition, when $\omega(t) \neq 0$, the system in (4.82) has the disturbance attenuation degree γ, which is proved below.

Premultiply and multiply (4.87) by matrix $diag(P_i, I, I, I)$, so that the following inequality results can be obtained:

$$\begin{bmatrix} -P_i & 0 & A_i^T \Omega_i & C_i^T \\ * & -\gamma^2 I & G_i^T \Omega_i & L_i^T \\ * & * & -\Sigma & 0 \\ * & * & * & -I \end{bmatrix} < 0 \tag{4.107}$$

According to the Schur complement theorem, equation (4.87) is equivalent to

$$\begin{bmatrix} B_{11} & B_{12} \\ * & B_{22} \end{bmatrix} < 0 \tag{4.108}$$

where

$$B_{11} = A_i^T \Omega_i \sum^{-1} \Omega_i^T A_i - P_i + C_i^T C_i \tag{4.109}$$

$$B_{12} = A_i^T \Omega_i \sum^{-1} \Omega_i^T G_i + C_i^T L_i \tag{4.110}$$

$$B_{22} = G_i^T \Omega_i \sum^{-1} \Omega_i^T G_i + L_i^T L_i - \gamma^2 I \tag{4.111}$$

Construct the following Lyapunov functional:

$$V_k(x_k, s_k) = x_k^T P_i x_k \quad (P_i > 0, i \in S) \tag{4.112}$$

and define the index function:

$$J_N = E\left[\sum_{k=0}^{N} \left(\|z_k\|^2 - \gamma^2 \|\omega_k\|^2\right)\right] \tag{4.113}$$

According to (4.113),

$$\begin{aligned} J_N &= E\left\{\sum_{k=0}^{N}\left(\|z_k\|^2 + E\left[V_{k+1}\left(x_{k+1}, s_{k+1}\right)\middle| x_k, s_k\right]\right.\right. \\ &\quad \left.\left. - V_k\left(x_k, s_k\right) - \gamma^2\|\omega_k\|^2\right)\right\} - E\left[V_{N+1}\left(x_{N+1}, s_{N+1}\right)\right] \\ &\le E\left\{\sum_{k=0}^{N}\left(\|z_k\|^2 + E\left[V_{k+1}\left(x_{k+1}, s_{k+1}\right)\middle| x_k, s_k\right]\right.\right. \\ &\quad \left.\left. - V_k\left(x_k, s_k\right) - \gamma^2\|\omega_k\|^2\right)\right\} \end{aligned} \tag{4.114}$$

For $i \in S$, according to (4.108),

$$\begin{aligned} J_N &\le E\left\{\sum_{k=0}^{N}[C_i x_k + L_i\omega_k]^T[C_i x_k + L_i\omega_k]\right. \\ &\quad + E\left[(A_i x_k + G_i\omega_k)^T P(s_{k+1})(A_i x_k + G_i\omega_k)\right. \\ &\quad \left.\left. - x_k^T P(s_k)x_k \middle| x_k, s_k = i\right] - \gamma^2\omega_k^T\omega_k\right\} \\ &= \begin{bmatrix} x_k^T & \omega_k^T \end{bmatrix}\begin{bmatrix} B_{11} & B_{12} \\ * & B_{22} \end{bmatrix}\begin{bmatrix} x_k \\ \omega_k \end{bmatrix} < 0 \end{aligned} \tag{4.115}$$

$$\sum_{k=0}^{N} E\left[\|z_k\|^2\right] \le \gamma^2 \sum_{k=0}^{N} E\left[\|\omega_k\|^2\right] \tag{4.116}$$

That is the system in (4.82) has the disturbance attenuation degree γ.

The stability criterion of a power system with multiple operating conditions described in (4.86) can be transformed to the following optimization problem:

$$\min_{Q_i, \gamma} t \quad s.t. \quad (4.87)-(4.89) \tag{4.117}$$

Use the feasibility problem-solving method to solve the LMI. If there exist $t < 0$ and positive definite matrix $P_i > 0$ that satisfy Theorem 4.4, then the system in (4.82) is robust stochastic stable and satisfies the disturbance attenuation degree γ. Otherwise, the system is unstable.

After the system angle stability is identified, the relationship between line fault probability and system stability is further studied. According to the stability criterion in (4.87) and (4.88), the factors that determine the stability of the Markov jump system not only include different operating conditions that form the jump system but also the transfer probability between the operating conditions (p_{ij}). It can be seen from (4.74) to (4.81) that, neglecting the influence of relay protection devices, the transfer probability between different operating conditions p_{ij} is dependent on four parameters: α_{ij}, β_{ij}, γ_{ij}, and D_i. Among the four parameters, β_{ij} and D_i are dependent on the transmission capacity of transmission lines $S_{i\max}$ and α_{ij} and γ_{ij} are dependent on flow distribution on the line. Thus, by adjusting line transmission capacity or line flow distribution, p_{ij} could be adjusted and then system stability could be improved.

Denote the matrix transformed from (4.82) to (4.87) as **R** and its eigenvalues as λ_i (i = 1, 2,..., n). Define the following stability degree index:

$$\mu = \frac{\sum_{i=1}^{n} imag(\lambda_i)}{n} \tag{4.118}$$

where the function *imag* means extracting the imaginary part of the variable. The stability degree index μ reflects the negative definite degree of a matrix. The smaller μ is, the farther the frequency-domain distribution of the matrix eigenvalues in the negative half-plane is from the imaginary axis, and thus the more stable the system represented by the matrix is. This index is a quantification of the influence of line fault probability on system stability, which could provide a solid reference for power system construction and for operation personnel to improve system stability by changing the transmission capacity of transmission lines or changing the system operation mode.

4.3.2.3 Simulation Analysis

Test System Take the IEEE 16-machine 68-bus New England–New York interconnected system shown in Figure 3.2, for example. The inter-area modes are obtained from the eigenvalue analysis and are displayed in Table 4.16. There are four critical inter-area modes. Generators 14, 15, and 16 are oscillating against generators 1–13 under mode 1, generator 14 is oscillating against generator 16 under mode 2, generators 12 and 13 are oscillating against generators 1–9 under mode 3, and generators 14 and 16 are oscillating against generator 15 under mode 4.

The WECC load model is assumed at all load buses. The real power is 80% constant current, the reactive power is 80% constant admittance, and all loads support the remaining 20% as an induction motor load. The simulations environment with a WECC load model is stricter than that with a constant load model. A reduced-order model based on selective modal analysis (SMA) is used to reduce the problem size. The size of **A** in

$$\mathbf{A}_t = \begin{bmatrix} \mathbf{0} & \omega_0 & \mathbf{I} \\ \frac{-k_{11t}}{M_1} & \frac{-k_{12t}}{M_1} & \frac{-K_{D1t}}{M_1} \\ \frac{-k_{21t}}{M_2} & \frac{-k_{22t}}{M_2} & \frac{-K_{D2t}}{M_2} \end{bmatrix}$$

Table 4.16 Inter-area modes of the study system.

Mode	Frequency (Hz)	Damping ratio	Dominant participated generator
1	0.2777	−0.0047	G1–G13 G14 G15 G16
2	0.4171	0.0142	G14 G16
3	0.6565	0.0283	G1–G9 G13 G12
4	0.7697	0.0323	G14 G15 G16

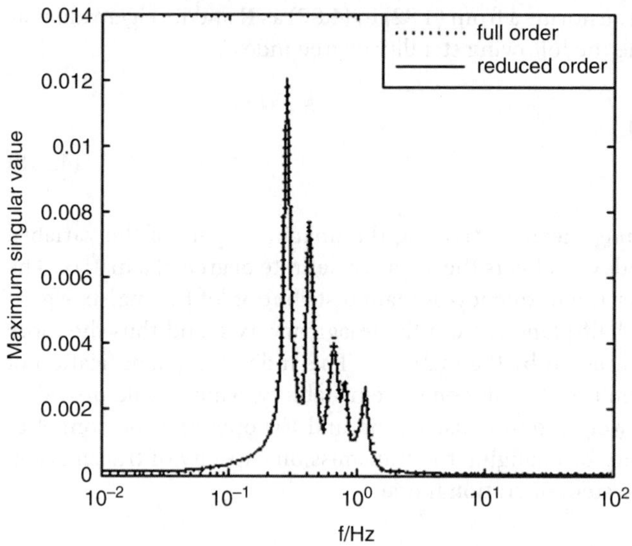

Figure 4.31 Multivariable frequency response plots of the input–output characteristics of a power system.

is reduced from 123 to 30 using SMA. When SMA converges, a reduced-order model preserving low-frequency eigenvalues has been obtained [14].

The test system has 141 states in the open loop; thus the matrix operation process will be rather time-consuming. For a large-scale system, the system order needs to be reduced in depth, on the premise that the input–output characteristics remain the same. The original system is reduced to a 10-order plant using the robust control toolbox, with Matlab performing the necessary computations. Schur model reduction [15] is used for order reduction of the model, which could reserve the outputs and inputs of the original system. The multivariable frequency responses are shown in Figure 4.31, which demonstrates that the full-order system could be represented by the reduced-order system.

It can be seen from Figure 4.31 that the reduced-order model has similar input–output characteristics in the desired frequency range to that of the full-order system. An order of 10 has proved to be satisfactory.

Stochastic Stability Analysis of Power System with Multiple Operating Conditions Suppose that the initial contingency is a line 4–14 fault and is cut out from service (outage). The load of line 4–14 is then transferred to the remaining lines. Calculate the state transfer probability of each remaining line p_{ij} according to (4.74) to (4.81), which is also the probability that the next stage of failure (in the cascading failure) occurs on each line. Shown in Table 4.17 is the state transfer probability of some lines after line 4–14 outage. If the previous fault path is the same, the bigger the transfer probability of a line, the more probable that the line will be the next fault line in the cascading failure. Therefore, the system operating conditions corresponding to a fault occurring on lines with relatively large state transfer probabilities are chosen to form the state set of the discrete Markov

Table 4.17 State transfer probability of some lines after line 4–14 outage.

Line number	State transfer probability
4–5	0.2542
14–15	0.1642
6–11	0.1410
10–11	0.1161
5–6	0.0906
3–18	0.0850
15–16	0.0103

Table 4.18 System operating condition set after line 4–14 outage.

Operating condition	Operating condition set
Condition 1	Line 4–14 outage
Condition 2	Line 4–14 and line 4–5 outage
Condition 3	Line 4–14 and line 14–15 outage
Condition 4	Line 4–14 and line 6–11 outage
Condition 5	Line 4–14 and line 10–11 outage
Condition 6	Line 4–14 and line 5–6 outage
Condition 7	Line 4–14 and line 3–18 outage
Condition 8	Line 4–14 and line 15–16 outage

system. Lines with relatively small state transfer probabilities are taken as nonfault lines.

To analyze the stability of a system when cascading fault occurs on the next line after line 4–14 outage, the state set shown in Table 4.18 is selected, denoted as $S = \{1, 2, \ldots, 8\}$.

Use the modal analysis method to obtain the state matrix of the 16-machine system in different operating conditions. Then use the SMA method to reduce the system order. The minimum damping ratio of the reduced-order system in different operating conditions is shown in Table 4.19. It can be seen that the system is unstable in two single operating conditions – Condition 3 and Condition 6. However, when the system varies from normal operating condition to Condition 1, and then to Condition 3 or Condition 6, instability does not result. As shown in Figure 4.32, the relative power angle between G1 and G13 gradually becomes stable after swinging for a short time. According to the feasibility problem-solving method, the LMI constructed in Theorem 4.4 is satisfied and a positive definite matrix $P_i > 0$ exists. Thus the system varying between multiple operating conditions is robust stochastic stable, which is consistent with the simulation result. On the other hand, stability analysis methods only dealing

Table 4.19 Minimum damping ratio of different operating conditions.

Operating condition	Minimum damping ratio
Condition 1	0.0237
Condition 2	0.0265
Condition 3	−0.0102
Condition 4	0.0253
Condition 5	0.0228
Condition 6	−0.0128
Condition 7	0.0259
Condition 8	0.0272

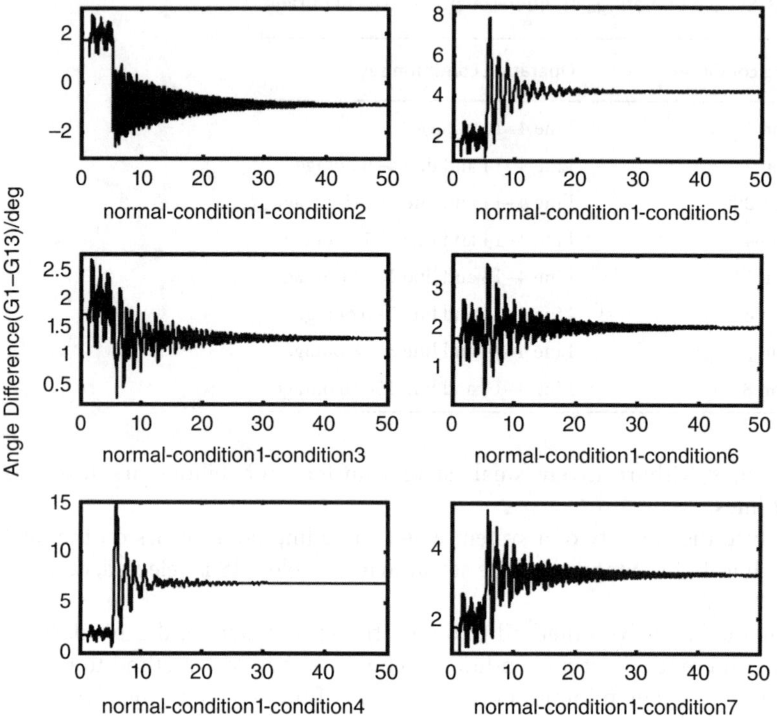

Figure 4.32 Dynamic responses of G1–G13 relative power angle in the 16-machine system.

with single operating conditions (such as the eigenvalue analysis method) may not correctly identify the stability of a system varying between multiple operating conditions due to cascading failure, even coming to the opposite result.

Stochastic Instability Cases of Power System with Multiple Operating Conditions Suppose that the initial contingency is a line 1–31 fault and is cut out from service. The load of line 1–31 is then transferred to the remaining

Table 4.20 State transfer probability of some lines after line 1–31 outage.

Fault type	Line number	State transfer probability
N-2	1–30	0.5956
N-2	30–31	0.3784
N-2	31–38	0.0105

Table 4.21 State transfer probability of some lines after line 1–31 and line 1–30 outage.

Fault type	Line number	State transfer probability
N-3	30–9	0.6784
N-3	1–47	0.2732
N-3	31–38	0.0392

lines. The state transfer probability of each remaining line p_{ij} is calculated according to (4.74) to (4.81), and the state transfer probability of some lines after line 1–31 outage (*N*-2) is shown in Table 4.20. Furthermore, the state transfer probability of some lines after line 1-31 and line 1-30 cascading outage (*N*-3) is shown in Table 4.21.

Select the system operating conditions corresponding to a fault occurring on lines with relatively large state transfer probabilities to form the state set of the discrete Markov system. Lines with relatively small state transfer probabilities are taken as non-fault lines. The selection results of the system operating conditions are shown in Table 4.22.

To analyze the stability of a system when a cascading fault occurs on the next line after line 1–31 outage (*N*-2 and *N*-3), their system state transfer probability matrices are calculated and shown as follows:

$$p_{N\text{-}2} = \begin{bmatrix} 0.0155 & 0.5956 & 0.3784 & 0.0105 \\ 0 & 1 & 0 & 0 \\ 0 & 0 & 1 & 0 \\ 0 & 0 & 0 & 1 \end{bmatrix} \tag{4.119}$$

$$p_{N\text{-}3} = \begin{bmatrix} 0.0092 & 0.6784 & 0.2732 & 0.0392 \\ 0 & 1 & 0 & 0 \\ 0 & 0 & 1 & 0 \\ 0 & 0 & 0 & 1 \end{bmatrix} \tag{4.120}$$

Use the modal analysis method to obtain the state matrix of the 16-machine system in different operating conditions. Then use the Schur method to reduce

Table 4.22 System operating condition set after line 1–31 outage

Operating condition	Fault type	Operating condition set
Condition 1	N-1	Line 1–31 outage
Condition 2	N-2	Line 1–31 and line 1–30 outage
Condition 3	N-2	Line 1–31 and line 30–31 outage
Condition 4	N-3	Line 1–31, line 1–30– and line 30-9 outage

Table 4.23 Minimum damping ratio of different operating conditions.

Operating condition	Minimum damping ratio
Condition 1	0.0545
Condition 2	0.0672
Condition 3	0.0552
Condition 4	0.0412

the system order. The minimum damping ratio of the reduced-order system in different operating conditions is shown in Table 4.23.

It can be seen that the system is stable in all single operating conditions. However, when the system varies from normal operating conditions to Condition 1, and then to the other conditions, instability results. As shown in Figure 4.33, the G1–G13 relative power angle and G1–G11 relative power angle become unstable. According to the feasibility problem-solving method, the LMI constructed in Theorem 4.4 is satisfied. However, there exists the nonpositive-definite matrix P_i; thus the system varying between multiple operating conditions is robust stochastic unstable, which is consistent with the simulation result. Therefore, a power system that is stable in all single operating conditions may become unstable when the system varies between multiple operating conditions due to cascading failure.

For normal dynamic simulations/time-domain simulations, when an initial fault is detected, permutations and combinations of all possible fault scenarios need to be conducted to form a series of fault chains. Time-domain simulation also needs to be carried out for all the fault chains. On this basis, the stability of a system in the case of cascading failure could be identified. As for the proposed method, permutations and combinations of all possible fault scenarios are not necessary. By using the Markov model to model and analyze the fault scenarios uniformly, the stability of a system in the case of cascading failure could be identified quickly. Thus, the main advantage of the proposed method is analysis speed.

When applied to the 16-machine system, consider the N-3 cascading outage. The computation time based on Matlab of normal dynamic simulations and the proposed method is compared in Table 4.24. The CPU of the computer used is Intel i5-4460 (3.2GHZ).

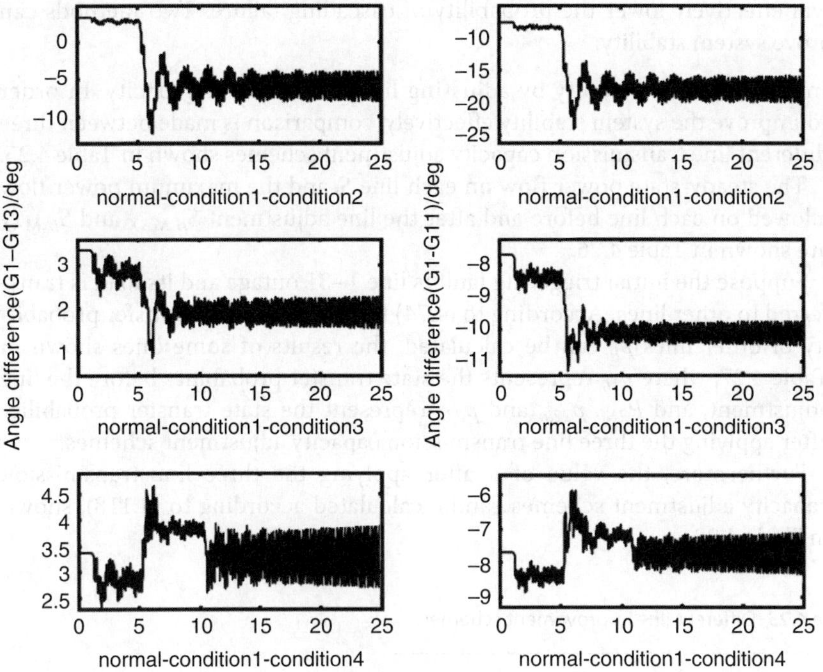

Figure 4.33 Dynamic responses of the relative power angle between generators in the 16-machine system.

Table 4.24 Time cost of the different analyze method.

Analyze method	Time cost
Normal dynamic simulations	18.2041 s
Proposed method	3.1520 s

Optimal Schemes to Improve System Stability Consider that the load transfer is the main cause of line protection maloperation and cascading failure. Then the line improvement or operation mode adjustment schemes used in this section are mainly based on the loaded degree of each line, that is, the ratio of the line power flow to the line maximum transmission capacity β_{ij}. As shown in (4.75),

$$\beta_{ij} = \left| \frac{S_{j(t_b)}}{S_{j\max}} \right|$$

which characterizes the loaded degree of the line. The larger that β_{ij} is, the more probable it is that the line protection will maloperate. Thus, after identifying the possible fault scenario where the system may undergo cascading failure according to (4.87), transforming or adjusting the operation mode of the lines with bigger

β_{ij} will effectively lower the probability of cascading failure. Two methods can improve system stability:

1) Improve system stability by adjusting line transmission capacity. In order to improve the system stability effectively, comparison is made between three different line transmission capacity adjustment schemes shown in Table 4.25.

   The steady-state power flow on each line S_j and the maximum power flow allowed on each line before and after the line adjustment $S_{j1,\mathrm{MAX}}$ and $S_{j2,\mathrm{MAX}}$ are shown in Table 4.26.

   Suppose the initial triggering fault is line 1–31 outage and its load is transferred to other lines. According to (4.74) to (4.81), the state transfer probability of other lines p_{ij} can be calculated, the results of some lines shown in Table 4.27, where p_{1ij} represents the state transfer probability before the line adjustment, and P_{21ij}, p_{22ij}, and p_{23ij} represent the state transfer probability after applying the three line transmission capacity adjustment schemes.

   Furthermore, the value of μ after applying the three line transmission capacity adjustment schemes can be calculated according to (4.118), shown in Table 4.28.

Table 4.25 Different line improvement schemes.

Scheme	Line improvement
1	$S_{j\mathrm{MAX}}$ of line 1–47 increases 50%
2	$S_{j\mathrm{MAX}}$ of line 30–31 increases 50%
3	$S_{j\mathrm{MAX}}$ of line 38–31 increases 50%

Table 4.26 Power flow of different line improvement schemes.

Line	S_j(MVA)	$S_{j1,\mathrm{MAX}}$(MVA)	$S_{j2,\mathrm{MAX}}$(MVA)
1–47	69.07	87	130.5
30–31	399.55	480	720
38–31	257.24	375	562.5

Table 4.27 State transfer probability of some lines after applying different line improvement schemes.

Line number	p_{1ij}	p_{21ij}	p_{22ij}	p_{23ij}
9–30	0.2357	0.2314	0.2294	0.2335
47–48	0.1214	0.1128	0.1113	0.1184
46–38	0.0875	0.0746	0.0685	0.0816

Table 4.28 The value of μx after applying different line improvement schemes.

Scheme	μ
1	−0.0875
2	−0.0942
3	−0.0834

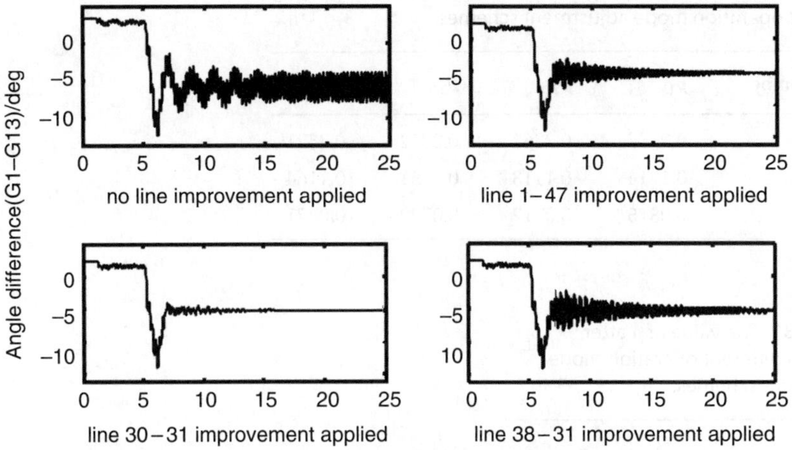

Figure 4.34 Dynamic responses of the relative power angle between generators in the 16-machine system.

Since the value of μ reflects the system stability degree, the smaller μ is, the higher is the stability degree. Thus, it can be seen from Table 4.28 that Scheme 2 is the most effective in improving system stability among the three schemes. To further verify this result, time-domain simulation tests were conducted based on Condition 2 in Table 4.22, with the results shown in Figure 4.34. It can be seen that the generator power angle instability caused by a system jump could be suppressed most quickly by increasing the transmission capacity of line 30–31, and the scheme with smaller μ has a better damping effect on the system. Based on the above analysis, power system construction staff could effectively improve system stability by selecting the optimal line transmission capacity adjustment scheme.

2) Improve system stability by adjusting the system operation mode. In order to improve the system stability effectively, a comparison is made between three different operation mode adjustment schemes shown in Table 4.29.

Suppose the initial triggering fault is line 1–31 outage and its load is transferred to other lines. According to (4.74) to (4.81), the state transfer probability of other lines p_{ij} can be calculated, the results of some lines shown in Table 4.30, where p_{1ij} represents the state transfer probability in the initial

Table 4.29 Different operation mode adjustment schemes.

Scheme	Operation mode adjustment
1	G16 and load at bus 1 decrease 20 MW
2	G15 and load at bus 30 decrease 20 MW
3	G14 and load at bus 47 decrease 20 MW

Table 4.30 State transfer probability of some lines after applying different operation mode adjustment schemes.

Line number	p_{1ij}	p_{21ij}	p_{22ij}	p_{23ij}
9–30	0.2357	0.2262	0.2312	0.2321
47–48	0.1214	0.1113	0.1151	0.1164
46–38	0.0875	0.0712	0.0732	0.0771

Table 4.31 The value of μ after applying different operation mode adjustment schemes.

Scheme	μ
1	−0.0983
2	−0.0952
3	−0.0926

operation mode, and P_{21ij}, p_{22ij}, and p_{23ij} represent the state transfer probability after applying different operation mode adjustment schemes.

Furthermore, the value of μ after applying the three operation mode adjustment schemes can be calculated according to (4.118), shown in Table 4.31.

Since the value of μ reflects the system stability degree, the smaller μ is, the higher is the stability degree. Thus, it can be seen from Table 4.31 that Scheme 1 has the optimal effect of improving system stability. In order to further verify this result, time-domain simulation tests were conducted based on Condition 2 in Table 4.22, the results shown in Figure 4.35. It can be seen that the generator power angle instability caused by a system jump could be suppressed most quickly by using operation mode adjustment Scheme 1, and the scheme with a smaller μ has a better damping effect. Therefore, power system operation staff could effectively improve system stability by selecting the optimal operation mode adjustment scheme.

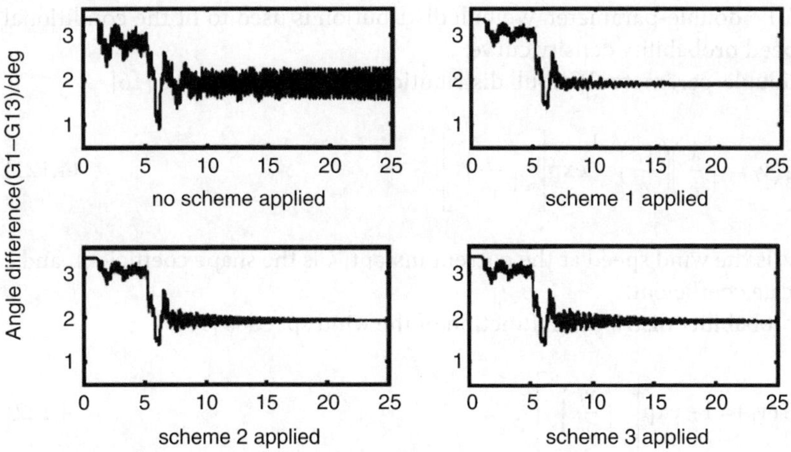

Figure 4.35 Dynamic responses of the relative power angle between generators in the 16-machine system.

4.4 Stability Analysis of Stochastic Excitation System

The development of wind power is an important part of the energy strategies that countries worldwide will apply in the future. However, the wind generator output will fluctuate constantly due to the stochastic characteristic of wind speed, which will endanger the safe, stable, and economic operation of a power system. In this situation, it is the most challenging issue to systematically and effectively evaluate the effects of wind power on power system dynamics.

A new method to analyze the stability of a power system with multiple operating conditions when considering the stochastic characteristic of wind speed is proposed in this section, which could identify the stability of a power system with multiple operating conditions without obtaining the operating trajectory of the system. Compared with the traditional time-domain simulation method, the computation amount is reduced and the identification efficiency is improved.

4.4.1 Model of Multiple Operating Conditions System Considering the Stochastic Characteristic of Wind Speed

The wind generator power output manifests various operation modes due to the influence of the stochastic characteristic of wind speed. If the stochastic characteristic of wind speed is analyzed properly, the operating condition of a power system containing the influence of wind generator power output could be simulated, and the stability of a power system with multiple operating conditions considering the stochastic characteristic of wind speed could be analyzed effectively.

4.4.1.1 Conditional Wind Speed Probability Density and Wind Generator Power Output Model

The conditional wind speed probability density $f(v_i|v_j)$ can be defined as the wind speed probability density at the next sampling instant or after a few sampling instants (the sampling interval is 10 min) under a certain conditional featured wind speed. The double-parameter Weibull distribution is used to fit the conditional wind speed probability density curve.

The double-parameter Weibull distribution can be described as [16]

$$g(v_i) = \left(\frac{k}{c}\right)\left(\frac{v_i}{c}\right)^{k-1} \exp\left[-\left(\frac{v_i}{c}\right)^k\right] \tag{4.121}$$

where v_i is the wind speed at the current instant, k is the shape coefficient, and c is the scale coefficient.

The probability distribution function of the wind speed is [16]

$$G(v_i) = 1 - \exp\left[-\left(\frac{v_i}{c}\right)^k\right] \tag{4.122}$$

In equation (4.123), the linear power model is used to describe the relationship between the wind generator output power and wind speed variation from the cut-in speed to the rated speed. This is a widely applied model to describe the wind power curve, which assumes that in the interval from the cut-in wind speed to the cut-out wind speed, the relationship between wind generator output power and wind speed variation is linear [17], that is,

$$P_1(v) = \frac{v_i - v_{ci}}{v_R - v_{ci}} P_R \tag{4.123}$$

Neglecting the electrical losses and the wind farm wake effect, the approximate function relationship between the wind generator power output and the wind speed can be expressed as

$$P(v_i) = \begin{cases} 0 & 0 \le v_i < v_{ci} \quad v_i \ge v_{co} \\ \dfrac{v_i - v_{ci}}{v_R - v_{ci}} P_R & v_{ci} < v_i < v_R \\ P_R & v_R \le v_i < v_{co} \end{cases} \tag{4.124}$$

where v_{ci} is the cut-in wind speed of wind generator, v_{co} is the cut-out wind speed of wind generator, v_R is the rated wind speed of wind generator, $P_1(v)$ is the power output of wind generator when the wind speed is between v_{ci} and v_{co}, and P_R is the rated power output of wind generator.

According to (4.121) to (4.124), the piecewise probability function of the wind generator power output can be gained:

$$F(P) = \begin{cases} G(v_{ci}) + 1 - G(v_{co}) & P = 0 \\ G(v_i) - G(v_{ci}) & 0 < P < P_R \\ G(v_{co}) - G(v_R) & P = P_R \end{cases} \tag{4.125}$$

4.4.1.2 Model of Power System with Multiple Operating Conditions Based on the Continuous Markov Process

The model of a power system with multiple operating conditions can be expressed using the continuous Markov jump system as

$$\begin{cases} \dot{x}_t = \left(A(s_t) + \Delta A(s_t, t)\right)x_t + B(s_t)u_t + G(s_t)w_t \\ z_t = C(s_t)x_t + D(s_t)u_t + L(s_t)w_t \end{cases} \tag{4.126}$$

where $x_t \in R^n$ is the state vector, $u_t \in R^p$ is the control input vector, w_t is white noise, which satisfies $E\{dw_t\} = 0, E\{dw_t^2\} = W\,dt(W > 0)$, and $\{s_t, t \geq 0\}$ is a Markov process defined in the finite space $S = \{1, 2, \ldots, l\}$, which corresponds to the system operating conditions under different wind speed modes and describe the variation of the system operating condition with the change of wind generator power output.

This section mainly deals with the stability of the relative angle between synchronous generators, which is affected by system operating condition variation caused by generator output power fluctuation. It could be conservative to use the second-order rotor motion equation to analyze the angle stability of the power system. The sixth-order detailed model considering the damping effect of rotor windings and the electromechanical transient of rotor motion is a widely applied generator model in engineering practice [5]. Thus this section applies the sixth-order model to describe the dynamic behavior of the generator rotor, shown below:

$$\begin{cases} \dfrac{d\delta}{dt} = \omega - 1 \\ M\dfrac{d\omega}{dt} = P_m - P_e + \sigma w \\ T'_{d0}\dfrac{dE'_q}{dt} = E_f - \dfrac{X_d - X_l}{X'_d - X_l}E'_q + \dfrac{X_d - X'_d}{X'_d - X_l}E''_q - \dfrac{(X_d - X'_d)(X''_d - X_l)}{X'_d - X_l}i_d \\ T'_{q0}\dfrac{dE'_d}{dt} = -\dfrac{X_q - X_l}{X'_q - X_l}E'_d + \dfrac{X_q - X'_q}{X'_q - X_l}E''_d + \dfrac{(X_q - X'_q)(X''_q - X_l)}{X'_q - X_l}i_q \\ T''_{d0}\dfrac{dE''_q}{dt} = \dfrac{X''_d - X_l}{X'_d - X_l}T''_{d0}pE'_q - E''_q + E'_q - (X'_d - X''_d)i_d \\ T''_{q0}\dfrac{dE''_d}{dt} = \dfrac{X''_q - X_l}{X'_q - X_l}T''_{q0}pE'_d - E''_d + E'_d + (X'_q - X''_q)i_q \end{cases} \tag{4.127}$$

where δ and ω is the generator power angle and rotating speed, P_m and P_e are the mechanical power and electromagnetic power, M is the inertia time constant, E_f is the excitation emf, x_l is the generator leakage reactance, E' and E'' are the generator transient emf and subtransient emf, T'_0 and T''_0 are the generator open-loop transient and subtransient time constant, x, x', and x'' are generator steady-state reactance, transient reactance, and subtransient reactance, and i is the generator output current. The subscripts d and q represent the direct axis and cross axis of the synchronous rotating coordinate system and σ is the external stochastic excitation intensity.

As for the DFIG model, Pena *et al.* have a detailed description of this model [18], where the electromagnetic part is as follows:

1) Stator voltage equations:

$$\dot{\psi}_{sd} = -r_s i_{sd} + \omega_s \psi_{sq} + u_{sd} \tag{4.128}$$

$$\dot{\psi}_{sq} = -r_s i_{sq} + \omega_s \psi_{sd} + u_{sq} \tag{4.129}$$

2) Rotor current equations:

$$\dot{i}_{rd} = \frac{1}{L_1}\left[r_s L_m i_{sd} - \omega_r L_m L_s i_{sq} - \omega_r L_r L_s i_{rq} + L_1 \omega_s i_{rq} - r_r L_s i_{rd} - L_m u_{sd} + L_s u_{rd}\right] \tag{4.130}$$

$$\dot{i}_{rq} = \frac{1}{L_1}\left[r_s L_m i_{sq} + \omega_r L_m L_s i_{sd} + \omega_r L_r L_s i_{rd} + L_1 \omega_s i_{rq} - r_r L_s i_{rq} - L_m u_{sq} + L_s u_{rq}\right] \tag{4.131}$$

3) Rotor current control equations:

$$\dot{x}_1 = i_{rd}^{ref} - i_{rd} \tag{4.132}$$

$$\dot{x}_2 = i_{rq}^{ref} - i_{rq} \tag{4.133}$$

where u_{sd} and u_{sq} are the DFIG stator d-axis voltage and q-axis voltage, ψ_{sd} and ψ_{sq} are the DFIG stator d-axis flux and q-axis flux, i_{sd} and i_{sq} are the DFIG stator d-axis current and q-axis current, ω_s is the rotating speed of the coordinate system, r_s is the armature resistance, i_{rd} and i_{rq} are the DFIG rotor d-axis current and q-axis current, u_{rd} and u_{rq} are the DFIG rotor d-axis voltage and q-axis voltage, ω_r is the rotor speed, r_r is the rotor resistance, L_s is the stator self-inductance, L_r is the rotor self-inductance, and L_m is the mutual inductance between the stator and the rotor, i_{rd}^{ref} and i_{rq}^{ref} are the reference values of the DFIG rotor d-axis current and q-axis current, and x_1 and x_2 are the accumulated tracking errors of the DFIG rotor d-axis current and q-axis current. In (4.130), $L_1 = L_s L_r - L_m^2$.

The algebra part of the DFIG is shown in (4.134) to (4.139):

1) Stator current equations:

$$i_{sd} = \frac{\psi_{sd} - L_m i_{rd}}{L_s} \tag{4.134}$$

$$i_{sq} = \frac{\psi_{sq} - L_m i_{rq}}{L_s} \tag{4.135}$$

2) Rotor flux equations:

$$\psi_{rd} = L_m i_{sd} + L_r i_{rd} \tag{4.136}$$

$$\psi_{rq} = L_m i_{sq} + L_r i_{rq} \tag{4.137}$$

3) Rotor voltage equations:

$$u_{\mathrm{rd}} = K_{\mathrm{p1}}\left(i_{\mathrm{rd}}^{\mathrm{ref}} - i_{\mathrm{rd}}\right) + r_{\mathrm{r}} i_{\mathrm{rd}} + K_{\mathrm{i1}} x_1 - (\omega_{\mathrm{s}} - \omega_{\mathrm{r}})\psi_{\mathrm{rq}} \tag{4.138}$$

$$u_{\mathrm{rq}} = K_{\mathrm{p2}}\left(i_{\mathrm{rq}}^{\mathrm{ref}} - i_{\mathrm{rq}}\right) + r_{\mathrm{r}} i_{\mathrm{rq}} + K_{\mathrm{i2}} x_2 - (\omega_{\mathrm{s}} - \omega_{\mathrm{r}})\psi_{\mathrm{rd}} \tag{4.139}$$

where ψ_{rd} and ψ_{rq} are the DFIG rotor d-axis flux and q-axis flux and K_{p1}, K_{p2}, K_{x1}, and K_{x2} are parameters of the rotor current controller.

The mechanical part considers the influence of the mechanical transmission device (i.e., the shaft) and the pitch angle control system. The shaft model uses the two mass model, which includes three parts: the low-speed axis, the gearbox and the high-speed axis. The detailed model of the mechanical part is shown below:

1) The shaft model:

$$\begin{cases} 2H_{tur}\dfrac{d\omega_{tur}}{dt} = T_M - K\theta_s - D\left(\omega_{tur} - \omega_{gen}\right) \\ 2H_{gen}\dfrac{d\omega_{gen}}{dt} = K\theta_s + D\left(\omega_{tur} - \omega_{gen}\right) - T_e \\ \dfrac{d\theta_s}{dt} = \omega_s\left(\omega_{tur} - \omega_{gen}\right) \end{cases} \tag{4.140}$$

2) The pitch angle control system model:

$$\frac{d\beta}{dt} = \frac{1}{T_{servo}}(\beta_0 - \beta) \tag{4.141}$$

where H_{tur} is the wind turbine inertia time constant of the paddle, hub, and low-speed axis, H_{gen} is the inertia time constant of the gearbox and high-speed axis, θ_s is the torsion angle of the low-speed axis from the high-speed axis, K is the stiffness coefficient of the shaft, D is the equivalent damping coefficient, T_M and T_e are the mechanical torque and electromagnetic torque of the wind turbine, ω_{tur} and ω_{gen} are the electrical angular speed of the wind turbine and the generator rotor, ω_s is the rotating speed of the stator magnetic field, and when the stator is directly connected to the grid, $\omega_s = 1$, T_{servo} is the inertia parameter of the pitch angle control system, and β_0 is the initial value of the pitch angle.

For each system operating condition corresponding to a certain wind speed model, linearize the above equations at the equilibrium point of the operating condition so that the coefficient matrix of the state vector $A(s_t) + \Delta A(s_t, t)$ can be gained. $A(s_t)$ is the certain part of the state matrix and $\Delta A(s_t, t)$ is the uncertain part of the state matrix, which could reflect the uncertainty of parameter measurement and estimation in a power system model. The stochastic disturbance coefficient matrix is $G(s_t) = \begin{bmatrix} 0 & \dfrac{\sigma}{M} & 0 & 0 & 0 & 0 \end{bmatrix}^T$. In this section the control effect is not considered; thus the coefficient matrices $C(s_t)$, $D(s_t)$, and $L(s_t)$, which correspond to the control input matrix, coefficient $B(s_t)$, and the control output matrix z_t, respectively, are all 0. Therefore, the Markov model of a time-varying power system can be gained.

The state transfer probability density of the Markov process is

$$p_{ij} = P(\eta_{t+\Delta} = j \mid \eta_t = i) = \begin{cases} \pi_{ij}\Delta + o(\Delta), & i \neq j \\ 1 + \pi_{ii}\Delta + o(\Delta), & i = j \end{cases} \tag{4.142}$$

where

$$\lim_{\Delta \to \infty} \frac{o(\Delta)}{\Delta} = 0 \qquad (\Delta > 0)$$

In (4.142), $\boldsymbol{\pi}$ is the Markov transfer probability density matrix, the element π_{ij} in matrix $\boldsymbol{\pi}$ represents the transfer probability density that renders the system operating condition at i when $t = t$ and renders the system operating condition at j when time $t = t + \Delta$:

$$\pi_{ii} = -\sum_{i \neq j} \pi_{ij} \quad (\pi_{ij} \geq 0, i \neq j) \tag{4.143}$$

The state probability P_{ij} and probability density π_{ij} can be calculated according to (4.121) to (4.125).

For each $s_t = i \in S$, denote $A(s_t)$, $B(s_t)$, $G(s_t)$, $C(s_t)$, $D(s_t)$, and $L(s_t)$ as A_i, B_i, G_i, C_i, D_i, and L_i, respectively. The uncertain parameters satisfy the following matching conditions:

$$\Delta A(s_t, t) = \Delta A(i, t) = H_i F(i, t) M_i \tag{4.144}$$

where H_i and M_i are given matrices. The real matrix $F(i, t)$ reflects the structure information of system uncertain parameters, and satisfies

$$F^T(i,t) F(i,t) \leq I \tag{4.145}$$

4.4.1.3 Stability Analysis of Power System with Multiple Operating Conditions Considering the Stochastic Characteristic of Wind Power

Definition 4.4 [19].For the system in (4.126), if $u(t) \equiv 0$, and for all initial conditions $x_0 \in R^n$ and $s_0 \in S$,

$$\int_0^\infty E\left\{ \left\| x(t, x_0, s_0) \right\|^2 \right\} dt < \infty \tag{4.146}$$

then the continuous Markov system in (4.126) is stochastically stable.

Theorem 4.5 For a given positive constant $\gamma > 0$, if there exists a set of symmetric positive definite matrices $P_i > 0$ $(i \in S)$, which satisfy the following matrix inequalities:

$$\begin{bmatrix} J_i & C_i^T L_i + P_i G_i \\ * & L_i^T L_i - \gamma^2 I \end{bmatrix} < 0 \tag{4.147}$$

$$J_i = A_i^T P_i + P_i A_i + \sum_{j=1}^{l} \pi_{ij} P_j + C_i^T C_i \tag{4.148}$$

then when $u(t) \equiv 0$, the system in (4.126) is robust stochastic stable, and satisfies the disturbance attenuation degree γ, that is,

$$\|z\|_2 \leq \gamma \left[\|w\|_2^2 + x_0^T P(s_0) x_0 \right]^{\frac{1}{2}} \tag{4.149}$$

Proof. According to the Schur Theorem, (4.147) could be transformed to

$$A_i^T P_i + P_i A_i + \sum_{j=1}^{l} \pi_{ij} P_j + C_i^T C_i < 0 \tag{4.150}$$

For $i \in S$, $C_i^T C_i \geq 0$ and thus

$$A_i^T P_i + P_i A_i + \sum_{j=1}^{l} \pi_{ij} P_j < 0 \tag{4.151}$$

Therefore, if it could be proved that the system in (4.126) is robust stochastic stable when (4.151) is satisfied, it could then be proved that the robust stochastic stability criterion in Theorem 4.5 is true.

Construct the following Lyapunov functional:

$$V(x_t, s_t = i) = V(x_t, i) = x_t^T P_i x_i \quad (P_i > 0, i \in S) \tag{4.152}$$

where $x_i = \begin{bmatrix} \delta & \omega & E_q' & U_R & U_F & E_f \end{bmatrix}^{\mathrm{T}}$ is the column vector of the system state variable of the ith operating condition, including the generator power angle, generator angular speed, generator postreactance emf, the excitation system voltage regulator output voltage, excitation system output voltage, excitation system excitation feedback voltage, etc.

When $w(t) = 0$, the weak infinitesimal operator of functional $V(x_t, i)$ is

$$\begin{aligned} \ell V(x_t, i) &= \sum_{j=1}^{l} \pi_{ij} V(x_t, i) + x_t^T A_i^T \frac{\partial}{\partial x} V(x_t, i) \\ &= x_t^T \left[\sum_{j=1}^{l} \pi_{ij} P_j + A_i^T P_j + P_j A_i \right] x_t < 0 \end{aligned} \tag{4.153}$$

where A_i is the linearized state matrix at the equilibrium point of the ith system operating condition. That is, there exists a set of positive definite matrices $Q_i > 0 (i \in S)$ that satisfy

$$\ell V(x_t, i) = -x_t^T Q_i x_t$$

Thus

$$\ell V(x_t, i) \leq -\min_{i \in S} \lambda_{\min} [Q_i] x_t^T x_t < 0 \tag{4.154}$$

According to the Dynkin Lemma,

$$E\left[V(x_t,i)\right]-E\left[V(x_0,s_0)\right]=E\left[\int_0^t \ell V(x_r,s_r)dr\right]$$
$$\leq -\min_{i\in S}\lambda_{\min}\left[Q_i\right]E\left[\int_0^t x_r^T x_r dr\bigg|(x_0,s_0)\right] \quad (4.155)$$

which is equivalent to

$$\begin{aligned}&\min_{i\in S}\lambda_{\min}\left[Q_i\right]E\left[\int_0^t x_r^T x_r dr\bigg|(x_0,s_0)\right]\leq E\left[V(x_0,s_0)\right]\\&-E\left[V(x_t,i)\right]\leq E\left[V(x_0,s_0)\right]\end{aligned} \quad (4.156)$$

Therefore,

$$E\left[\int_0^t x_r^T x_r dr\bigg|(x_0,s_0)\right]\leq \frac{E\left[V(x_0,s_0)\right]}{\min_{i\in S}\lambda_{\min}\left[Q_i\right]} \quad (4.157)$$

When $t\to\infty$,

$$E\left[\int_0^\infty x_r^T x_r dr\bigg|(x_0,s_0)\right]\leq \frac{E\left[V(x_0,s_0)\right]}{\min_{i\in S}\lambda_{\min}\left[Q_i\right]}<\infty \quad (4.158)$$

Thus, according to (4.151) to (4.158), the system in (4.126) is stochastically stable.

When $w(t)\neq 0$, the weak infinitesimal operator of functional $V(x_t, i)$ is

$$\ell V(x_t,i)=x_t^T\left[\sum_{j=1}^{l}\pi_{ij}P_j+A_i^T P_j+P_j A_i\right]x_t+2x_t^T P_i G_i w_t \quad (4.159)$$

Define the index function:

$$J_T=E\left[\int_0^T\left(z_t^T z_t-\gamma^2 w_t^T w_t\right)dt\right] \quad (4.160)$$

Since

$$\begin{aligned}&z_t^T z_t-\gamma^2 w_t^T w_t+\ell V(x_t,i)\\&=x_t^T C_i^T C_i x_t-\gamma^2 w_t^T w_t+x_t^T C_i^T L_i w_t+w_t^T L_i^T C_i x_t+w_t^T L_i^T L_i w_t\\&\quad+x_t^T\left[\sum_{j=1}^{l}\pi_{ij}P_j+A_i^T P_j+P_j A_i\right]x_t+2w_t^T G_i^T P_i x_t+x_t^T P_i G_i w_t\\&=\begin{bmatrix}x_t^T & w_t^T\end{bmatrix}\begin{bmatrix}J_i & C_i^T L_i+P_i G_i\\ * & L_i^T L_i-\gamma^2 I\end{bmatrix}\begin{bmatrix}x_t\\ w_t\end{bmatrix}\end{aligned}$$

then

$$J_T = E\left[\int_0^T \left(z_t^T z_t - \gamma^2 w_t^T w_t + \ell V(x_t, i)\right) dt\right] - E\left[\int_0^T \left(\ell V(x_t, i)\right) dt\right] \quad (4.161)$$

According to the Dynkin Lemma,

$$E\left[\int_0^T \left(\ell V(x_t, i)\right) dt\right] = E\left[V(x_T, s_T)\right] - E\left[V(x_0, s_0)\right]$$

so that

$$J_T = E\left\{\int_0^T \begin{bmatrix} x_t^T & w_t^T \end{bmatrix} \begin{bmatrix} J_i & C_i^T L_i + P_i G_i \\ * & L_i^T L_i - \gamma^2 I \end{bmatrix} \begin{bmatrix} x_t \\ w_t \end{bmatrix} dt\right\} - E\left[V\left(x_T, s_T\right)\right] + E\left[V\left(x_0, s_0\right)\right] \quad (4.162)$$

According to (4.151) and $E\left[V(x_T, s_T)\right] \geq 0$,

$$J_T \leq E\left[V\left(x_0, s_0\right)\right]$$

When $T \to \infty$,

$$\|z\|_2 \leq \gamma\left[\|w\|_2^2 + x_0^T P(s_0) x_0\right]^{\frac{1}{2}}$$

Thus according to (4.159) to (4.162), (4.147) renders the system in (4.126) with disturbance attenuation degree γ.

Finally, the stability of the system in (4.160) with multiple operating conditions can be expressed with the following optimization problem:

$$\min_{P_i, \gamma} t \quad s.t. \quad (4.147)(4.148) \quad (4.163)$$

The system with operating condition variation due to wind generator output power fluctuation is a jump system. The jump parameters evolve according to a given Markov chain with limited states and the stability of such a system is jointly determined by each varying state and the law of transfer between different states. Thus, the dynamic variation of system structure and parameter caused by wind generator output power fluctuation is decoupled to linearized equations at the equilibrium points of different operating conditions. Combined with the transfer probability matrix of each operating condition, the continuous Markov model for a power system with multiple operating conditions considering the stochastic characteristic of wind speed is established. On this basis, construct the Lyapunov functional containing the continuous Markov power system model with multiple operating conditions and transform the robust stochastic stability to the feasibility problem of LMI. Then use the feasibility problem-solving method to solve the LMI.

If there exist $t < 0$ and the positive definite matrix $P_i > 0$ that satisfy Theorem 4.5, then the system in (4.126) is robust stochastic stable and satisfies the disturbance attenuation degree γ. Otherwise, the system is unstable.

4.4.2 Simulation Analysis

4.4.2.1 Calculation of the Conditional Wind Speed Probability Density Matrix

Take the wind speed measurements of a wind farm in one month, for example. The cut-in wind speed is 3 m/s, the rated wind speed is 12 m/s, and the cut-out wind speed is 22 m/s. The wind generator power output curve is shown in Figure 4.36 [20]. With the interval width of 1 m/s, divide the wind speed into 10 modes, which form a continuous Markov model state set denoted by $S = \{1, 2, \ldots, 10\}$. The information of the wind speed modes is shown in Table 4.32.

According to the mode division in Table 4.32, Weibull distribution fitting of the conditional wind speed probability density curve in each wind speed mode is conducted, with the results shown in Table 4.33. Then, according to Table 4.33 and (4.155), the conditional wind speed probability density matrices can be calculated as follows:

$$\pi = \begin{bmatrix} -0.2414 & 0.2413 & 0.0001 & 0 & 0 & 0 & 0 & 0 & 0 & 0 \\ 0.2061 & -0.4255 & 0.2184 & 0.001 & 0 & 0 & 0 & 0 & 0 & 0 \\ 0.0469 & 0.2205 & -0.5102 & 0.2388 & 0.004 & 0 & 0 & 0 & 0 & 0 \\ 0.0079 & 0.0520 & 0.2102 & -0.5481 & 0.2694 & 0.0086 & 0 & 0 & 0 & 0 \\ 0.0012 & 0.0103 & 0.0553 & 0.202 & -0.5736 & 0.2901 & 0.0147 & 0 & 0 & 0 \\ 0.0001 & 0.0013 & 0.01 & 0.0523 & 0.1953 & -0.5678 & 0.2975 & 0.0113 & 0 & 0 \\ 0 & 0.0001 & 0.0011 & 0.0086 & 0.0483 & 0.1979 & -0.5319 & 0.2728 & 0.0031 & 0.0001 \\ 0 & 0 & 0.0001 & 0.0009 & 0.0073 & 0.045 & 0.2057 & -0.4779 & 0.2189 & 0.0003 \\ 0 & 0 & 0 & 0 & 0.0005 & 0.0046 & 0.0357 & 0.204 & -0.3848 & 0.14 \\ 0 & 0 & 0 & 0 & 0 & 0 & 0 & 0 & 0.0049 & -0.0049 \end{bmatrix} \tag{4.164}$$

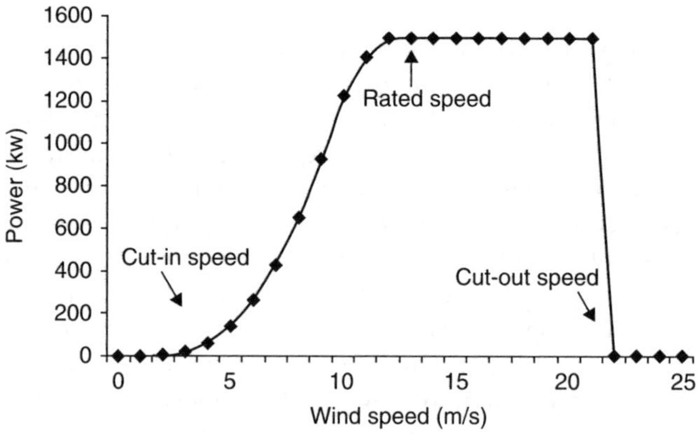

Figure 4.36 Wind generator power curve.

Table 4.32 Patterns of conditional wind speed.

Mode number	Wind speed interval (m/s)	Featured wind speed (m/s)
Mode 1	3–4	3.5
Mode 2	4–5	4.5
Mode 3	5–6	5.5
Mode 4	6–7	6.5
Mode 5	7–8	7.5
Mode 6	8–9	8.5
Mode 7	9–10	9.5
Mode 8	10–11	10.5
Mode 9	11–12	11.5
Mode 10	>12	12.5

Table 4.33 Weibull fitting coefficient of conditional wind speed probability density in different wind speed modes.

Mode number	Shape coefficient	Scale coefficient (m/s)
Mode 1	6.9168	3.8533
Mode 2	7.0870	4.7546
Mode 3	7.6680	5.7612
Mode 4	8.6074	6.8192
Mode 5	9.4138	7.8676
Mode 6	11.0615	8.8791
Mode 7	12.1481	9.8236
Mode 8	13.7081	10.7550
Mode 9	13.0406	11.7599
Mode 10	8.4070	14.6598

4.4.2.2 Four-Machine System

Take the IEEE four-machine 11-bus system for example, as shown in Figure 3.1. The generators apply the sixth-order detailed model, the excitation system applies the IEEE-DC1 excitation, and the load applies the WECC load model – 80% constant active power load, 80% constant reactive power impedance load, and 20% motor load. In the simulation tests, G4 in Area 2 is substituted with a wind farm with equal capacity, so that the variation of wind generator power output could be simulated.

First, use the modal analysis method to obtain the state matrix of the four-generator system in each wind generator power output mode. Second, reduce the order of the state matrix with the SMA (selective modal analysis) method so that the calculation amount is reduced and at the same time maintain the state variables related to oscillation, such as the power angle and rotating speed. Then apply the reduced-order state matrix and the state transfer probability density matrix in (4.164) to (4.163) (the disturbance attenuation degree γ is 0.1), so that the positive definite matrix P_i that satisfies the LMI could be obtained according to the feasibility problem-solving method. Therefore, the four-generator system is robust stochastic stable. The values of some positive definite matrices P_i are shown in the following equations:

$$P_1 = \begin{bmatrix} 1.062 & -0.945 & 0.083 & 2.552 & 1.408 & 10.81 \\ -0.945 & 1.162 & -0.145 & -2.065 & 1.058 & -13.30 \\ 0.083 & -0.145 & 1.837 & -12.01 & 12.54 & 0.772 \\ 2.552 & -2.065 & -12.01 & 4184.3 & -938.6 & -1333.5 \\ 1.407 & 1.058 & 12.53 & -938.6 & 4012.6 & -1766.2 \\ 10.81 & -13.31 & 0.772 & -1333.5 & -1766.2 & 6190.9 \end{bmatrix} \tag{4.165}$$

$$P_3 = \begin{bmatrix} 1.117 & -0.891 & -0.012 & 1.807 & 0.635 & 3.632 \\ -0.891 & 1.182 & -0.171 & -1.736 & 0.886 & -4.96 \\ -0.012 & -0.171 & 2.047 & -4.119 & 4.671 & 0.562 \\ 1.807 & -1.736 & -4.119 & 3588.8 & -1091.1 & -1054.9 \\ 0.635 & 0.886 & 4.671 & -1091.1 & 4152.4 & -1684.2 \\ 3.632 & -4.961 & 0.562 & -1054.9 & -1684.2 & 6430.9 \end{bmatrix} \tag{4.166}$$

Based on the above calculation results, the four-generator system meets the robust stochastic stability conditions in Theorem 4.5, that is the system is robust stochastic stable when the stochastic characteristic of wind speed is considered. Concerning the possible system operating condition changes in real operation, time-domain simulation tests are carried out. When the system operating condition varies (only once) from one mode to another mode, the dynamic responses of the G2–G3 relative power angle are shown in Figure 4.37. When the system operating condition varies (more than once) between multiple modes, the dynamic responses of the G2–G3 relative power angle are shown in Figure 4.38. The profile of the wind speed following a pattern/profile, which consists of different types of variation, has been considered in this section. In the simulation tests, the wind speed jumping from Mode 2 to Mode 7 to Mode 2 corresponds to the gusty wind case; the wind speed jumping from Mode 1 to Mode 3 to Mode 5 corresponds to the ramp wind case; and the small-scale fluctuations of wind speed from Mode 5 to Mode 4 and from Mode 7 to Mode 8 correspond to the random wind case.

It can be seen that the real system simulation results are consistent with the identification results of the proposed method. Thus the effectiveness and correctness of the proposed method is proved. In addition, the proposed method could identify the stability of a system correctly without obtaining the system operating trajectory, thus making the computation amount smaller.

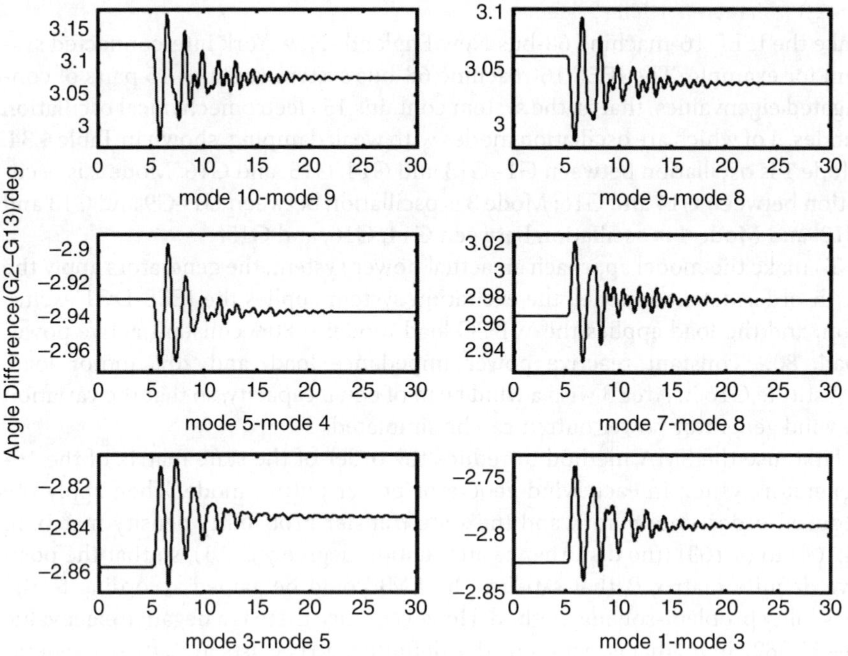

Figure 4.37 Dynamic responses of the G2–G3 relative power angle in the four-generator system when the system operating condition varies from one mode to another mode.

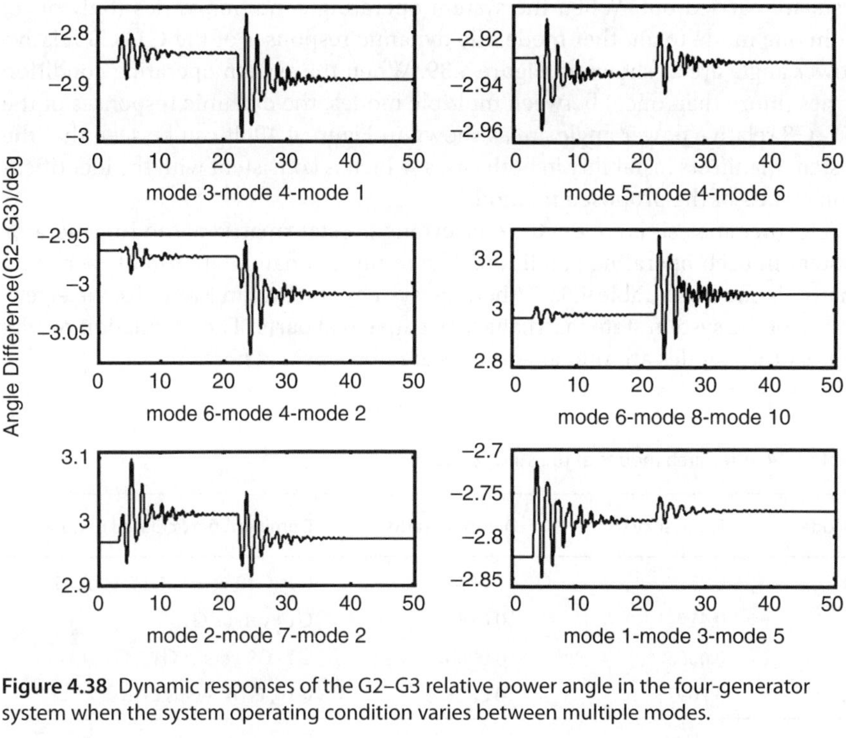

Figure 4.38 Dynamic responses of the G2–G3 relative power angle in the four-generator system when the system operating condition varies between multiple modes.

4.4.2.3 Sixteen-Machine System

Take the IEEE 16-machine 68-bus New England–New York interconnected system for example. The IEEE 16-machine 68-bus system contains 15 pairs of conjugated eigenvalues, that is, the system contains 15 electromechanical oscillation modes, 4 of which are oscillation modes with weak damping, shown in Table 4.34. Mode 1 is oscillation between G1–G13 and G14, G15, and G16; Mode 2 is oscillation between G14 and G16; Mode 3 is oscillation between G1–G9 and G13 and G12; and Mode 4 is oscillation between G14, G16, and G15.

To make the model approach an actual power system, the generators apply the sixth-order detailed model, the excitation system applies the IEEE-DC1 excitation, and the load applies the WECC load model – 80% constant active power load, 80% constant reactive power impedance load, and 20% motor load. Substitute G16 in Area 3 with a wind farm of equal capacity, so that the variation of wind generator power output can be simulated.

First, use the SMA method to reduce the order of the state matrix of the 16-generator system in each wind generator power output mode. Then apply the reduced-order state matrix and the state transfer probability density matrix in (4.164) to (4.163) (the disturbance attenuation degree γ is 0.1), so that the positive definite matrix P_i that satisfies the LMI could be gained according to the feasibility problem-solving method. However, since there is a negative eigenvalue of −15.3827 in P_6, that is, a nonpositive definite matrix exists in P_i, the 16-generator system does not meet the robust stochastic stability conditions when the variation of wind generator power output is considered. Concerning the possible system operating condition changes in real operation, time-domain simulation tests are carried out. When the system operating condition varies (only once) from one mode to another mode, the dynamic responses of the G1–G8 relative power angle are as shown in Figure 4.39. When the system operating condition varies (more than once) between multiple modes, the dynamic responses of the G1–G8 relative power angle are as shown in Figure 4.40. It can be seen that the system manifests instability in both cases, which is consistent with the identification results of the proposed method.

A further analysis is conducted concerning the state matrix of the 16-generator system in each operating condition. The minimum damping ratio of each state matrix is shown in Table 4.35. When the system operates in Mode 10, the eigenvalues of the system state matrix have negative real parts. The calculation results of the other modes are similar and thus are not repeated here.

Table 4.34 Inter-area modes of the study system.

Mode	Frequency (Hz)	Damping ratio	Dominant participated generator
1	0.2777	−0.0047	G1–G13 versus G14, G15, G16
2	0.4171	0.0142	G14 versus G16
3	0.6565	0.0283	G1–G9 versus G13, G12
4	0.7697	0.0323	G14, G16 versus G15

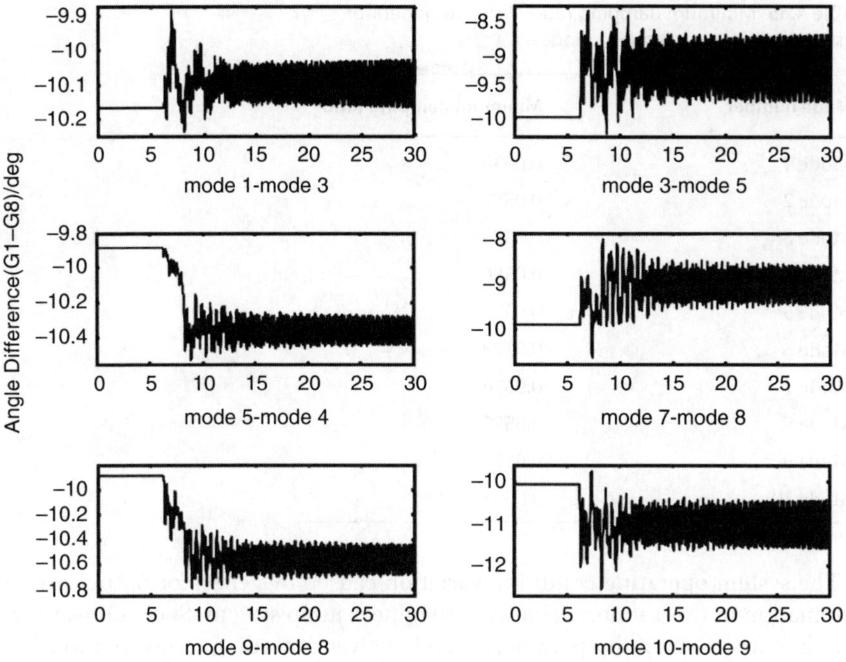

Figure 4.39 Dynamic responses of the G1–G8 relative power angle in the 16-generator system when the system operating condition varies from one mode to another mode.

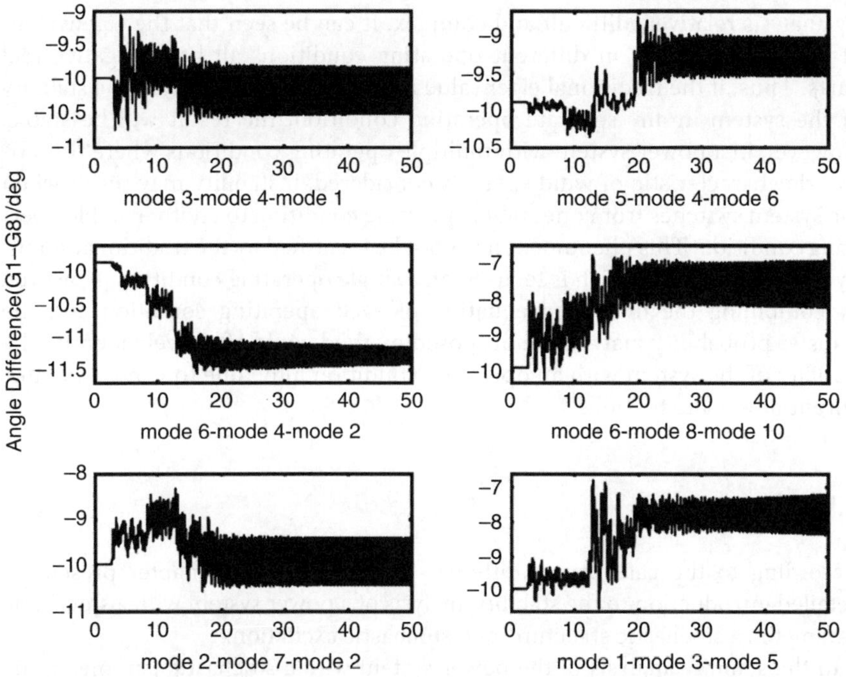

Figure 4.40 Dynamic responses of the G1–G8 relative power angle in the 16-generator system when the system operating condition varies between multiple modes.

Table 4.35 Minimum damping ratio of the 16-generator system in different operation modes.

Mode number	Minimum damping ratio
Mode 1	0.0630
Mode 2	0.0588
Mode 3	0.0583
Mode 4	0.0580
Mode 5	0.0579
Mode 6	0.0578
Mode 7	0.0576
Mode 8	0.0575
Mode 9	0.0574
Mode 10	0.0573

The system operating condition variation caused by generator output power fluctuation in this section belongs to the linear jump system. Such a linear system has stochastic jump parameters and evolves according to a given Markov chain with limited states. Due to the state switch, the dynamic characteristics of such a system experience fundamental change, that is, the stability of the system at the equilibrium points of different operation modes cannot reflect the system stability after the state jumps from one mode to another. This makes the stability analysis relatively difficult and complex. It can be seen that the eigenvalues of the state matrices in different operating conditions all have negative real parts. Thus, if the traditional eigenvalue method is used to analyze the stability of the system in any separate operating condition, the result will be stable. However, in a power system with multiple operating conditions where the stochastic characteristic of wind speed is considered, instability may result when the system switches from one stable operating condition to another stable operating condition. This phenomenon cannot be identified by the traditional stability analysis method, which is focused on a single operating condition. However, by combining the linearized equation of each operating condition and the transfer probability matrix, the proposed method could effectively identify the stability of the system with an operating condition jump due to wind generator output power fluctuations.

4.5 Summary

According to the category of different disturbances, this chapter presents a detailed introduction to the stability analysis of a power system with a stochastic parameter, a stochastic structure, and stochastic excitation.

In the stability analysis of the power system with a stochastic parameter, the low-frequency oscillation modal analysis method and the small-signal stability region calculation method are introduced. For the low-frequency oscillation modal analysis method based on the second-order perturbation theory, even when multiple parameters in the system vary, the small-signal stability of the system can still be estimated accurately. Compared with the first-order perturbation theory, the low-frequency oscillation modal obtained according to the second-order perturbation theory is closer to the real value. For the improved small-signal stability region (ISSSR) calculation method based on the Guardian map, problems that are not considered in traditional small-signal stability region calculations such as continuous oscillation and overly low damping ratio in the system are solved. Furthermore, this method also analyzes the influence of the generator excitation system parameter on the improved small-signal stability region (ISSSR), which is of great instructive value for power grid dispatch and operation staff in monitoring the system stability region and preventing small-signal instability.

In the stability analysis of a power system with stochastic structure, methods to analyze the stability of the power system in the case of a single fault and a cascading fault are introduced, respectively. The model-trajectory-based method for analyzing the time-varying oscillation characteristics of a fault system is able to track the eigenvalue change of the system more accurately compared to the equilibrium method. It is also able to avoid the window choosing problem when calculating the time-varying eigenvalue of the system using the trajectory eigenvalue method. The method to analyze the power angle stability of a power system with multiple operating conditions when considering cascading failure could accurately identify the system stability status when the system structure jumps. In addition, this method transforms the power angle stability identification to a feasibility problem, with no need to acquire the real system operation trajectory. Thus the computation amount is small and the identification process is simple and effective.

In the stability analysis of a power system with stochastic excitation, a method to analyze the stability of the power system with multiple operating conditions considering the stochastic characteristic of wind speed is introduced. This method classifies the operating conditions of the power system according to the stochastic characteristic of wind speed, and then constructs the stability criterion based on the continuous Markov model. By this method, the stability of the power system when the wind generator output power varies could be analyzed quickly and effectively, with a small computation amount and good feasibility.

References

1 Ma, J., Wang, S., *et al.* (2014) A power system small signal stability region (SSSR) calculation method based on the guardian map theory. *IET Generation Transmission and Distribution*, **8**(8), 1479–1488.

2 Moore, R.E. (1966) *Interval Analysis*, Prentice-Hall, New York.

3 Chen, S.H. (1999) *Matrix Perturbation Theory in Structural Dynamic Designs*, Science Press, Beijing.

4 Ma, J., Li, Y., Liu, C., *et al.* (2015) A model-trajectory-based method for analyzing the time-varying oscillation characteristics of a fault system. *Electric Power Components and Systems*, **43**(7), 781–791.
5 Kundur, P. (1994) *Power System Stability and Control*, McGraw-Hill, New York.
6 Zhang, X., Tsiotras, P., and Lanzon, A. (2006) An approach for computing the exact stability domain for a class of LTI parameter dependent systems. *International Journal of Control*, **79**(9), 1046–1061.
7 Barmish, B.R. (1994) *New Tools for Robustness of Linear Systems*, Macmillan Publishing Company, London.
8 Zhou, K., Doyle, J.C., and Glover, K. (1966) *Robust and Optimal Control*, Prentice-Hall, New York.
9 Ma, J., Wang, S., *et al.* (2017) Angle stability analysis of power system with multiple operating conditions considering cascading failure. *IEEE Transactions on Power Systems*, **32**(2), 873–882.
10 Pan, X., Xue, Y., Zhang, X., *et al.* (2008) Analytical calculation of power system trajectory eigenvalues and its error analysis. *Automation of Electric Power Systems*, **32**(19), 10–14.
11 Dobson, I. Chen, J. Throp, J., *et al.* (2002) Examining criticality of blackouts in power system models with cascading events. *35th Hawaii International Conference on System Sciences*, Hawaii.
12 Horowitz, S.H. and Phadke, A.G. (2004) *Power System Relaying*, John Wiley & Sons, Inc., New York.
13 Wang, A., Luo, Y., Tu, G., and Liu, P. (2011) Vulnerability assessment scheme for power system transmission networks based on the fault chain theory. *IEEE Transactions on Power Systems*, **26**(1), 442–450.
14 Verghese, G.C., Perez-Arriage, I.J., Schweppe, F.C. (1982) Selective modal analysis with applications to electric power systems, Part II: The dynamic stability problem. *IEEE Transactions on Power Apparatus and Systems*, **101**(9), 3126–3134.
15 Skogestad, S. and Postlethwaite, I. (2001) *Multivariable Feedback Control*, John Wiley & Sons, Inc., New York.
16 Tai-Her, Y. and Li, W. (2008) A study on generator capacity for wind turbines under various tower heights and rated wind speeds using Weibull distribution. *IEEE Transactions on Energy Conversion*, **23**(2), 592–602.
17 Mathew, S. (2006) *Wind Energy: Fundamentals, Resource Analysis and Economics*, Springer, Berlin.
18 Pena, R., Clare, J., and Asher, G. (1996) Doubly-fed induction generators using back-to-back PWM converters and its applications to variable-speed wind-energy generation. *IEE Proceedings, Part B, Electric Power Applications*, **153**(3), 231–241.
19 Feng, X.B., Loparo, K.A., Yuandong, J., and Chizeck, H.J. (1992) Stochastic stability properties of jump linear systems. *IEEE Transactions on Automatic Control*, **37**(1), 38–53.
20 Wan, Y.-H., Ela, E., and Orwig, K. (2010) Development of an equivalent wind plant power-curve. Preprint[J]. Nrel.

5

Stability Analysis of Time-Delay System

5.1 Introduction

With the rapid development of PMU technology, control technology, big data analysis and calculation technology, and the communication network, when considering the stability requirement of modern power grid, wide-area stability control based on PMU has been the trend of stability control research for an interconnected power grid. However, due to the non-negligible communication delay of WAMS information in the transmission process, a wide-area power system is inevitably a time-delay power system. Apart from the WAMS signal transmission delay, other time delays also exist in the power system, such as the time delay of the switch signal and the time delay of the EMS/SCADA system data. Research on the WAMS signal transmission delay demonstrates that the time delays in communication systems vary with different communication media. However, even a small time delay can render a controller, which has excellent performance in a power system, that does not consider the influence of time delay ineffective. Therefore, it is of great importance to study the effect of a time delay on power system stability in depth, so that the stable operation of a wide-area power system can be guaranteed.

According to whether the system state and parameter jumps or not, a time-delay system could be classified into a non-jump time-delay system and a jump time-delay system for stability analysis:

1) For the non-jump system, considering the influence of a stochastic disturbance on system stability, a method to analyze the stochastic time-delay stability of a power system based on the *Itô* differential is introduced. First, the objective functional based on the *Itô* differential and Lyapunov stability theory is constructed. Then free weighting items formed according to the Newton–Leibniz formula are added to the weak infinitesimal generating operators of the objective functional. Finally, the time-delay stability upper bound of the power system under different operating conditions can be calculated accurately using the GEVP method.
2) For the jump system, when considering the stochastic switch between the subsystems, a power system time-delay stability analysis method based on the Markov process is introduced. First, a class of Lyapunov–Krasovskii is

Power System Wide-Area Stability Analysis and Control, First Edition. Jing Ma.

constructed, which is functional considering the Markov transfer probability matrix. Then, free weighting items formed according to the transfer probability matrix and the Newton–Leibniz formula, respectively, are added to the weak infinitesimal generating operators of the objective functional and the whole time-varying time-delay interval is decomposed into two subintervals. Finally, the maximum time delay that the power system could allow for as a system parameter or structure varies is calculated quantitatively using the GEVP method.

5.2 Stability Analysis of a Non-Jump Time-Delay System

With the rapid development of new energy power generation and the wide application of various stochastic loads such as electric vehicles to the power system, it is no longer feasible to use the certainty differential equation theory to analyze the time-delay stability of a power system. To solve this problem, a novel method to analyze the stochastic time-delay stability of a power system is proposed in this section based on the *Itô* differential. The maximum time delay that power system could bear in unfavorable operation states, such as overload, weak damping, and high excitation magnification, is solved by considering different stochastic excitation intensities.

5.2.1 Stochastic Stability Analysis of Power System with Time Delay Based on *Itô* Differential

5.2.1.1 Stochastic Stability Analysis of Power System with Time Delay

Stochastic Power System with Time Delay In power system generation control, there are external stochastic excitation factors with similar characteristics to the Gauss white noise that can cause dynamic behavior in the system and exert considerable influence on the stability of the power system. Stochastic disturbances in power generation control mainly include three categories [1]: (1) input mechanical power shake caused by stochastic variation in the motive power, such as the water level/pressure change, steam pressure fluctuation, etc.; (2) stochastic disturbance on the grid side, such as the stochastic load fluctuation; and (3) control loop measurement noise, especially where modern digital controllers are used. In this section, the stochastic excitation factors are described as a Gauss process. Therefore, the generator rotor motion equations considering the external stochastic excitation are

$$\begin{cases} \dfrac{d\delta}{dt} = \omega - 1 \\ M\dfrac{d\omega}{dt} = P_m - P_e + \vartheta_0 G(t) \end{cases} \tag{5.1}$$

where $P_m, P_e, \delta, \omega, M$ are the generator mechanical power, electromagnetic power, power angle, rotating speed, and inertia time constant, respectively. $G(t)$ is the

Gauss process and ϑ_0 is the stochastic excitation intensity, which represents the maximum possible value of the stochastic excitation.

According to (5.1), the time-delay system model considering the stochastic factors can be expressed as

$$\begin{cases} dx(t) = f\left(x(t), x(t-h_t)\right)dt + g\left(x(t)\right)d\omega(t) \\ x(\alpha) = \varphi(\alpha), \forall \alpha \in \left[-\bar{h}, 0\right] \\ f\left(x(t), x(t-h_t)\right) = Ax(t) + A_d x(t-h_t) \end{cases} \tag{5.2}$$

where $x(t) \in \mathbf{R}^n$ is the state vector of the system. The time delay h_t meets $0 \le h_t \le \bar{h}, \dot{h}_t < \mu$, and A, A_d are given real matrices. $W(t)$ is a Wiener process defined in the complete probability space (Ω, F, P) and $\varepsilon\left(dW(t)\right) = 0, \varepsilon\left(dW^2(t)\right) = dt$.

Theorem 5.1 If there exists a Lyapunov–Krasovskii functional $V(t, x(t)) > 0$, whose weak infinitesimal generating operator meets $\zeta V(t, x(t)) < 0$, then the stochastic system in (5.2) is asymptotically mean square stable.

Proof: According to the *Itô* differential formula, for $V(t, x(t))$,

$$\begin{aligned} dV\left(t, x(t)\right) &= \left[V_t + V_x f + \frac{1}{2} tr\left(g V_{xx} g^T\right)\right]dt + V_x g d\omega(t) \\ &= \zeta V dt + V_x g d\omega(t) \end{aligned} \tag{5.3}$$

In (5.3), the weak infinitesimal generating operator is defined as

$$\zeta = \frac{\delta}{\delta x} + \left(\frac{\delta}{\delta x} f\right)^T + \frac{1}{2} tr\left(g \frac{\delta^2}{\delta x \delta x^T} g^T\right) \tag{5.4}$$

Calculate the mathematical expectation of (5.3), and according to $\mathrm{E}\{dw(t)\} = 0$,

$$\mathrm{E}\left\{\frac{dV\left(t, x(t)\right)}{dt}\right\} = \mathrm{E}\left\{\zeta V\left(t, x(t)\right)\right\} \tag{5.5}$$

Therefore, for any vector $x(t)$, $t \ge 0$, if there exists $V(t, x(t)) > 0$ which makes $\zeta V(t, x(t)) < 0$, then

$$\mathrm{E}\left\{\frac{dV\left(t, x(t)\right)}{dt}\right\} < 0$$

that is, the stochastic system in (5.2) is asymptotically mean square stable.

Lemma 5.1 [2].For any vectors $x, y \in R^n$ and symmetric positive definite matrix P with appropriate dimensions, the following inequality is true:

$$\pm 2x^T y \le x^T P^{-1} x + y^T P y \tag{5.6}$$

Combining Theorem 5.1 and Lemma 5.1, the following conclusion can be made.

Theorem 5.2 For the stochastic time-delay system described in (5.2), with given scalar quantities $\bar{h} > 0$ and μ, if there exist $P = P^T > 0$, $Q = Q^T \geq 0$, $R = R^T \geq 0$, $K = K^T > 0$, $Z = Z^T \geq 0$, and appropriately dimensioned matrices L, M, and S, which meet the following LMI, then the stochastic time-delay system in (5.2) is mean square stable:

$$\begin{bmatrix} \Omega & L \\ * & -T_1 - Z/\bar{h} \end{bmatrix} < 0 \tag{5.7}$$

where

$$\Omega = \Omega_1 + \Omega_2 + \Omega_2^T$$

$$\Omega_2 = \begin{bmatrix} L + SA & SA_d & -L & 0 & -S \end{bmatrix}$$

$$\Omega_1 = \begin{bmatrix} Q + K + G_1^T P G_1 - R/\bar{h} & R/\bar{h} & 0 & A^T M & P \\ * & G_2^T P G_2 - (1-\mu)Q - R/\bar{h} & 0 & A_d^T M & 0 \\ * & * & -K & 0 & 0 \\ * & * & * & \bar{h}R - M - M^T & 0 \\ * & * & * & * & \bar{h}Z \end{bmatrix}$$

Proof: Define the vector domain $y(t) \in R_{n'}$ which meets

$$y(t)dt = dx(t), y(\alpha) = \varphi(\alpha), \forall \alpha \in [-\bar{h}, 0] \tag{5.8}$$

First, form the following Lyapunov-Krasovskii functional:

$$\begin{aligned} V(x_t, t) = {} & x^T(t)Px(t) + \int_{t-h_t}^{t} x^T(s)Qx(s)ds \\ & + \int_{t-\bar{h}}^{t} x^T(s)Kx(s)ds \\ & + \int_{-\bar{h}}^{0}\int_{t+\theta}^{t} y^T(s)Ry(s)dsd\theta \\ & + \int_{-\bar{h}}^{0}\int_{t+\theta}^{t} f^T(s)Zf(s)dsd\theta \end{aligned} \tag{5.9}$$

According to the *Itô* differential formula, the stochastic differential of $V(x_t, t)$ along the system in (5.2) is

$$dV(t, x(t)) = \zeta V dt + 2x^T(t)Pgd\omega(t) \tag{5.10}$$

In (5.10), the weak infinitesimal generating operator is

$$\begin{aligned}\zeta V(t,x(t)) = {} & 2x^T(t)Pf(x(t)) + x^T(t)(Q+K)x(t) \\ & -(1-\dot{h}_t)x^T(t-h_t)Qx(t-h_t) \\ & -x^T(t-\bar{h})Kx(t-\bar{h}) \\ & +\bar{h}y^T(t)Ry(t) - \int_{t-\bar{h}}^{t} y_s^T Ry_s ds \\ & +\bar{h}f^T(t)Zf(t) - \int_{t-\bar{h}}^{t} f_s^T Zf_s ds \\ & +tr\{g^T Pg\}\end{aligned} \tag{5.11}$$

According to the Newton–Leibniz formula, for any matrices $\boldsymbol{L}$ and $\boldsymbol{M}$ with appropriate dimensions, the following equations are true:

$$2\xi^T(t)L\left[x(t) - x(t-\bar{h}) - \int_{t-\bar{h}}^{t} dx_s\right] = 0 \tag{5.12}$$

$$2y^T(t)M\left[(Ax + A_d x_{t-h_t} - y(t))dt + gd\omega(t)\right] = 0 \tag{5.13}$$

where

$$\xi(t) = \begin{bmatrix} x^T & x^T(t-h_t) & x^T(t-\bar{h}) & y^T(t) & f^T(t) \end{bmatrix}^T \tag{5.14}$$

Meanwhile, according to $f(t) = Ax(t) + A_d x(t-h_t)$, for any appropriately dimensioned matrix S, the following equation is constantly true:

$$2\zeta^T(t)S\left[Ax(t) - A_d x(t-h_t) - f(t)\right] = 0 \tag{5.15}$$

Furthermore, according to the Jensen inequality [3] and Lemma 5.1, the following inequalities are true:

$$-\int_{t-\bar{h}}^{t} y_s^T Ry_s ds \le \begin{bmatrix} x \\ x_{t-h_t} \end{bmatrix} \begin{bmatrix} -R/\bar{h} & R/\bar{h} \\ R/\bar{h} & -R/\bar{h} \end{bmatrix} \begin{bmatrix} x & x_{t-h_t} \end{bmatrix} \tag{5.16}$$

$$\begin{aligned}-2\xi^T(t)N\int_{t-h_t}^{t} dx_s = {} & -2\xi^T(t)N\int_{t-h_t}^{t} f(s)ds \\ & -2\xi^T(t)N\int_{t-h_t}^{t} g(s)d\omega(s) \\ \le {} & -2\zeta^T(t)N\int_{t-h_t}^{t} f(s)ds \\ & +\xi^T(t)NT_1^{-1}N^T\xi(t) \\ & +\left(\int_{t-h_t}^{t} g(s)d\omega(s)\right)^T T_1 \int_{t-h_t}^{t} g(s)d\omega(s)\end{aligned} \tag{5.17}$$

Based on the above analysis, apply (5.12), (5.13), (5.15), (5.16), and (5.17) to (5.11), so that

$$\begin{aligned}\zeta V(t,x(t)) \le \xi^T(t)\left(\Omega + LT_1^{-1}L^T + \bar{h}\cdot LZ^{-1}L^T\right)\xi(t) \\ -\int_{t-h_t}^{t}\left[\xi^T(t)L + f^T(t)Z\right]Z^{-1}\left[L^T\xi(t) + Zf(t)\right]ds\end{aligned} \tag{5.18}$$

Since $Z > 0$, the following inequality is constantly true:

$$\Omega + LT_2^{-1}L^T + \bar{h}\cdot LZ^{-1}L^{TT} < 0 \tag{5.19}$$

If for any $\xi(t) \neq 0, \mathrm{E}\{\zeta V(t,x(t))\} < 0$ is true, then the stochastic system in (5.1) is asymptotically mean square stable in the probability space. According to the Schur complement theorem [4], (5.19) is equivalent to (5.7) and thus Theorem 5.2 is proved.

Time-Delay Stability Upper Bound The matrix inequality in (5.7) could only be used to identify whether the system is stable or not; the time-delay stability upper bound of the system is still unknown. Consider that the solving of a time-delay stability upper bound is a convex optimization problem with a linear inequality constraint and manifests itself in the form of GEVP (generalized eigenvalue). In this section, the GEVP method is used to calculate the time-delay stability upper bound of the system. Since (5.7) is not a standard GEVP form and cannot be solved with the GEVP method directly, appropriate transformation is necessary. According to the nature of the Schur complement, (5.7) can be transformed to

$$\begin{bmatrix} \Omega & L & 0 & 0 \\ * & -T_1 - Z/\bar{h} & R & 0 \\ * & * & -R/\bar{h} & Z \\ * & * & * & -Z/\bar{h} \end{bmatrix} < 0 \tag{5.20}$$

Set $-Z/\bar{h} < Z_1, -R/\bar{h} < -R_1$ and $d = 1/\bar{h}$, so that

$$\begin{bmatrix} T_1 & 0 & 0 \\ * & T_2 & 0 \\ * & * & T_3 \end{bmatrix} < d\cdot \begin{bmatrix} Z_1 & 0 & 0 \\ * & Z_2 & 0 \\ * & * & K \end{bmatrix} \tag{5.21}$$

In (5.21), substitute $-Z/\bar{h}, -R/\bar{h}$ with $-Z_1, -R_1$, respectively; thus

$$\begin{bmatrix} \Omega & L & 0 & 0 \\ * & -T_1 - Z_1 & R & 0 \\ * & * & -R_1 & Z \\ * & * & * & -Z_1 \end{bmatrix} < 0 \tag{5.22}$$

According to (5.20) to (5.22), the solving of the time-delay stability upper bound $\bar{h}$ could be transformed to the following optimization problem:

$$\min\ \tau \quad s.t.(5.21),(5.22) \tag{5.23}$$

By solving (5.23), the minimum τ can be gained. Then, according to $\bar{h}_{\max} = 1/\tau$, the time-delay stability upper bound of the stochastic system can be gained.

5.2.1.2 Simulation Analysis

The IEEE 16-machine 68-bus New England–New York interconnected system is applied. The generator applies the classical model, the excitation system applies the IEEE-DC2 excitation, and the load model is a 50% constant active power load with a 50% constant reactive power load. Taking this system as the test system, the effectiveness of the proposed method is verified through time-domain simulation and the time-delay stability upper bound of the stochastic power system is calculated.

The scale of a real power system is very large, with various regulation and control devices such as PSS, FACTS, and AGC, etc. Thus the system model is usually high-ordered. To improve the computation speed and avoid the "curse of dimensionality," order reduction is necessary, that is, the high-order system model needs to be transformed to a lower-order model. In this section, the Schur order reduction method is used to reduce the order of the system, on the premise that the input–output characteristic of the frequency band concerned is not affected. The reduced-order system still has the same form as that shown in (5.2). Based on the test system, considering different stochastic excitation intensities, the stochastic time-delay stability of the power system in unfavorable operation states such as overload, weak damping, and high excitation magnification is analyzed successively.

Overload Operation State Suppose the other parameters are fixed and the load in the test system is set to be 1.2 times the rated value. Set the stochastic excitation intensity ϑ_0 to be 0.03 and 0.06, respectively. Apply the state matrix $\boldsymbol{A}$ and time-delay matrix $\boldsymbol{A}_\mathbf{d}$ of the reduced-order system and the value of ϑ_0 to (5.21) and (5.22), so that the time-delay stability upper bound of the stochastic system when $\vartheta_0 = 0.03$ and $\vartheta_0 = 0.06$ can be calculated respectively: $h_{0.03} = 111.5\,\text{ms}$ and $h_{0.06} = 98.7\,\text{ms}$. In order to verify the feasibility and effectiveness of the proposed method, the damping controller is designed in the test system based on the H_2/H_∞ control method [5]. When $\vartheta_0 = 0.03$, set the time delay to be 0 ms, 50 ms, 111.5 ms, and 150 ms, respectively, and observe the dynamic responses of power angle difference between G1 and G13. The results are shown in Figure 5.1.

When $\vartheta_0 = 0.06$, set the time delay to be 0 ms, 50 ms, 98.7 ms and 130 ms, respectively, and observe the dynamic responses of power angle difference between G8 and G16. The results are shown in Figure 5.2.

It can be seen from Figure 5.1 and Figure 5.2 that, when the time delay is below the time-delay stability upper bound, the stochastic system is able to damp the inter-area oscillation quickly (within 10 s) and remains stable after $t = 10\,\text{s}$.

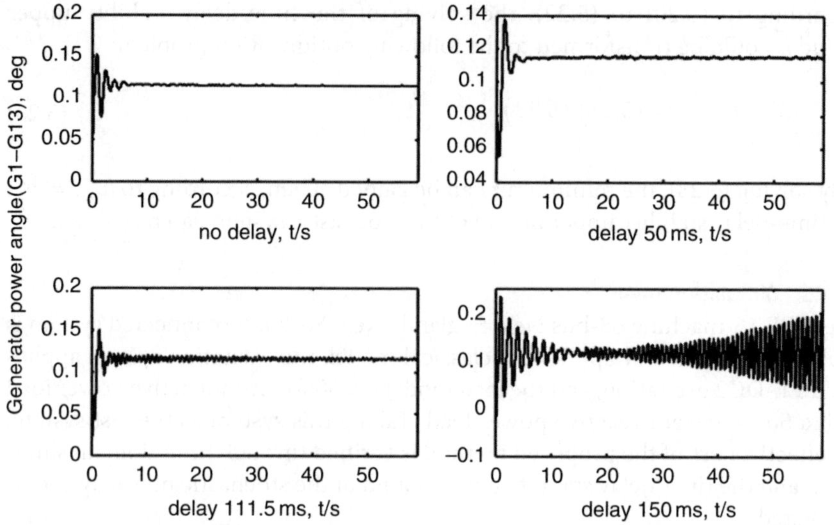

Figure 5.1 Dynamic responses of the G1–G13 relative power angle with different time delays when $\vartheta_0 = 0.03$.

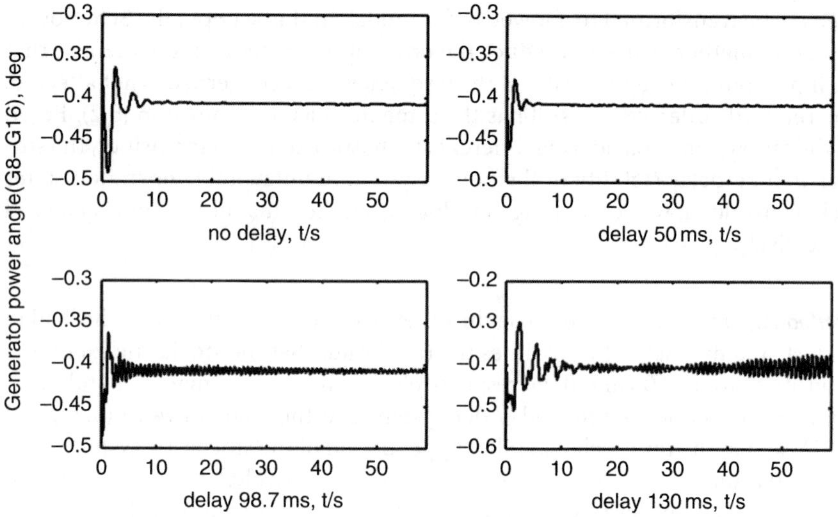

Figure 5.2 Dynamic responses of G8–G16 relative power angle with different time delays when $\vartheta_0 = 0.06$.

When the time delay reaches the time-delay stability upper bound, the generator relative power angle fluctuates slightly, which means the stochastic system is in a weak damping state. When the time delay rises above the time-delay stability upper bound, the generator relative power angle curve oscillates with constant amplitude or even increasing amplitude, which means the system turns from a weak damping state to an unstable state under the stochastic excitation.

Table 5.1 Damping ratio of the power angle difference between G1 and G13 with different time delays when $\vartheta_0 = 0.03$.

Time delay	Damping ratio
0 ms	0.2627
50 ms	0.2028
111.5 ms	0.1235
150 ms	0.0213

Table 5.2 Damping ratio of the power angle difference between G8 and G16 with different time delays when $\vartheta_0 = 0.06$.

Time delay	Damping ratio
0 ms	0.2353
50 ms	0.1861
98.7 ms	0.0945
130 ms	0.0600

The prony algorithm [6] is used to calculate the damping ratios of the G1–G13 relative power angle curve and the G8–G16 relative power angle curve with different time delays. The results are shown in Table 5.1 and Table 5.2, respectively.

It can be seen from Tables 5.1 and 5.2 that, when $\vartheta_0 = 0.03$ and the time delay reaches 111.5 ms, the damping ratio of the relative power angle curve is 12.35%. When $\vartheta_0 = 0.06$ and the time delay reaches 98.7 ms, the damping ratio of the relative power angle curve is 12.35%. In these two cases, the damping ratios are both below 13% and thus the controller will not meet the control requirements. In addition, a comparison between the time-delay stability upper bound of the stochastic system when $\vartheta_0 = 0.03$ and $\vartheta_0 = 0.06$ shows that the time-delay stability upper bound decreases as the stochastic excitation intensity increases from 0.03 to 0.06. This means that the time-delay stability of the system becomes worse with the increase in the stochastic excitation. Meanwhile, the simulation results in Tables 5.1 and 5.2 demonstrate that, when the system is in the overload operation state, the proposed method is effective in solving the time-delay stability upper bound of a power system under stochastic excitation.

Weak Damping Operation State Suppose the other parameters are fixed and the damping coefficients of 16 generators in the test system are set to be 50% of the rated value. Set the stochastic excitation intensity ϑ_0 to be 0.03 and 0.06, respectively. Apply the state matrix $\boldsymbol{A}$ and time-delay matrix $\boldsymbol{A}_{\mathbf{d}}$ of the reduced-order system and the value of ϑ_0 to (5.21) and (5.22), so that the time-delay stability upper bound of the stochastic system when $\vartheta_0 = 0.03$ and $\vartheta_0 = 0.06$ can be calculated, respectively: $h_{0.03} = 97.1$ ms and $h_{0.06} = 83.6$ ms. When $\vartheta_0 = 0.03$, set the time delay to be 0 ms, 50 ms, 97.1 ms and 130 ms respectively, and observe the dynamic responses of the power angle difference between G1 and G16. The results are shown in Figure 5.3. When $\vartheta_0 = 0.06$, set the time delay to be 0 ms, 50 ms, 83.6 ms, and 120 ms, respectively, and observe the dynamic responses of the power angle difference between G6 and G15. The results are shown in Figure 5.4.

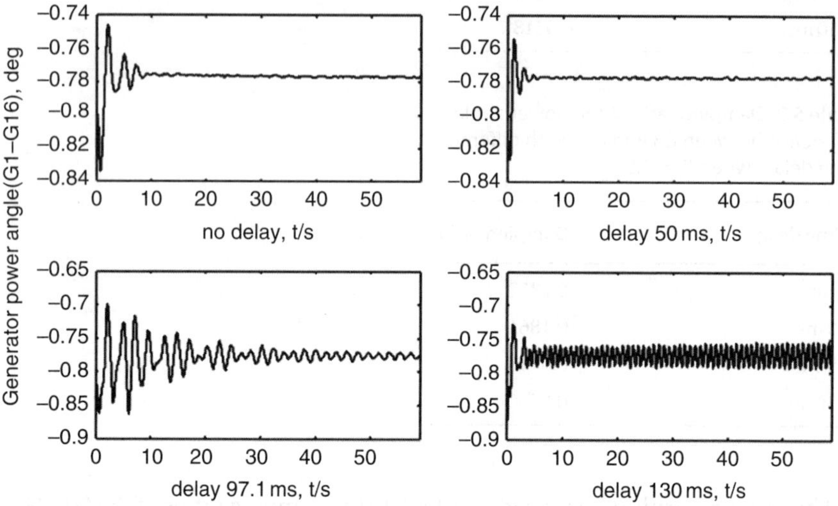

Figure 5.3 Dynamic responses of the G1–G16 relative power angle with different time delays when $\vartheta_0 = 0.03$.

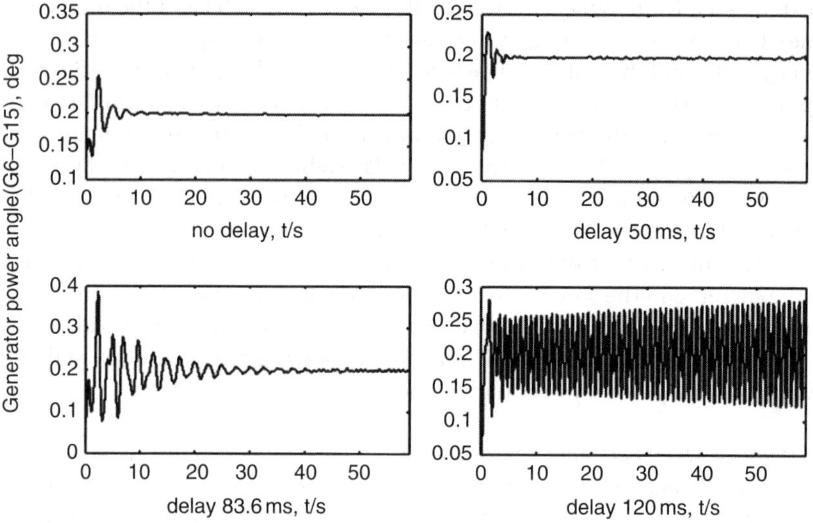

Figure 5.4 Dynamic responses of the G6–G15 relative power angle with different time delays when $\vartheta_0 = 0.06$.

It can be seen from Figures 5.3 and 5.4 that, when the time delay is below the time-delay stability upper bound, the stochastic system is able to damp the inter-area oscillation quickly (within 10 s) and remains stable after $t = 10$ s. When the time delay reaches the time-delay stability upper bound, the generator relative power angle fluctuates slightly, that is, the stochastic system is in a weak damping state. When the time delay rises above the time-delay stability upper bound, the generator relative power angle curve oscillates with a constant amplitude or an even increasing amplitude, that is, the system turns from a weak damping state to an unstable state under the stochastic excitation.

The damping ratios of the G1–G16 relative power angle curve and the G6–G15 relative power angle curve with different time delays are shown in Tables 5.3 and 5.4, respectively. It can be seen that, when the time delay is below the time-delay stability upper bound, the damping ratio of the stochastic system is relatively large and the system could operate in the steady state. However, when the time delay reaches the time-delay stability upper bound, the damping ratio decreases to 10.89% and 11.46%, respectively, which are below 12%, and thus the controller will not meet the control requirements. Analysis of Figures 5.3 and 5.4 and Tables 5.1 and 5.2 demonstrates that the proposed method is feasible when the system is in a weak damping operation state and is effective in solving the time-delay stability upper bound of a weak damping power system under stochastic excitation.

Table 5.3 Damping ratio of the power angle difference between G1 and G16 with different time delays when $\vartheta_0 = 0.03$.

Time delay	Damping ratio
0 ms	0.3544
50 ms	0.1943
97.1 ms	0.1089
130 ms	0.0789

Table 5.4 Damping ratio of the power angle difference between G6 and G15 with different time delays when $\vartheta_0 = 0.06$.

Time delay	Damping ratio
0 ms	0.3574
50 ms	0.2473
83.6 ms	0.1146
120 ms	0.0048

High Excitation Magnification Operation State Suppose the other parameters are fixed and the excitation magnification K_a of 16 generators in the test system is set to be 110% of the rated value. Set the stochastic excitation intensity ϑ_0 to be 0.03 and 0.06, respectively. Apply the state matrix $\boldsymbol{A}$ and time-delay matrix $\boldsymbol{A_d}$ of the reduced-order system and the value of ϑ_0 to (5.21) and (5.22), so that the time-delay stability upper bound of the stochastic system when $\vartheta_0 = 0.03$ and $\vartheta_0 = 0.06$ can be calculated, respectively: $h_{0.03} = 126.3$ ms and $h_{0.06} = 99.4$ ms. When $\vartheta_0 = 0.03$, set the time delay to be 0 ms, 50 ms, 126.3 ms and 150 ms, respectively, and observe the dynamic responses of the power angle difference between G9 and G15. The results are shown in Figure 5.5. When $\vartheta_0 = 0.06$, set the time delay to be 0 ms, 50 ms, 99.4 ms, and 130 ms, respectively, and observe the dynamic responses of the power angle difference between G4 and G13. The results are shown in Figure 5.6.

It can be seen from Figures 5.5 and 5.6 that, when the time delay is below the time-delay stability upper bound, the system is able to damp the inter-area oscillation quickly (within 10 s) and after $t = 10$ s the generator relative power angle remains constant, which means the stochastic system is stable. When the time delay reaches the time-delay stability upper bound, the generator relative power angle fluctuates slightly around the constant value, which means the stochastic system is in a weak damping state. When the time delay rises above the time-delay stability upper bound, the generator relative power angle curve oscillates with a constant amplitude or an even increasing amplitude, which means the system turns from a weak damping state to an unstable state under the stochastic excitation.

The damping ratios of the G9–G15 relative power angle curve and the G4–G13 relative power angle curve with different time delays are shown in Tables 5.5 and 5.6, respectively. It can be seen that, when $\vartheta_0 = 0.03$ and the time delay reaches 126.3 ms, the damping ratio of the relative power angle curve is 8.88%. When $\vartheta_0 = 0.06$ and the time delay reaches 99.4 ms, the damping ratio of the

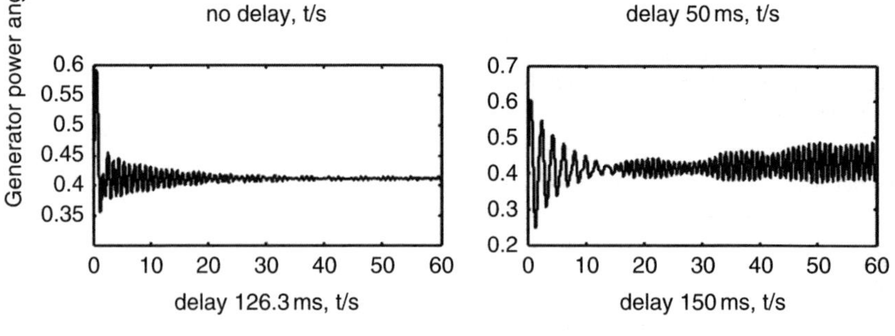

Figure 5.5 Dynamic responses of the G9–G15 relative power angle with different time delays when $\vartheta_0 = 0.03$.

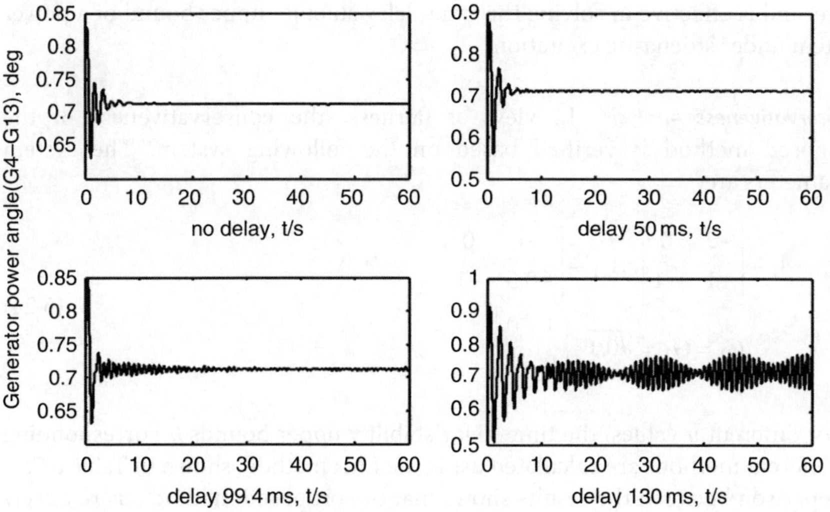

Figure 5.6 Dynamic responses of the G4–G13 relative power angle with different time delays when $\vartheta_0 = 0.06$.

Table 5.5 Damping ratio of the power angle difference between G9 and G15 with different time delays when $\vartheta_0 = 0.04$.

Time delay	Damping ratio
0 ms	0.3483
50 ms	0.2078
126.3 ms	0.0888
150 ms	0.0079

Table 5.6 Damping ratio of the power angle difference between G4 and G13 with different time delays when $\vartheta_0 = 0.06$.

Time delay	Damping ratio
0 ms	0.3325
50 ms	0.1828
99.4 ms	0.1084
130 ms	0.0637

relative power angle curve is 10.84%. In these two cases, the damping ratios are both below 11% and thus the system is in a weak damping state. Analysis of Figures 5.5 and 5.6 and Tables 5.5 and 5.6 demonstrates that the proposed method is feasible when the system is in a high excitation magnification operation state, and is effective in solving the time-delay stability upper bound of a power system under stochastic excitation.

Conservativeness Analysis In view of fairness, the conservativeness of the proposed method is verified based on the following system. The system parameters are

$$A_1 = \begin{bmatrix} -2 & 0 \\ -1 & -1 \end{bmatrix}, A_{d1} = \begin{bmatrix} -1 & 0 \\ -0.5 & -1 \end{bmatrix}$$
$$G_1 = G_2 = \sqrt{0.1} * \begin{bmatrix} 1 & 0 \\ 0 & 1 \end{bmatrix} \tag{5.24}$$

For different μ values, the time-delay stability upper bounds $\bar{h}$ corresponding to different methods are calculated using the LMI method, shown in Table 5.7. A comparison between the results shows that the proposed method has relatively low conservativeness.

5.2.2 Stochastic Time-Delay Stability Analysis of a Power System with Wind Power Connection

In recent years, with the capacity and voltage level of fluctuating new energy power generation ever increasing and various stochastic loads being connected to the power system, conventional methods to analyze the time-delay stability of a power system based on the certainty differential equation theory are facing serious challenges. In view of this problem, a method is proposed in this section to analyze the influence of stochastic excitations on the time-delay stability of the power system by directly modeling the state equations of the interconnected system under stochastic excitations in the stochastic space. The variation of the system time-delay stability upper bound with the stochastic excitation intensity of wind power, excitation magnification, and damping coefficient is calculated and the correctness and effectiveness of the proposed method in different system operating conditions is verified.

Table 5.7 Time-delay stability upper bound $\bar{h}$ obtained using different methods.

μ	**0**	**0.5**	**0.9**
Theorem 5.1 in this section	6.2103	6.2408	3.5606
Reference [7]	4.1322	1.0679	0.6471
Theorem 1 in reference [8]	4.1491	1.2956	0.8180
Corollary 1 in reference [8]	4.3564	1.9013	1.0812

5.2.2.1 Modeling of a Power System with DFIG Connection

DFIG Model Pena *et al.* have made a detailed description of the DFIG model [9], where the electromagnetic part is as follows:

1) Stator voltage equations:

$$\dot{\psi}_{sd} = -r_s i_{sd} + \omega_s \psi_{sq} + u_{sd} \tag{5.25}$$

$$\dot{\psi}_{sq} = -r_s i_{sq} + \omega_s \psi_{sd} + u_{sq} \tag{5.26}$$

2) Rotor current equations:

$$\dot{i}_{rd} = \frac{1}{L_1}\left[r_s L_m i_{sd} - \omega_r L_m L_s i_{sq} - \omega_r L_r L_s i_{rq} + L_1 \omega_s i_{rq} - r_r L_s i_{rd} - L_m u_{sd} + L_s u_{rd}\right] \tag{5.27}$$

$$\dot{i}_{rq} = \frac{1}{L_1}\left[r_s L_m i_{sq} + \omega_r L_m L_s i_{sd} + \omega_r L_r L_s i_{rd} + L_1 \omega_s i_{rq} - r_r L_s i_{rq} - L_m u_{sq} + L_s u_{rq}\right] \tag{5.28}$$

3) Rotor current control equations:

$$\dot{x}_1 = i_{rd}^{ref} - i_{rd} \tag{5.29}$$

$$\dot{x}_2 = i_{rq}^{ref} - i_{rq} \tag{5.30}$$

4) Phase-lock loop tracking equation:

$$\dot{x}_3 = -u_{sd} \tag{5.31}$$

5) Motion equation under a d–q coordinate system:

$$\dot{\theta} = \omega_n - \omega_s \tag{5.32}$$

where u_{sd} and u_{sq} are the DFIG stator d-axis voltage and q-axis voltage, ψ_{sd} and ψ_{sq} are the DFIG stator d-axis flux and q-axis flux. i_{sd} and i_{sq} are the DFIG stator d-axis current and q-axis current, ω_s is the rotating speed of the coordinate system, r_s is the armature resistance, i_{rd} and i_{rq} are the DFIG rotor d-axis current and q-axis current, u_{rd} and u_{rq} are the DFIG rotor d-axis voltage and q-axis voltage, ω_r is the rotor speed, r_r is the rotor resistance, L_s is the stator self-inductance, L_r is the rotor self-inductance, and L_m is the mutual inductance between stator and rotor, i_{rd}^{ref} and i_{rq}^{ref} are the reference values of the DFIG rotor d-axis current and q-axis current, x_1 and x_2 are the accumulated tracking errors of the DFIG rotor d-axis current and q-axis current, x_3 is the accumulated tracking error of the DFIG stator d-axis voltage, θ is the angular difference between the synchronously rotating x–y coordinate system and d–q coordinate system, and ω_n is the grid frequency. In (5.27), $L_1 = L_s L_r - L_m^2$.

The algebra part of the eighth-order detailed model of the DFIG is shown in (5.33) to (5.39).

6) Stator current equations:

$$i_{sd} = \frac{\psi_{sd} - L_m i_{rd}}{L_s} \tag{5.33}$$

$$i_{sq} = \frac{\psi_{sq} - L_m i_{rq}}{L_s} \tag{5.34}$$

7) Rotor flux equations:

$$\psi_{rd} = L_m i_{sd} + L_r i_{rd} \tag{5.35}$$

$$\psi_{rq} = L_m i_{sq} + L_r i_{rq} \tag{5.36}$$

8) Rotor voltage equations:

$$u_{rd} = K_{p1}\left(i_{rd}^{ref} - i_{rd}\right) + r_r i_{rd} + K_{i1} x_1 - \left(\omega_s - \omega_r\right)\psi_{rq} \tag{5.37}$$

$$u_{rq} = K_{p2}\left(i_{rq}^{ref} - i_{rq}\right) + r_r i_{rq} + K_{i2} x_2 - \left(\omega_s - \omega_r\right)\psi_{rd} \tag{5.38}$$

9) Rotating speed equation under d–q coordinate system:

$$\omega_s = -K_{p3} u_{sd} + K_{i3} x_3 \tag{5.39}$$

where ψ_{rd} and ψ_{rq} are the DFIG rotor d-axis flux and q-axis flux, K_{p1}, K_{p2}, K_{x1}, and K_{x2} are parameters of the rotor current controller, and K_{p3} and K_{i3} are parameters of the PI controller.

To reach a general conclusion, the DFIG eighth-order electromagnetic model needs to be reduced in order, so that the simplified model for stability analysis can be obtained. For DFIG, transient attenuation of the stator flux is much slower than that of the rotor current [10]. Thus, the transient process of the DFIG stator flux has little influence on the electromechanical transient of the system. Meanwhile, numerical calculation demonstrates that the simplified second-order dynamic model retaining only the phase-lock loop has a similar stability boundary to the eighth-order electromagnetic model [11]. In addition, the shaft of wind generators will result in relevant oscillation modes. When a fault occurs in the power system, the wind mechanical system will undergo oscillation and swing with certain frequencies [12]. Therefore, on the basis of the phase-lock loop simplified second-order dynamic model, the mechanical model of a wind generator is introduced in this section, so that the error due to negligence of the mechanical part can be avoided. The electromechanical kinetic equation of the wind generator is

$$2H\left(1-s\right)\dot{s} = P_m - P_e \tag{5.40}$$

where H is the inertia coefficient, P_e is the electromagnetic power, s is the slip, P_m is the input mechanical power to the wind generator, and P_e and P_m can be calculated according to (5.41) and (5.42), respectively:

$$P_m = P_0 + P_\Delta \tag{5.41}$$

$$P_e = \mathrm{Re}\left\{u_s i_s^*\right\} \tag{5.42}$$

where P_0 is the definite part of the wind generator mechanical power and P_Δ is the mechanical power fluctuation caused by the stochastic variation of the wind speed, which can be described by the Wiener process as

$$dP_\Delta = \vartheta d\omega(t) \tag{5.43}$$

where $\omega(t)$ is the Wiener process, ϑ is the stochastic excitation intensity coefficient, which represents the intensity of power stochastic fluctuation, and (5.22), (5.23), and (5.40) form the third-order dynamic model of DFIG. Suppose that (x_{3_0}, θ_0, s_0) is an equilibrium point of DFIG. Linearize the third-order dynamic model at the equilibrium point so that the state equations can be expressed as:

$$d\begin{bmatrix}\Delta x_3\\ \Delta\theta\\ \Delta s\end{bmatrix} = \begin{bmatrix} -k_2 L_1 i_{sq0} & k_3\cos\theta_0 & 0\\ -k_2 & -k_1\cos\theta_0 & 0\\ \dfrac{k_2 L_1 i_{sd0} i_{sq0}}{2H(1-s_0)} & \dfrac{-k_4\cos\theta_0}{2H(1-s_0)} & 0\end{bmatrix}\cdot\begin{bmatrix}\Delta x_3\\ \Delta\theta\\ \Delta s\end{bmatrix} dt + \begin{bmatrix}0\\ 0\\ \dfrac{\vartheta}{2H(1-s_0)}\end{bmatrix} d\omega(t) \tag{5.44}$$

where

$$\begin{aligned} K_1 &= \frac{K_{p3}U_0}{K_{p3}L_1 i_{sq0}-1}, \quad K_2 = \frac{-K_{i3}}{K_{p3}L_1 i_{sq0}-1}\\ K_3 &= \frac{U_0}{K_{p3}L_1 i_{sq0}-1}, \quad K_4 = \frac{-U_0 i_{sd0}}{K_{p3}L_1 i_{sq0}-1}\end{aligned} \tag{5.45}$$

Interconnected Power System Modeling and Analysis Suppose that DFIG and synchronous generator are connected to the infinite system in parallel. A diagram of the two-machine infinite-bus system is shown in Figure 5.7. The stochastic time-delay stability of the system is analyzed as follows.

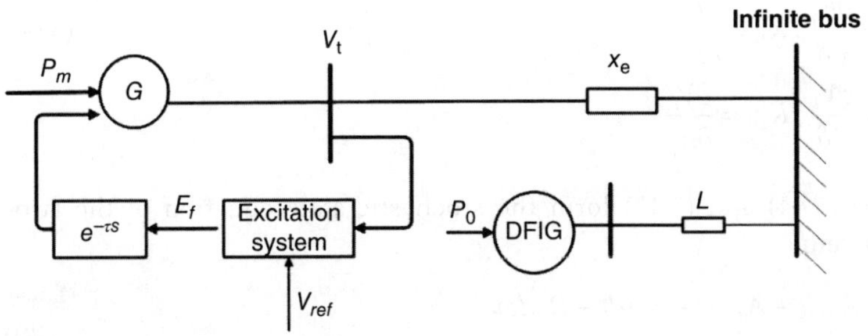

Figure 5.7 Two-machine infinite-bus system.

Consider that the excitation system output voltage has a time delay and the system model is a fourth-order differential equation. Then the state equations of the time-delay system can be expressed as

$$\begin{aligned}
&\dot{\delta} = \omega_s \omega \\
&M\dot{\omega} = P_m - P_e - D\omega \\
&\dot{E}'_q = \frac{1}{T'_d}\left[E_f(t-h_t) - E'_q - (x_d - x'_d)i_d\right] \\
&\dot{E}_f = \frac{1}{T_A}\left[K_A(V_{ref} - V_t) - E_f\right]
\end{aligned} \tag{5.46}$$

where δ, ω, E'_q, and E_f are the generator power angle, angular speed, the postreactance emf, and the excitation emf. P_m is the prime mover output power, ω_s is the system rated angular speed, D is the generator damping coefficient, T'_d and T_A are time constants of the generator stator loop and excitation loop, K_A is the magnification coefficient of the excitation loop, V_{ref} is the reference value of the terminal voltage, x_d and x'_d are the steady-state reactance and transient reactance of the generator, x_e is the line reactance, V_0 is the infinite bus voltage, P_e is the generator output power, and V_t and i_d are the generator terminal voltage and pilot output current.

Linearize the fourth-order dynamic model at the equilibrium point $(\delta_0, \omega_0, E'_{q0}, E_{f0})$ so that the state equations are

$$\begin{bmatrix}\Delta\delta \\ \Delta\omega \\ \Delta E'_q \\ \Delta E_f\end{bmatrix} = \begin{bmatrix} 0 & \omega_s & 0 & 0 \\ \frac{-K_5}{M} & \frac{-D}{M} & \frac{-K_6}{M} & 0 \\ \frac{-K_7}{T'_d} & 0 & \frac{-K_8}{T'_d} & 0 \\ \frac{-K_9}{T_A} & 0 & \frac{-K_{10}}{T_A} & \frac{-1}{T_A}\end{bmatrix} \cdot \begin{bmatrix}\Delta\delta \\ \Delta\omega \\ \Delta E'_q \\ \Delta E_f\end{bmatrix} dt + \begin{bmatrix} 0 & 0 & 0 & 0 \\ 0 & 0 & 0 & 0 \\ 0 & 0 & 0 & \frac{1}{T'_d} \\ 0 & 0 & 0 & 0\end{bmatrix}\begin{bmatrix} 0 \\ 0 \\ \Delta E_f(t-h_t) \\ 0\end{bmatrix} \tag{5.47}$$

where

$$\begin{aligned}
K_5 &= \frac{\partial P_e}{\partial \delta}, K_6 = \frac{\partial P_e}{\partial E'_q} \\
K_7 &= \frac{\partial E_q}{\partial \delta}, K_8 = \frac{\partial E_q}{\partial E'_q} \\
K_9 &= \frac{\partial V_t}{\partial \delta}, K_{10} = \frac{\partial V_t}{\partial E'_q}
\end{aligned} \tag{5.48}$$

Equations (5.44) and (5.47) form the stochastic state equation of the two-machine system:

$$\begin{aligned}
dx &= Axdt + A_d x(t-h_t)dt + B_\omega d\omega_t \\
&= f(t)dt + B_\omega d\omega_t
\end{aligned} \tag{5.49}$$

where

$$x=\left[x_3,\theta,s,\delta,\omega,E'_q,E_f\right]^T$$

$$A=\begin{bmatrix} A_1 & 0 \\ 0 & A_2 \end{bmatrix}$$

$$A_d=\begin{bmatrix} & & & 0_{6\times 7} & & & \\ 0 & 0 & 0 & 0 & 0 & 1/T'_d & 0 \\ & & & 0_{1\times 7} & & & \end{bmatrix}$$

$$B_\omega=\begin{bmatrix} & & & 0_{2\times 7} & & & \\ 0 & 0 & \vartheta/2H(1-s_0) & 0 & 0 & 0 & 0 \\ & & & 0_{4\times 7} & & & \end{bmatrix}$$

It can be seen from (5.49) that stochastic factors exist in the time-delay system and therefore traditional methods to analyze the time-delay stability are not applicable. In this section, the *Itô* differential theory is used to analyze the time-delay stability of the stochastic system in (5.49).

Theorem 5.3 For the stochastic time-delay power system described in (5.49), with given scalar quantities $\bar{h}>0$ and μ, if there exist $P=P^T>0$, $Q=Q^T\geq 0$, $R=R^T\geq 0$, $K=K^T>0$, $Z=Z^T\geq 0$, scalar quantity $\varepsilon>0$, and appropriately dimensioned matrices L, M, and S that meet the LMI in (5.50), then the stochastic time-delay system in (5.49) is mean square stable:

$$\begin{bmatrix} \Omega & L \\ * & -T_1-Z/\bar{h} \end{bmatrix}<0 \tag{5.50}$$

where

$$\Omega=\Omega_1+\Omega_2+\Omega_2^T$$

$$\Omega_2=\left[L+SA \quad SA_d \quad -L \quad 0 \quad -S\right]$$

$$\Omega_1=\begin{bmatrix} Q+K+\varepsilon B_\omega^T B_\omega - R/\bar{h} & R/\bar{h} & 0 & A^T M & P \\ * & \varepsilon B_\omega^T B_\omega-(1-\mu)Q-R/\bar{h} & 0 & A_d^T M & 0 \\ * & * & -K & 0 & 0 \\ * & * & * & \bar{h}R-M-M^T & 0 \\ * & * & * & * & \bar{h}Z \end{bmatrix}$$

Proof: Define vector domain $y(t) \in R_n$ which meets

$$y(t)dt = dx(t), y(\alpha) = \varphi(\alpha), \forall \alpha \in \left[-\bar{h}, 0\right] \tag{5.51}$$

Form the following Lyapunov–Krasovskii functional:

$$\begin{aligned} V(x_t, t) &= x^T(t)Px(t) + \int_{t-h_t}^{t} x^T(s)Qx(s)ds + \int_{t-\bar{h}}^{t} x^T(s)Kx(s)ds \\ &+ \int_{-\bar{h}}^{0}\int_{t+\theta}^{t} y^T(s)Ry(s)dsd\theta + \int_{-\bar{h}}^{0}\int_{t+\theta}^{t} f^T(s)Zf(s)dsd\theta \end{aligned} \tag{5.52}$$

According to the *Itô* differential formula, the stochastic differential of $V(x_t, t)$ along the system in (5.49) is

$$dV(t, x(t)) = \zeta V dt + 2x^T(t)PB_\omega d\omega(t) \tag{5.53}$$

In (5.53), the weak infinitesimal generating operator is

$$\begin{aligned} \zeta V(t, x(t)) &= 2x^T(t)Pf(x(t)) + x^T(t)(Q+K)x(t) - (1-\dot{h}_t)x^T(t-h_t)Qx(t-h_t) \\ &- x^T(t-\bar{h})Kx(t-\bar{h}) + \bar{h}y^T(t)Ry(t) - \int_{t-\bar{h}}^{t} y_s^T R y_s ds \\ &+ \bar{h}f^T(t)Zf(t) - \int_{t-\bar{h}}^{t} f_s^T Z f_s ds + tr\left\{B_\omega{}^T PB_\omega\right\} \end{aligned} \tag{5.54}$$

According to the Newton-Leibniz formula, for any matrices **L** and **M** with appropriate dimensions, the following equations are true:

$$2\xi^T(t)L\left[x(t) - x(t-\bar{h}) - \int_{t-\bar{h}}^{t} dx_s\right] = 0 \tag{5.55}$$

$$2y^T(t)M\left[\left(Ax + A_d x_{t-h_t} - y(t)\right)dt + B_\omega d\omega(t)\right] = 0 \tag{5.56}$$

where

$$\xi(t) = \left[x^T \quad x^T(t-h_t) \quad x^T(t-\bar{h}) \quad y^T(t) \quad f^T(t)\right]^T \tag{5.57}$$

Meanwhile, according to $f(t) = Ax(t) + A_d x(t-h_t)$, for any appropriately dimensioned matrix S, the following equation is constantly true:

$$2\zeta^T(t)S\left[Ax(t) - A_d x(t-h_t) - f(t)\right] = 0 \tag{5.58}$$

Furthermore, according to the Jensen inequality [3], the following inequalities are true:

$$-\int_{t-\bar{h}}^{t} y_s^T R y_s ds \le \begin{bmatrix} x \\ x_{t-h_t} \end{bmatrix} \begin{bmatrix} -R/\bar{h} & R/\bar{h} \\ R/\bar{h} & -R/\bar{h} \end{bmatrix} \begin{bmatrix} x & x_{t-h_t} \end{bmatrix} \tag{5.59}$$

$$\begin{aligned} -2\xi^T(t)N\int_{t-h_t}^{t} dx_s &= -2\xi^T(t)N\int_{t-h_t}^{t} f(s)ds - 2\xi^T(t)N\int_{t-h_t}^{t} g(s)d\omega(s) \\ &\le -2\zeta^T(t)N\int_{t-h_t}^{t} f(s)ds + \xi^T(t)NT_1^{-1}N^T\xi(t) \\ &\quad + \left(\int_{t-h_t}^{t} g(s)d\omega(s)\right)^T T_1 \int_{t-h_t}^{t} g(s)d\omega(s) \end{aligned} \tag{5.60}$$

$$\begin{aligned} tr\left\{B_\omega{}^T PB_\omega\right\} &= tr(P)^* tr\left\{B_\omega{}^T B_\omega\right\} \\ &\le \lambda_{\max}(P)^* tr\left\{B_\omega{}^T B_\omega\right\} \\ &< \varepsilon B_\omega{}^T B_\omega \end{aligned} \tag{5.61}$$

Based on the above analysis, apply (5.45) to (5.50) to (5.44), so that

$$\begin{aligned} \zeta V(t,x(t)) &\le \xi^T(t)\left(\Omega + LT_1^{-1}L^T + \bar{h}\cdot LZ^{-1}L^T\right)\xi(t) \\ &\quad - \int_{t-h_t}^{t} \left[\xi^T(t)L + f^T(t)Z\right]Z^{-1}\left[L^T\xi(t) + Zf(t)\right]ds \end{aligned} \tag{5.62}$$

Since $Z > 0$, if the following inequality is true,

$$\Omega + LT_2^{-1}L^T + \bar{h}LZ^{-1}L^{TT} < 0 \tag{5.63}$$

Then, for any $\xi(t) \ne 0$, $\mathrm{E}\{\zeta V(t,x(t))\} < 0$ is true and the stochastic system in (5.25) is asymptotically mean square stable in the probability space. According to the Schur complement theorem [4], (5.63) is equivalent to (5.50), and thus Theorem 5.3 is proved.

Time-Delay Stability Upper Bound The matrix inequality in (5.50) could only be used to identify whether the system is stable or not as the time-delay stability upper bound of the system is still unknown. Consider that solving the time-delay stability upper bound is a convex optimization problem with a linear inequality constraint and manifests the form of the GEVP (generalized eigenvalue). In this section the GEVP method is used to calculate the time-delay stability upper bound of the system. Since (5.50) is not a standard GEVP form and cannot be solved with the GEVP method directly, appropriate transformation is necessary.

According to the nature of the Schur complement, (5.50) can be transformed to

$$\begin{bmatrix} \Omega & L & 0 & 0 \\ * & -T_1 - Z/\bar{h} & R & 0 \\ * & * & -R/\bar{h} & Z \\ * & * & * & -Z/\bar{h} \end{bmatrix} < 0 \tag{5.64}$$

Set $-Z/\bar{h} < Z_1, -R/\bar{h} < -R_1$ and $d = 1/\bar{h}$ so that

$$\begin{bmatrix} T_1 & 0 & 0 \\ * & T_2 & 0 \\ * & * & T_3 \end{bmatrix} < d \begin{bmatrix} Z_1 & 0 & 0 \\ * & Z_2 & 0 \\ * & * & K \end{bmatrix} \tag{5.65}$$

In (5.65), substitute $-Z/\bar{h}$ and $-R/\bar{h}$ with $-Z_1$ and $-R_1$, respectively; thus

$$\begin{bmatrix} \Omega & L & 0 & 0 \\ * & -T_1 - Z_1 & R & 0 \\ * & * & -R_1 & Z \\ * & * & * & -Z_1 \end{bmatrix} < 0 \tag{5.66}$$

According to (5.64) to (5.66), solving the time-delay stability upper bound $\bar{h}$ could be transformed to the following optimization problem:

$$\min\ \tau \qquad s.t.(5.65),(5.66) \tag{5.67}$$

By solving (5.67), the minimum τ can be gained. Then, according to $\bar{h}_{\max} = 1/\tau$, the time-delay stability upper bound of the stochastic system can be gained.

5.2.2.2 Simulation Analysis

Interconnected Power System Modeling and Analysis The diagram of the two-machine infinite-bus system is shown in Figure 5.7, where the synchronous generator parameters are: $P_m = 1.3$; $\omega_s = 377$; $D = 10$; $T_d' = 10$, $T_A = 1$; $K_A = 190$; $V_{ref} = 1.05$; $x_d = 1$, $x_d' = 0.4$, $x_e = 0.5$; and $V_t = 1.0$. The DFIG parameters are: $r_r = 0.006298$; $H = 3.025$; $L_s = 0.004872$; $L_r = 0.005098$; $L_m = 0.004603$; $L = 0.00075$; $K_{p1} = 0.155$, $K_{p2} = 0.155$, $K_{x2} = 15$; $K_{p3} = 1$, and $K_{i3} = 345$. With the above parameters, the coefficient matrices of the linearized state equations of the stochastic time-delay system in (5.49) at the equilibrium point when the stochastic excitation intensity $\vartheta = 0.01$ can be calculated:

$$A = \begin{bmatrix} A_1 & 0 \\ 0 & A_2 \end{bmatrix} \tag{5.68}$$

$$A_1 = \begin{bmatrix} 32.4204 & -0.7769 & 0 \\ -377.4204 & 0.7769 & 0 \\ -22.8965 & 0.5460 & 0 \end{bmatrix} \tag{5.69}$$

$$A_2 = \begin{bmatrix} 0 & 377 & 0 & 0 \\ -0.07615 & -1 & -0.09587 & 0 \\ -0.05752 & 0 & -0.16667 & 0 \\ 49.5017 & 0 & -93.0975 & -1 \end{bmatrix} \tag{5.70}$$

$$A_d = \begin{bmatrix} & & & 0_{6\times 7} & & & \\ 0 & 0 & 0 & 0 & 0 & 0.1 & 0 \\ & & & 0_{1\times 7} & & & \end{bmatrix} \tag{5.71}$$

$$B_\omega = \begin{bmatrix} 0 & 0 & 0 & 0 & 0 & 0 & 0 \\ 0 & 0 & 0 & 0 & 0 & 0 & 0 \\ 0 & 0 & 0.0016 & 0 & 0 & 0 & 0 \\ 0 & 0 & 0 & 0 & 0 & 0 & 0 \\ 0 & 0 & 0 & 0 & 0 & 0 & 0 \\ 0 & 0 & 0 & 0 & 0 & 0 & 0 \\ 0 & 0 & 0 & 0 & 0 & 0 & 0 \end{bmatrix} \tag{5.72}$$

To verify the correctness and effectiveness of the proposed method, apply (5.68) to (5.72) to the GEVP solver in the LMI toolbox. It can be calculated that the time-delay stability upper bound of the stochastic system when the time-delay variation rate $\mu = 0.1$ is $\bar{h} = 328.1$ ms. In other words, when $0 \le h \le 328.1$ ms, the stochastic system is time-delay stable and when $h > 328.1$ ms, the system will go unstable (feasible solutions of the system in this case can be seen in Appendix A (A1) to (A6) at the end of the chapter). Set the time delay of the stochastic system to be 50 ms, 328.1 ms, and 360 ms, respectively, and observe the variation of the generator power angle. The results are shown in Figure 5.8. It can be seen in Figure 5.8 that, when $0 \le h \le 328.1$ ms, the stochastic system is time-delay stable and when $h > 328.1$ ms, the system becomes unstable. This is in line with the conclusions reached according to the proposed method.

Influence of Stochastic Excitation Intensity Variation on Time-Delay Stability Upper Bound Suppose the other parameters are fixed and the stochastic excitation intensity ϑ varies between 0.02 and 0.2. Calculate the corresponding time-delay stability upper bound of the system based on GEVP; the results are shown in Figure 5.9. It can be seen that the variation of the system time-delay stability upper bound with ϑ is relatively complex. When ϑ increases, the time-delay stability upper bound first decreases and then increases. Thus, the system stochastic time-delay stability region is a nonconvex set, with a distinct concave area. It should be noted that, in the $(\vartheta, \bar{h})$ space, the $\bar{h} - \vartheta$ curve corresponds to the upper boundary of the system stochastic time-delay stability region, and underneath the curve is the system stochastic time-delay stability region.

To verify the correctness of Figure 5.9, set $\vartheta = 0.06$, $\vartheta = 0.1$, and $\vartheta = 0.14$, respectively, when the time delay of the system in (5.49) is $h = 220$ ms. In the above operation modes, time-domain simulation tests are conducted on the system and the synchronous generator power angle variation curves are shown in

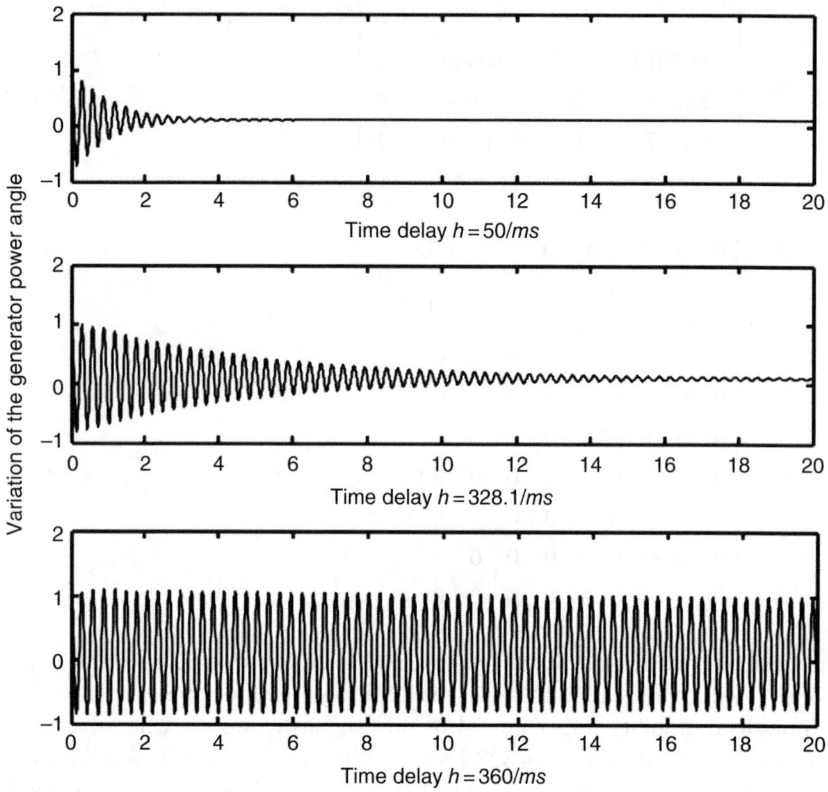

Figure 5.8 Power angle of a synchronous generator under different time delays.

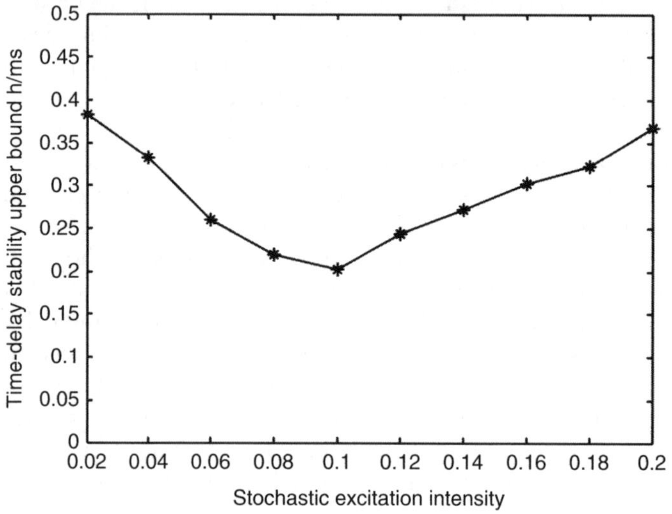

Figure 5.9 Variation of the time-delay stability upper bound with ϑ.

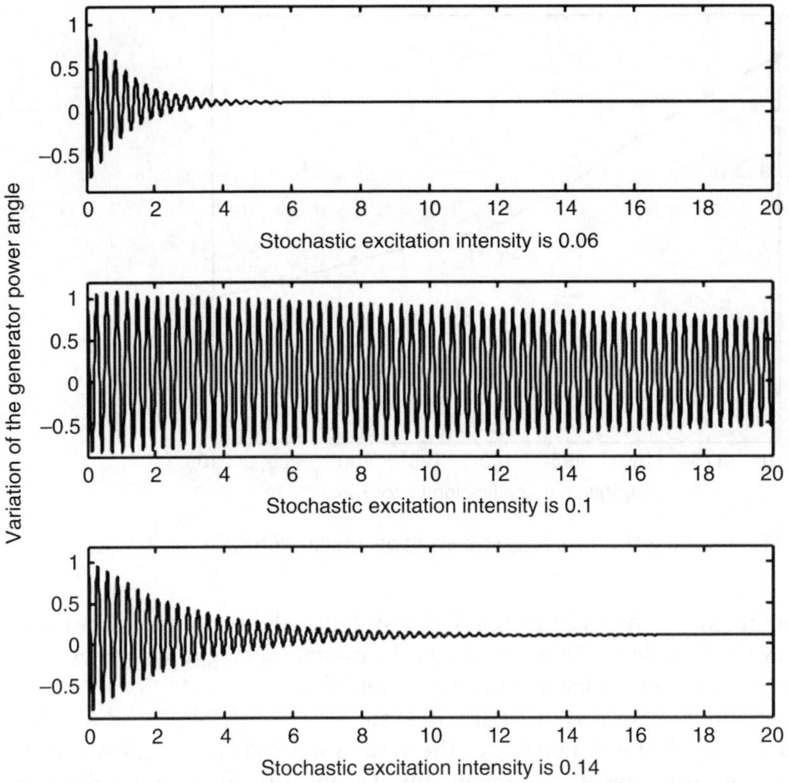

Figure 5.10 Variation of the synchronous generator power angle with ϑ.

Figure 5.10. It can be seen that, when the time delay $h = 250$ ms, the system is stable when $\vartheta = 0.06$ or $\vartheta = 0.14$ and is unstable when $\vartheta = 0.1$. This shows that when $\vartheta = 0.06$ or $\vartheta = 0.14$, the time-delay stability upper bound of the system is higher than 250 ms, and when $\vartheta = 0.1$, the time-delay stability upper bound is lower than 250 ms, which is in line with the variation of the time-delay stability upper bound with ϑ in Figure 5.9. According to Figure 5.9, when $\vartheta = 0.06$, $\vartheta = 0.1$, and $\vartheta = 0.14$, the time-delay stability upper bound is 259.81 ms, 202.61 ms, and 271.64 ms, respectively. Meanwhile, by comparing the generator power angle curves of $\vartheta = 0.06$ and $\vartheta = 0.14$, it can be seen that, though the system is stable in both cases, it takes longer for the generator power angle to reach the stability value when $\vartheta = 0.14$ than when $\vartheta = 0.06$. This demonstrates that, when the stochastic excitation intensity increases, the stability of the system is weakened.

Influence of Generator Excitation Magnification Factor Variation on Time-Delay Stability Upper Bound Suppose the other parameters are fixed and the excitation magnification factor K_A varies between 50 and 210. The calculation results are shown in Figure 5.11. It can be seen that the time-delay stability upper bound of the system decreases as K_A increases. This is because the bigger K_A is, the higher is the sensitivity of the generator excitation system, that is, the more sensitive the excitation voltage is to the difference between the generator terminal voltage and

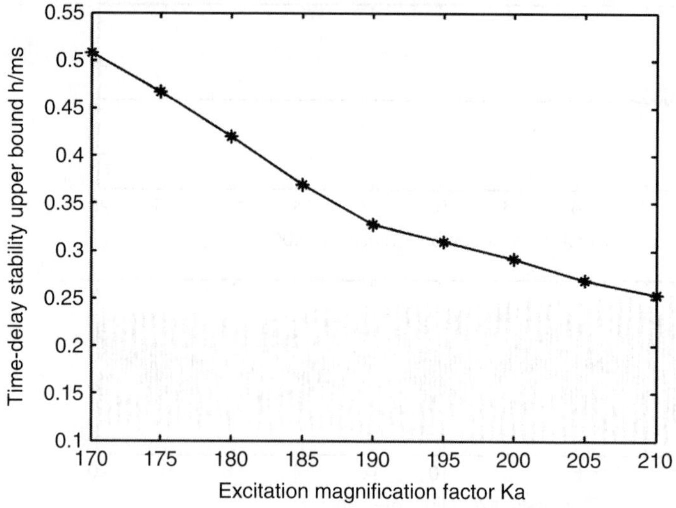

Figure 5.11 Variation of the time-delay stability upper bound with K_a.

the reference value and the more easily the system stability is affected by errors caused by the time delay. Thus, the bigger the excitation magnification factor is, the smaller is the time-delay stability upper bound.

To verify the correctness of Figure 5.11, set $K_a = 180$, $K_a = 190$, and $K_a = 200$, respectively, when the time delay of the system in (5.49) is $h = 320$ ms. In the above operation modes, time-domain simulation tests are conducted on the system and the synchronous generator power angle variation curves are shown in Figure 5.12. It can be seen that, when the time delay $h = 320$ ms, the system is stable when $K_a = 180$ or $K_a = 190$ and is unstable when $K_a = 200$. This shows that when $K_a = 180$ or $K_a = 190$, the time-delay stability upper bound of the system is higher than 320 ms and when $K_a = 200$, the time-delay stability upper bound is lower than 320 ms, which is in line with the variation of the time-delay stability upper bound with K_a in Figure 5.11. Thus the proposed method is proved correct.

Influence of Generator Damping Coefficient Variation on the Time-Delay Stability Upper Bound Suppose the other parameters are fixed and the generator damping coefficient D varies between 0 and 40. The calculation results are shown in Figure 5.13. According to the theoretical analysis, as the system damping increases, the stability of the power system will increase; thus the maximum time delay that the power system could allow for in order to remain stable will also increase. It can be seen from Figure 5.13 that, as the damping coefficient D increases from 0 to 40, the time-delay stability upper bound of the system increases monotonously. This is in line with the theoretical analysis result.

To further verify the correctness of Figure 5.13, set $D = 0$, $D = 10$, and $D = 20$ respectively, when the time delay of the system in (5.49) is $h = 320$ ms. In the above operation modes, the time-domain simulation tests are conducted on the system and the synchronous generator power angle curves are shown in

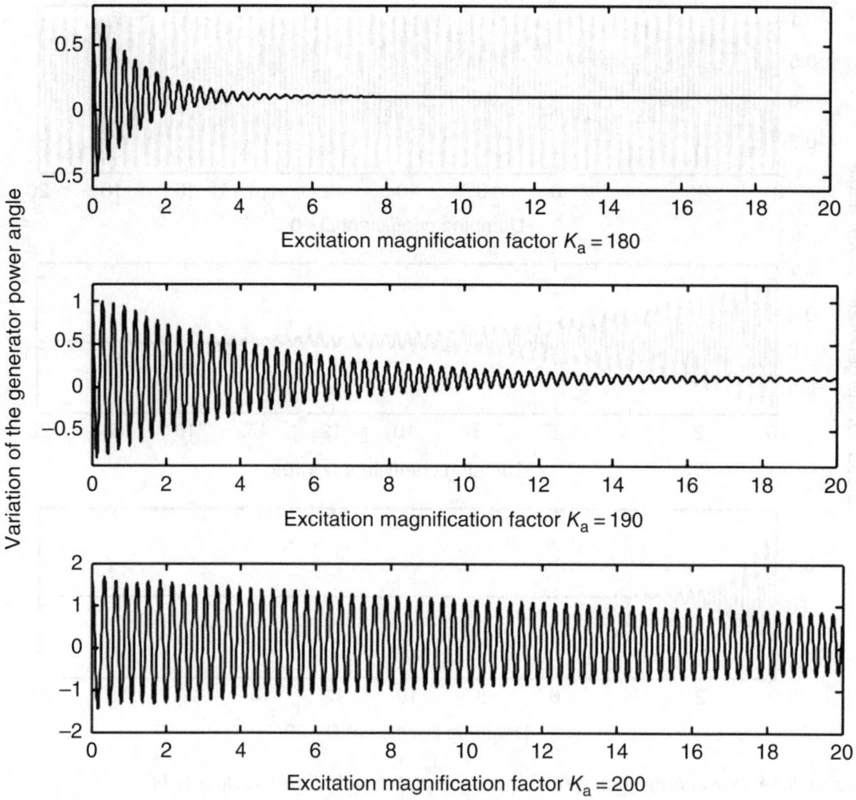

Figure 5.12 Power angle of a synchronous generator with different values of K_a.

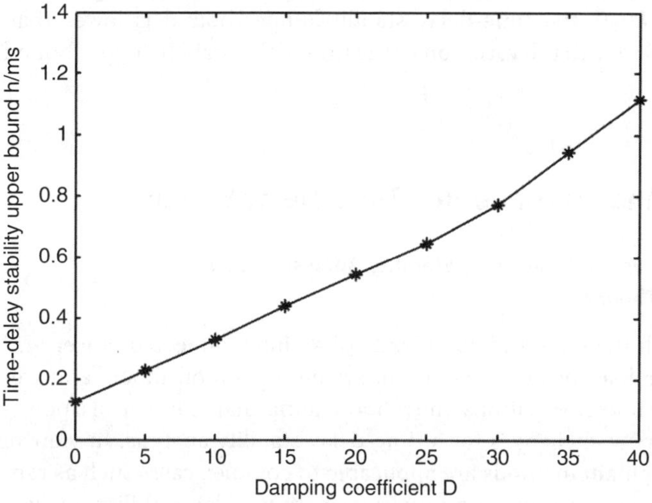

Figure 5.13 Variation of the time-delay stability upper bound with D.

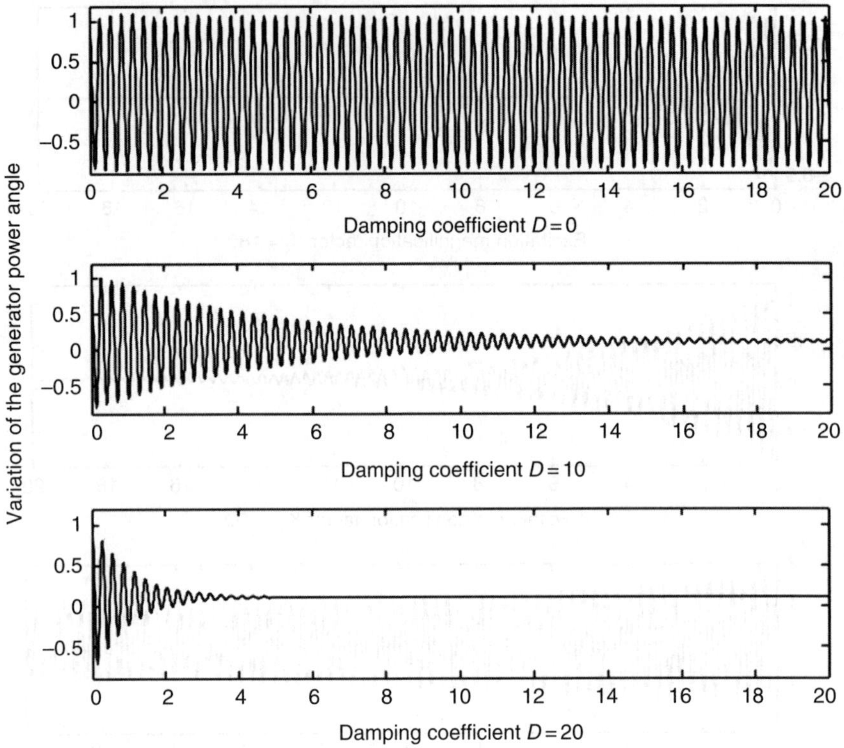

Figure 5.14 Power angle of a synchronous generator with different values of D.

Figure 5.14. It can be seen that, when the time delay $h = 320\,\text{ms}$, the system is stable when $D = 10$ or $D = 20$ and is unstable when $D = 0$. This shows that, when $D = 10$ or $D = 20$, the time-delay stability upper bound of the system is higher than 320 ms, and when $D = 0$, the time-delay stability upper bound is lower than 320 ms, which is in line with the variation of the time-delay stability upper bound in Figure 5.13.

5.3 Stability Analysis of a Jump Time-Delay System

5.3.1 Jump Power System Time-Delay Stability Analysis Based on Discrete Markov Theory

In recent years, with the scale and complexity of an interconnected power grid ever increasing, the load power fluctuation, system operation mode, and frequent operation parameter variations, the random jump characteristic of a power system has caused new challenges for a time-delay stability analysis. In view of the fact that time-domain methods are applicable to complex cases such as random jump, in this section a jump power system time-delay stability analysis method based on the discrete Markov theory is proposed. Then the maximum time-delay that a power system could allow for in different jump cases, such as a system operation mode, operation parameter variation, and load power fluctuation, are solved.

5.3.1.1 Time-Delay Power System Model

The closed-loop power system considering a time delay can be described as:

$$\begin{cases} \dot{\mathbf{x}}(t) = \boldsymbol{A}\mathbf{x}(t) + \boldsymbol{BK}\mathbf{u}(t - d(t)) \\ \mathbf{u}(t - d(t)) = \mathbf{x}(t - d(t)) \end{cases} \tag{5.73}$$

where $\mathbf{x} \in \mathbf{R}^n$ is the state vector, $\mathbf{u} \in \mathbf{R}^n$ is the control input vector, $\boldsymbol{A} \in \mathbf{R}^{n \times n}$ is the system state matrix, $\boldsymbol{B} \in \mathbf{R}^{n \times m}$ is the system control matrix, and $\boldsymbol{K} \in \mathbf{R}^{m \times n}$ is the system controller, and *d(t)* is the system time delay, which meets the inequalities below:

$$0 \le d(t) \le h, \quad \dot{d}(t) < \mu \tag{5.74}$$

Suppose the time delay matrix $\boldsymbol{A_d} = \boldsymbol{BK}$; then the state equation of the power system is

$$\begin{cases} \dot{\mathbf{x}}(t) = \boldsymbol{A}\mathbf{x}(t) + \boldsymbol{A_d}\mathbf{u}(t - d(t)) \\ \mathbf{u}(t - d(t)) = \mathbf{x}(t - d(t)) \end{cases} \tag{5.75}$$

When the power system operation mode/parameter changes, or the load power fluctuates, the state equation becomes [13]:

$$\begin{cases} \dot{\mathbf{x}}(t) = \boldsymbol{A}(\eta)\mathbf{x}(t) + \boldsymbol{A_d}(\eta)\mathbf{u}(t - d(t)) \\ \mathbf{u}(t - d(t)) = \mathbf{x}(t - d(t)) \end{cases} \tag{5.76}$$

where $\mathbf{x}(t) \in R^n$ is the state vector and η is a discrete Markov process defined in the finite state space S = {1,2, ..., N}. The state jump probability of η is

$$p_{ij} = P(\eta_{t+\Delta} = j \mid \eta_t = i) = \begin{cases} \pi_{ij}\Delta + \mathrm{o}(\Delta), & i \ne j \\ 1 + \pi_{ii}\Delta + \mathrm{o}(\Delta), & i = j \end{cases} \tag{5.77}$$

where

$$\lim_{\Delta \to \infty} \frac{\mathrm{o}(\Delta)}{\Delta} = 0 \qquad (\Delta > 0) \tag{5.78}$$

In (5.77), π_{ij} is the element in the Markov jump probability density matrix $\boldsymbol{\pi}$, which represents the jump probability density that renders the system modal at state *i* when time = *t*, and renders the system modal at state *j* when time = $t + \Delta$:

$$\pi_{ii} = -\sum_{i \ne j} \pi_{ij} \quad (\pi_{ij} \ge 0, i \ne j) \tag{5.79}$$

5.3.1.2 Jump Power System Time-Delay Stability Analysis Method

Jump Power System Time-Delay Stability Criterion

Theorem 5.4 For the jump system described in (5.75), with given scalar quantities $h_m > \mathbf{0}$ and μ, if there exist $\mathbf{P} = \mathbf{P}^{\mathrm{T}} > \mathbf{0}$, $\mathbf{Q} = \mathbf{Q}^{\mathrm{T}} \geq \mathbf{0}$, $\mathbf{R} = \mathbf{R}^{\mathrm{T}} \geq \mathbf{0}$, $\mathbf{Z}_i = \mathbf{Z}_i^{\mathrm{T}} \geq \mathbf{0}$ $(i = 1,\ 2)$, $\mathbf{N} = [\mathbf{N}_1\ \ \mathbf{N}_2\ \ \mathbf{N}_3]^{\mathrm{T}}$, $\mathbf{S} = [\mathbf{S}_1\ \ \mathbf{S}_2\ \ \mathbf{S}_3]^{\mathrm{T}}$, and $\mathbf{M} = [\mathbf{M}_1\ \ \mathbf{M}_2\ \ \mathbf{M}_3]^{\mathrm{T}}$, which satisfy the following LMI:

$$\begin{bmatrix} \boldsymbol{f} & h\mathbf{N} & h\mathbf{S} & h\mathbf{M} & h\mathbf{A}_{ci}^{\mathrm{T}}(\mathbf{Z}_1 + \mathbf{Z}_2) \\ * & -h\mathbf{Z}_1 & \mathbf{0} & \mathbf{0} & \mathbf{0} \\ * & * & -h\mathbf{Z}_1 & \mathbf{0} & \mathbf{0} \\ * & * & * & -h\mathbf{Z}_2 & \mathbf{0} \\ * & * & * & * & -h(\mathbf{Z}_1 + \mathbf{Z}_2) \end{bmatrix} < \mathbf{0} \tag{5.80}$$

then the jump system in (5.76) that meets the conditions in (5.74) is asymptotically stable, where

$$\mathbf{A}_{\mathrm{ci}} = \begin{bmatrix} \mathbf{A}_{\mathrm{i}} & \mathbf{A}_{\mathrm{di}} & \mathbf{0} \end{bmatrix}$$

$$\mathbf{\Phi} = \mathbf{\Phi}_1 + \mathbf{\Phi}_2 + \mathbf{\Phi}_2^{\mathrm{T}}$$

$$\mathbf{\Phi}_2 = \begin{bmatrix} \mathbf{N} + \mathbf{M} & -\mathbf{N} + \mathbf{S} & -\mathbf{M} - \mathbf{S} \end{bmatrix}$$

$$\mathbf{\Phi}_1 = \begin{bmatrix} \mathbf{P}_{\mathrm{i}} A_{\mathrm{i}} + A_{\mathrm{i}}^{\mathrm{T}} \mathbf{P}_{\mathrm{i}} + \mathbf{Q} + \mathbf{R} + \sum_{j=1}^{S} \pi_{\mathrm{ij}} \mathbf{P}_{\mathrm{j}} & \mathbf{P}_{\mathrm{i}} \mathbf{A}_{\mathrm{di}} & \mathbf{0} \\ * & -(1-\mu)\mathbf{Q} & \mathbf{0} \\ * & * & -\mathbf{R} \end{bmatrix}$$

$$i = \{1, 2, \ldots, N\}$$

Proof: First, the following Lyapunov–Krasovskii functional is formed:

$$\begin{aligned} \mathbf{V}(\mathbf{x}, \pi_t) &= \mathbf{x}^{\mathrm{T}}(t)\mathbf{P}(\pi_t)\mathbf{x}(t) \\ &+ \int_{t-d(t)}^{t} \mathbf{x}^{\mathrm{T}}(s)\mathbf{Q}\mathbf{x}(s)ds \\ &+ \int_{t-h}^{t} \mathbf{x}^{\mathrm{T}}(s)\mathbf{R}\mathbf{x}(s)ds \\ &+ \int_{-h}^{0}\int_{t+\theta}^{t} \dot{\mathbf{x}}^{\mathrm{T}}(s)(\mathbf{Z}_1 + \mathbf{Z}_2)\dot{\mathbf{x}}(s)dsd\theta \end{aligned} \tag{5.81}$$

According to the Newton–Leibniz formula, for any matrices **N**, **S**, and **M** that have the same dimension as the state matrix $\mathbf{A}_i$ and the time-delay matrix $\mathbf{A}_{di}$, the following equations are true:

$$2\varsigma_1^{\mathbf{T}}(t)\mathbf{N}\left[\mathbf{x}(t)-\mathbf{x}(t-d(t))-\int_{t-d(t)}^{t}\dot{\mathbf{x}}(s)ds\right]=\mathbf{0} \tag{5.82}$$

$$2\varsigma_1^{T}(t)\mathbf{S}\left[\mathbf{x}(t-d(t))-\mathbf{x}(t-h)-\int_{t-h}^{t-d(t)}\dot{\mathbf{x}}(s)ds\right]=0 \tag{5.83}$$

$$2\varsigma_1^{T}(t)\mathbf{M}\left[\mathbf{x}(t)-\mathbf{x}(t-h)-\int_{t-h}^{t}\dot{\mathbf{x}}(s)ds\right]=0 \tag{5.84}$$

where

$$\varsigma_1(t)=\left[\mathbf{x}^{\mathbf{T}}(t)\quad \mathbf{x}^{\mathbf{T}}(t-d(t))\quad \mathbf{x}^{\mathbf{T}}(t-h)\right]^{\mathbf{T}}$$

In (5.81), the derivative function of $\mathbf{V}(\mathbf{x},\pi_t)$ on t is

$$\begin{aligned}\dot{\mathbf{V}}(\mathbf{x},\pi_t)&=\mathbf{2x}^{\mathbf{T}}(t)\mathbf{P}_i\dot{\mathbf{x}}(t)+\mathbf{x}^{\mathbf{T}}(t)\mathbf{Qx}(t)\\&\quad-\left(1-\dot{d}(t)\right)\mathbf{x}^{\mathbf{T}}(t-d(t))\mathbf{Qx}(t-d(t))\\&\quad+\mathbf{x}^{\mathbf{T}}(t)\mathbf{Rx}(t)-\mathbf{x}^{\mathbf{T}}(t-h)\mathbf{Rx}(t-h)\\&\quad+h\dot{\mathbf{x}}^{\mathbf{T}}(t)(\mathbf{Z}_1+\mathbf{Z}_2)\dot{\mathbf{x}}(t)\\&\quad-\int_{t-h}^{t}\dot{\mathbf{x}}^{\mathbf{T}}(s)(\mathbf{Z}_1+\mathbf{Z}_2)\dot{\mathbf{x}}(s)ds\\&\quad+\sum_{j=1}^{S}\pi_{ij}\mathbf{x}^{\mathbf{T}}(t)\mathbf{P}_j\mathbf{x}(t)\end{aligned} \tag{5.85}$$

Apply $\dot{\mathbf{x}}(t)=A_{ci}\cdot\varsigma_1$ and (5.82) to (5.84) to the right side of (5.85), so that

$$\begin{aligned}\dot{\mathbf{V}}(\mathbf{x})&\le\mathbf{x}^{\mathbf{T}}(t)\left(\mathbf{P}_i\mathbf{A}_i+\mathbf{A}_i{}^{\mathbf{T}}\mathbf{P}_i+\mathbf{Q}+\mathbf{R}+\sum_{j=1}^{S}\pi_{ij}\mathbf{P}_j\right)\mathbf{x}(t)\\&\quad+2\mathbf{x}^{\mathbf{T}}(t)\mathbf{P}_i\mathbf{A}_{di}\mathbf{x}(t-d(t))\\&\quad-(1-\mu)\mathbf{x}^{\mathbf{T}}(t-d(t))\mathbf{Qx}(t-d(t))\\&\quad-\mathbf{x}^{\mathbf{T}}(t-h)\mathbf{Rx}(t-h)+h\dot{\mathbf{x}}^{\mathbf{T}}(t)(\mathbf{Z}_1+\mathbf{Z}_2)\dot{\mathbf{x}}(t)\\&\quad-\int_{t-d(t)}^{t}\dot{\mathbf{x}}^{\mathbf{T}}(s)\mathbf{Z}_1\dot{\mathbf{x}}^{\mathbf{T}}(s)ds\\&\quad-\int_{t-h}^{t-d(t)}\dot{\mathbf{x}}^{\mathbf{T}}(s)\mathbf{Z}_1\dot{\mathbf{x}}^{\mathbf{T}}(s)ds-\int_{t-h}^{t}\dot{\mathbf{x}}^{\mathbf{T}}(s)\mathbf{Z}_2\dot{\mathbf{x}}^{\mathbf{T}}(s)ds\end{aligned} \tag{5.86}$$

$$
\begin{aligned}
&+2\varsigma_1^{\mathrm{T}}(t)\mathbf{N}\left[\mathbf{x}(t)-\mathbf{x}(t-d(t))-\int_{t-d(t)}^{t}\dot{\mathbf{x}}(s)ds\right]\\
&+2\varsigma_1^{\mathrm{T}}(t)\mathbf{S}\left[\mathbf{x}(t-d(t))-\mathbf{x}(t-h)-\int_{t-d(t)}^{t}\dot{\mathbf{x}}(s)ds\right]\\
&+2\varsigma_1^{\mathrm{T}}(t)\mathbf{M}\left[\mathbf{x}(t)-\mathbf{x}(t-h)-\int_{t-h}^{t}\dot{\mathbf{x}}(s)ds\right]\\
&\le\varsigma_1^{\mathrm{T}}(t)[\Phi+\Phi_s]\varsigma_1(t)-\int_{t-d(t)}^{t}\Phi_1 ds\\
&-\int_{t-h}^{t-d(t)}\Phi_2 ds-\int_{t-h}^{t}\Phi_3 ds
\end{aligned}
$$

where

$$
\begin{aligned}
\Phi_s = {} & hA_{ci}^T(\mathbf{Z}_1+\mathbf{Z}_2)A_{ci}+h\mathbf{N}\mathbf{Z}_1^{-1}\mathbf{N}^T\\
& +h\mathbf{S}\mathbf{Z}_1^{-1}\mathbf{S}^T+h\mathbf{M}\mathbf{Z}_2^{-1}\mathbf{M}^T
\end{aligned}
\tag{5.87}
$$

$$
\Phi_1=\left[\varsigma_1^{\mathrm{T}}(t)\mathbf{N}+\dot{\mathbf{x}}(t)\mathbf{Z}_1\right]\mathbf{Z}_1^{-1}\left[\mathbf{N}^{\mathrm{T}}\varsigma_1(t)+\mathbf{Z}_1\dot{\mathbf{x}}(t)\right]
\tag{5.88}
$$

$$
\Phi_2=\left[\varsigma_1^{\mathrm{T}}(t)\mathbf{S}+\dot{\mathbf{x}}(t)\mathbf{Z}_1\right]\mathbf{Z}_1^{-1}\left[\mathbf{S}^{\mathrm{T}}\varsigma_1(t)+\mathbf{Z}_1\dot{\mathbf{x}}(t)\right]
\tag{5.89}
$$

$$
\Phi_3=\left[\varsigma_1^{\mathrm{T}}(t)\mathbf{M}+\dot{\mathbf{x}}(t)\mathbf{Z}_2\right]\mathbf{Z}_2^{-1}\left[\mathbf{M}^{\mathrm{T}}\varsigma_1(t)+\mathbf{Z}_1\dot{\mathbf{x}}(t)\right]
\tag{5.90}
$$

Since $\mathbf{Z}_i=\mathbf{Z}_i^{\mathrm{T}}\ge\mathbf{0}$ $(i=1, 2)$, (5.88) to (5.90) are all above 0. Thus the last three items in (5.86) are all below 0. If $\Phi+\Phi_s\le\mathbf{0}$, then, according to (5.86), $\dot{\mathbf{V}}(\mathbf{x})\le 0$, that is, the jump system in (5.76) is asymptotically stable.

Jump Power System Time-Delay Stability Upper Bound The LMI in (5.80) could only be used to identify whether the jump system is stable or not, the time-delay stability upper bound of the jump system is still unknown. Considering that the GEVP method is able to solve the global minimum value of an optimization problem, in this section the GEVP method is used to calculate the time-delay stability upper bound of the jump system. Since (5.80) is not a standard GEVP form and cannot be solved using the GEVP method directly, the following transformation is made, that is, multiplying both the left side and right side of (5.80) by the following:

$$
\begin{bmatrix}
\mathbf{I} & \mathbf{0} & \mathbf{0} & \mathbf{0} & \mathbf{0}\\
\mathbf{0} & \mathbf{I}/h & \mathbf{0} & \mathbf{0} & \mathbf{0}\\
\mathbf{0} & \mathbf{0} & \mathbf{I}/h & \mathbf{0} & \mathbf{0}\\
\mathbf{0} & \mathbf{0} & \mathbf{0} & \mathbf{I}/h & \mathbf{0}\\
\mathbf{0} & \mathbf{0} & \mathbf{0} & \mathbf{0} & \mathbf{I}/h
\end{bmatrix}>\mathbf{0}
\tag{5.91}
$$

so that

$$\begin{bmatrix} \Phi & \mathbf{N} & \mathbf{S} & \mathbf{M} & \mathbf{A}_{ci}^{T}(\mathbf{Z}_1+\mathbf{Z}_2) \\ \mathbf{N}^T & -\nu\mathbf{Z}_1 & \mathbf{0} & \mathbf{0} & \mathbf{0} \\ \mathbf{S}^T & \mathbf{0} & -\nu\mathbf{Z}_1 & \mathbf{0} & \mathbf{0} \\ \mathbf{M}^T & \mathbf{0} & \mathbf{0} & -\nu\mathbf{Z}_2 & \mathbf{0} \\ (\mathbf{Z}_1^T+\mathbf{Z}_2^T)A_c & \mathbf{0} & \mathbf{0} & \mathbf{0} & -\nu(\mathbf{Z}_1+\mathbf{Z}_2) \end{bmatrix} < \mathbf{0} \tag{5.92}$$

where $\nu = 1/h$.

In order to obtain the time-delay stability upper bound h, that is, the minimum ν, a supplementary matrix $\mathbf{Y}_i = \mathbf{Y}_i^T \geq \mathbf{0}$ $(i = 1,\ 2)$ is introduced, which meets the inequality in

$$\begin{bmatrix} \mathbf{Y}_1 & \mathbf{0} \\ \mathbf{0} & \mathbf{Y}_2 \end{bmatrix} < \nu \begin{bmatrix} \mathbf{Z}_1 & \mathbf{0} \\ \mathbf{0} & \mathbf{Z}_2 \end{bmatrix} \tag{5.93}$$

Apply (5.93) to (5.92) so that

$$\begin{bmatrix} \Phi & \mathbf{N} & \mathbf{S} & \mathbf{M} & \mathbf{A}_{ci}^{T}(\mathbf{Z}_1+\mathbf{Z}_2) \\ \mathbf{N}^T & -\mathbf{Y}_1 & \mathbf{0} & \mathbf{0} & \mathbf{0} \\ \mathbf{S}^T & \mathbf{0} & -\mathbf{Y}_1 & \mathbf{0} & \mathbf{0} \\ \mathbf{M}^T & \mathbf{0} & \mathbf{0} & -\mathbf{Y}_2 & \mathbf{0} \\ (\mathbf{Z}_1^T+\mathbf{Z}_2^T)\mathbf{A}_c & \mathbf{0} & \mathbf{0} & \mathbf{0} & -(\mathbf{Y}_1+\mathbf{Y}_2) \end{bmatrix} < \mathbf{0} \tag{5.94}$$

Solving the time-delay stability upper bound could be transformed to use in the following optimization problem:

$$\min \quad \nu\ \mathbf{P},\mathbf{Q},\mathbf{R},\mathbf{Z}_i,\mathbf{Y}_i \qquad s.t.(5.94),(5.93) \tag{5.95}$$

In (5.95), $\mathbf{P}$, $\mathbf{Q}$, $\mathbf{R}$, $\mathbf{Z}_i$, $\mathbf{Y}_i$ are variables and (5.94) and (5.93) are constraints. By solving (5.95), the minimum ν could be gained. Then, according to $h = 1/\nu$, the time-delay stability upper bound of the jump system could be found.

5.3.1.3 Simulation Analysis

Taking the IEEE 16-machine 68-bus system, for example, time-domain simulation tests are conducted to calculate the time-delay stability upper bound of the power system in different jump cases such as operation mode/parameter variation and load power fluctuation. The generators apply the sixth-order detailed model, the excitation system applies the IEEE-DC1 excitation, and the load applies the WECC load model – 80% constant active power load, 80% constant reactive power impedance load, and 20% dynamic load.

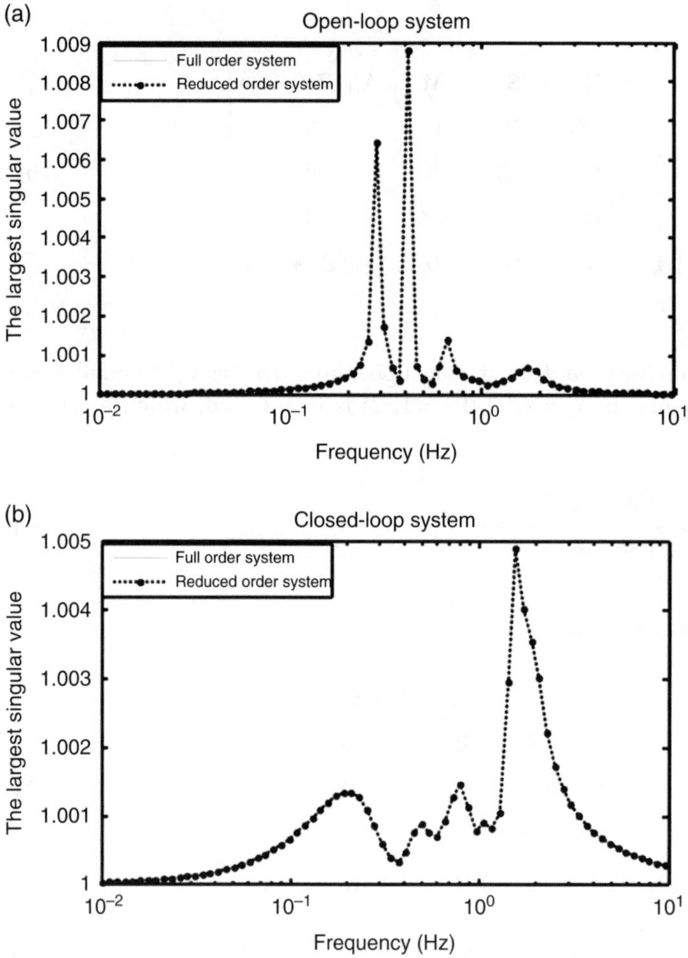

Figure 5.15 Frequency responses of a Schur-order reduction.

First, the Schur balance order reduction method is used to reduce the order of the system, on the premise that the input–output characteristic of the frequency band concerned is not affected. The frequency responses of the full-order system and reduced-order system are shown in Figure 5.15. In Figure 5.15(a), the solid line represents the frequency response of the open-loop full-order system and the dashed line represents the frequency response of the open-loop reduced-order system. It can be seen that the input–output characteristics of the full-order system and reduced-order system are approximately the same. Similarly, in Figure 5.15(b) the solid line represents the frequency response of the closed-loop full-order system and the dashed line represents the frequency response of the closed-loop reduced-order system.

It can be seen that the closed-loop reduced-order system retains the characteristic of the full-order system. Therefore, it is effective and feasible to use the reduced-order system matrix to calculate the time-delay stability upper bound of the full-order system.

Operation Mode Variation Suppose that the system originally operates in the baseline condition and then one of the lines in line 1–27 falls out of operation. When the 16-machine system operates in the baseline condition, the reduced-order state matrix $\mathbf{A}_{re}$ and the time-delay matrix $\mathbf{A}_{dre}$ are shown below in (5.96) and (5.97). When one of the lines in line 1–27 falls out of operation, the reduced-order state matrix $\mathbf{A}_{re}$ and time-delay matrix $\mathbf{A}_{dre}$ are shown in (5.98) and (5.99). Apply (5.96) to (5.99) to (5.95), so that the time-delay stability upper bound of the jump system can be calculated: $h = 61.5$ ms. Set the time-delay to 0 ms, 50 ms, 61.5 ms, and 80 ms, respectively, and observe the dynamic response of the power angle difference between G4 and G15. The result is shown in Figure 5.16. It can be seen that, when the signal time-delay is below the time-delay stability upper bound, the system is able to damp the inter-area oscillation within 10 s. However, as the time-delay increases, the damping effect decreases. Then the prony example is used to analyze the damping ratio of the power angle difference after a system jump, which is the result shown in Table 5.8. It can be seen that, when the system jumps, if the time-delay of control signals reaches 61.5 ms, the damping ratio of the power angle difference will be 7.90%, which is below 10%

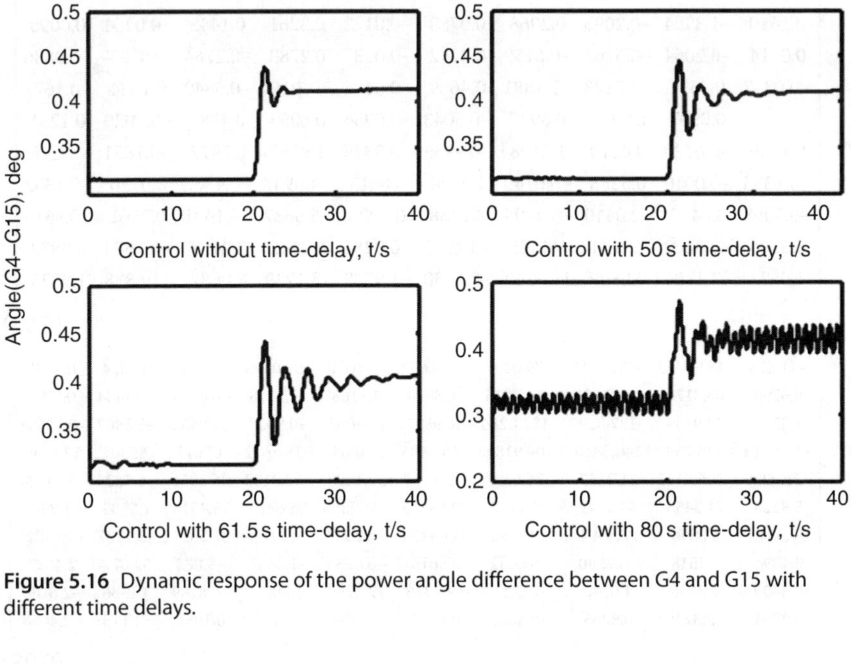

Figure 5.16 Dynamic response of the power angle difference between G4 and G15 with different time delays.

Table 5.8 Damping ratio of the power angle difference between G4 and G15 with different time delays.

Time delay	Damping ratio
0 ms	0.2964
50 ms	0.1365
61.5 ms	0.0790
80 ms	0.0113

and thus the controller will not meet the control requirements. Therefore, the proposed method is proved effective in solving the time-delay stability upper bound of the power system when the system operation mode jumps.

$$\mathbf{A}_{re} = \begin{bmatrix} -16.1234 & 191.6215 & -34.7234 & -93.6034 & -46.9813 & -40.7245 & 148.6905 & -9.4037 & 0.5458 & -0.1871 \\ -8.5822 & 73.8511 & 38.3796 & 28.2807 & -31.1118 & -3.5328 & 3.0915 & -0.8165 & 0.4670 & 0.3871 \\ 8.6901 & -56.6390 & -119.0312 & -154.2303 & 89.4859 & 8.3981 & 7.6545 & 47.8142 & -16.0572 & 20.4179 \\ -16.3933 & 128.3466 & 100.7138 & 91.8190 & -63.4074 & 1.0582 & -30.2333 & -5.4016 & 1.8643 & -2.0558 \\ -21.5589 & 183.0718 & 41.8903 & -25.8071 & -3.0065 & 15.5680 & -58.5954 & 54.2491 & -17.9821 & 24.6753 \\ 8.9164 & -85.9699 & 41.1758 & 105.8473 & -47.0510 & -8.1517 & 54.5981 & 35.2028 & -2.4424 & 8.9442 \\ -3.3296 & 27.0818 & 6.4952 & -4.4047 & 2.3259 & 5.3976 & -8.7728 & 33.8530 & -9.1379 & 13.1160 \\ 0.6362 & -4.3077 & -5.9066 & -6.7725 & 3.3248 & -1.2021 & -0.6273 & -11.9533 & 5.6377 & -0.8597 \\ 0.3450 & -3.4204 & 2.0537 & 4.4665 & -1.9128 & -0.9482 & 0.6125 & -6.2861 & 4.3435 & 6.7691 \\ 0.4673 & -4.1903 & 0.9437 & 3.9651 & -2.1654 & -0.7490 & 2.0926 & -2.3038 & -5.0688 & -4.4607 \end{bmatrix} \tag{5.96}$$

$$\mathbf{A}_{\mathbf{d}re} = \begin{bmatrix} -0.0168 & 0.1387 & 0.0401 & 0.0041 & -0.0139 & 0.0056 & -0.0470 & -0.0035 & 0 & 0.0023 \\ 0.1610 & -1.3994 & -0.2093 & 0.2968 & 0.0753 & -0.0122 & 0.3261 & 0.1445 & -0.0264 & 0.0095 \\ 0.0414 & -0.2064 & -0.5163 & -0.6159 & 0.2242 & -0.0837 & 0.2789 & -0.2764 & 0.0783 & -0.0886 \\ -0.0247 & 0.3531 & -0.7123 & -1.1881 & 0.4669 & -0.0484 & -0.1044 & -0.5840 & 0.1524 & -0.1677 \\ 0 & 0.0947 & 0.4236 & 0.6927 & -0.5043 & -0.0954 & 0.6054 & 0.4081 & -0.1020 & 0.1251 \\ 0.1159 & -0.6120 & 0.4210 & 1.2198 & -0.9955 & -0.3345 & 1.9667 & 0.7972 & -0.1851 & 0.2245 \\ -0.1383 & -0.0010 & 0.5268 & -0.0149 & 1.2620 & 0.8615 & -4.0937 & -0.4567 & 0.0720 & -0.1352 \\ -0.5098 & 3.6470 & -2.0410 & -5.4194 & 3.2488 & 0.7903 & -5.5637 & -3.1691 & 0.7561 & -0.8615 \\ 0.5506 & -3.8102 & 1.9174 & 5.4325 & -3.4277 & -0.9281 & 6.2442 & 3.2459 & -0.7673 & 0.8833 \\ 1.0523 & -7.1184 & 4.5809 & 11.8740 & -7.5930 & -1.9539 & 13.1229 & 7.0687 & -1.6838 & 1.9525 \end{bmatrix} \tag{5.97}$$

$$\mathbf{A}_{re} = \begin{bmatrix} -11.678 & -191.6462 & -70.3227 & -79.0497 & -48.3652 & -110.648 & 104.0195 & -5.3713 & 0.7074 & 0.4355 \\ 6.6790 & 63.3176 & -26.7956 & -14.2015 & 18.8844 & 4.1168 & -1.5178 & -0.6594 & -0.1434 & 0.7901 \\ 7.3569 & 33.9139 & -127.3597 & -121.0284 & 60.9821 & 11.4067 & -0.9038 & 31.9452 & -5.5867 & -23.7998 \\ -16.8713 & -125.7599 & 140.3816 & 109.9182 & -65.7495 & 5.4865 & -16.3672 & -17.0152 & 2.5307 & 12.6249 \\ -21.9229 & -206.0143 & 31.3457 & -26.6892 & -3.9202 & 52.1408 & -52.2132 & 36.3051 & -6.5625 & -28.1744 \\ 5.8223 & 71.3453 & 49.1749 & 72.1177 & -29.4323 & -31.7583 & 35.6815 & 13.4516 & 1.5783 & -1.3225 \\ -0.5412 & 5.3546 & 28.5494 & 29.9752 & -11.4154 & -2.2537 & 9.3097 & 37.5824 & -3.1737 & -19.3021 \\ 0.4894 & 2.4518 & -6.2580 & -5.6032 & 2.3616 & -0.8286 & -1.0832 & -9.083 & 5.8877 & 2.4837 \\ 0.3409 & 4.7596 & 4.8740 & 6.2723 & -2.5946 & -2.2759 & 1.2081 & -5.8189 & 3.1536 & -2.6165 \\ 0.2794 & -2.3226 & 0.8065 & -0.3805 & 0.4742 & 1.0909 & -1.102 & 0.1058 & 8.2175 & -3.4058 \end{bmatrix} \tag{5.98}$$

$$\mathbf{A}_{\mathbf{d}re} = \begin{bmatrix} -0.0082 & -0.7845 & 1.0689 & 0.3216 & -0.1436 & 0.3915 & -0.0972 & -0.2581 & -0.4216 & -0.1397 \\ -13.254 & -6.7371 & -1.1454 & 0.2471 & 3.4813 & 1.1416 & -2.9427 & 4.0879 & -4.3914 & 0.5616 \\ 5.634 & -6.5425 & -0.7037 & 1.3303 & 1.5530 & 0.8414 & 1.0332 & 0.8684 & -2.2289 & 1.3764 \\ -91.6852 & 1.3403 & -17.4012 & -4.281 & 5.7699 & -1.9927 & -3.6826 & -0.1831 & -1.5932 & -7.0902 \\ 1.1806 & -17.9896 & -4.1276 & 1.3708 & 7.0354 & 3.78 & -7.0897 & 2.8827 & -4.5897 & 1.9537 \\ 0.6597 & -4.3634 & -0.02 & 0.5344 & -1.3938 & -0.8509 & 0.2225 & -6.4847 & 2.5059 & -1.5712 \\ 16.5805 & -6.049 & -6.1162 & 0.7899 & 2.7477 & 1.0602 & -0.195 & -5.4722 & 4.0705 & -1.1533 \\ -0.7350 & 8.2783 & -2.3227 & -0.7543 & -5.1677 & 4.4179 & 3.0756 & -1.3717 & -0.455 & 0.7396 \\ -0.2877 & 22.766 & 3.4572 & -3.2913 & -5.3352 & 3.0256 & -2.1839 & 10.456 & -3.5807 & 4.8781 \\ 6.0941 & 5.8894 & -11.1380 & -1.247 & -0.8054 & -4.0081 & 1.3539 & 2.6537 & -3.8119 & -0.385 \end{bmatrix} \tag{5.99}$$

Operation Parameter Variation When the other parameters are fixed and the generator excitation magnification factor K_a varies between 60 and 200, the calculated time-delay stability upper bound is as shown in Figure 5.17. According to theoretical analysis, when the system time delay is fixed, the larger the excitation magnification factor, the more sensitive the excitation emf is to the difference of the terminal voltage from the reference voltage, and the system is more easily affected by the error caused by a signal time delay. Therefore, the time-delay stability upper bound of a power system will decrease as the excitation magnification factor increases. This is verified in the calculation result shown in Figure 5.17.

Suppose that the generator excitation magnification factor K_a jumps from 180 to 200; the conservativeness of the proposed method is examined. In this case, the time-delay stability upper bound of the jump system is $h = 94.6$ ms. Set the time delay to be 0 ms, 80 ms, 94.6 ms, and 120 ms, respectively, and observe the dynamic response of the power angle difference between G8 and G15. The result is shown in Figure 5.18. It can be seen that, when the signal time-delay is below the time-delay stability upper bound, the system is able to damp the inter-area

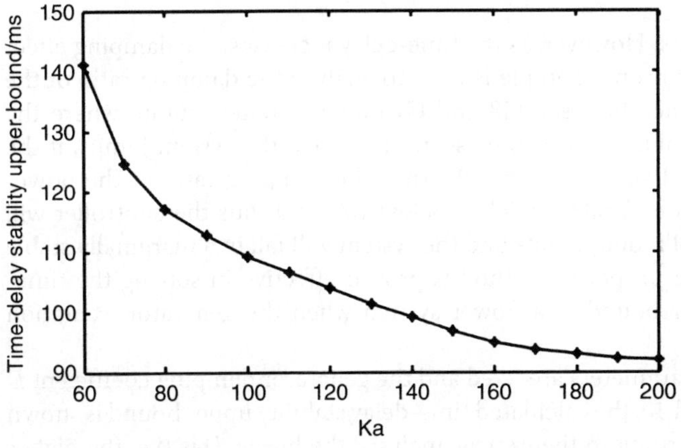

Figure 5.17 Time-delay stability upper bound of the 16-machine system with different K_a.

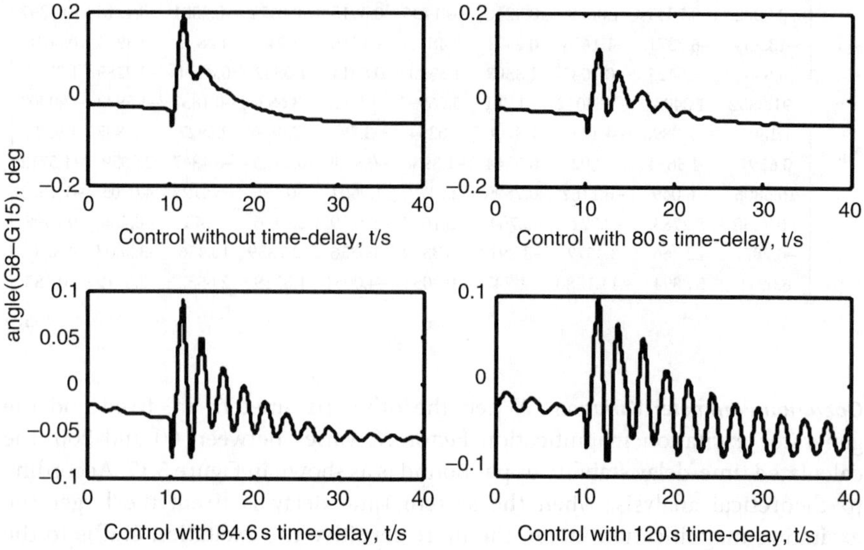

Figure 5.18 Dynamic response of the power angle difference between G8 and G15 with different time delays.

Table 5.9 Damping ratio of the power angle difference between G8 and G15 with different time delays.

Time delay	Damping ratio
0 ms	0.2741
80 ms	0.1328
94.6 ms	0.0753
120 ms	0.0236

oscillation within 10 s. However, as the time-delay increases, the damping effect decreases. Then the prony example is used to analyze the damping ratio of the power angle difference between G8 and G15 after a system jump, where the result is shown in Table 5.9. It can be seen that, when the system jumps, if the time delay of control signals reaches 94.6 ms, the damping ratio of the power angle difference will be 7.53%, which is below 10% and thus the controller will not meet the control requirements and the system will fall in a marginally stable state. Therefore, the proposed method is proved effective in solving the time-delay stability upper bound of a power system when the generator excitation parameter jumps.

When the other parameters are fixed and the generator damping coefficient D varies between 0 and 40, the calculated time-delay stability upper bound is shown in Figure 5.19. According to theoretical analysis, the bigger D is (i.e. the higher the system damping is), the system is more capable of damping oscillation with

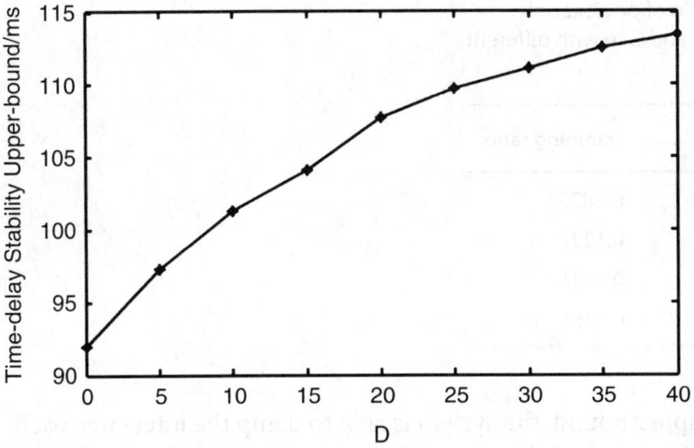

Figure 5.19 Time-delay stability upper bound of the 16-machine system with different *D* values.

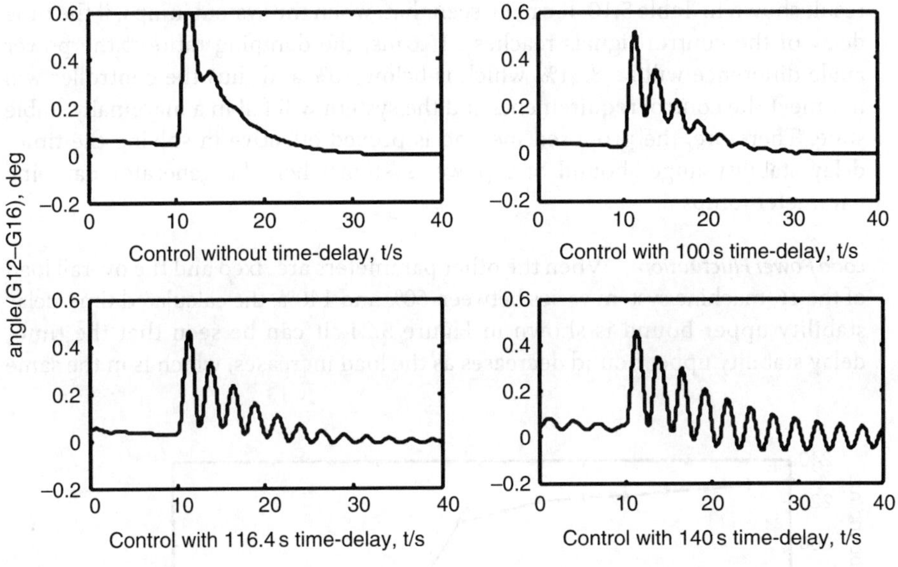

Figure 5.20 Dynamic response of the power angle difference between G12 and G16 with different time delays.

its own damping, and the system stability and safety is less easily affected by the time delay of control signals. Therefore, the time-delay stability upper bound of a power system will increase as the generator damping coefficient increases. This is verified in the calculation result shown in Figure 5.19.

Suppose that the generator damping coefficient D jumps from 30 to 40; the conservativeness of the proposed method is examined. In this case, the time-delay stability upper bound of the jump system is $h = 116.4$ ms. Set the time delay to be 0 ms, 100 ms, 116.6 ms, and 140 ms, respectively, and observe the dynamic response of the power angle difference between G12 and G16. The result is shown in Figure 5.20. It can be seen that, when the signal time delay is below the

Table 5.10 Damping ratio of power angle difference between G12 and G16 with different time delays.

Time delay	Damping ratio
0 ms	0.2875
100 ms	0.1227
116.4 ms	0.0841
140 ms	0.0182

time-delay stability upper bound, the system is able to damp the inter-area oscillation within 10 s. However, as the time delay increases, the damping effect decreases. Then the prony example is used to analyze the damping ratio of the power angle difference between G12 and G16 after a system jump, giving the result shown in Table 5.10. It can be seen that, when the system jumps, if the time delay of the control signals reaches 116.6 ms, the damping ratio of the power angle difference will be 8.41%, which is below 10% and thus the controller will not meet the control requirements and the system will fall in a marginally stable state. Therefore, the proposed method is proved effective in solving the time-delay stability upper bound of a power system when the generator damping parameter jumps.

Load Power Fluctuation When the other parameters are fixed and the overall load of the 16-machine system varies between 60% and 140%, the calculated time-delay stability upper bound is shown in Figure 5.21. It can be seen that the time-delay stability upper bound decreases as the load increases, which is in the same

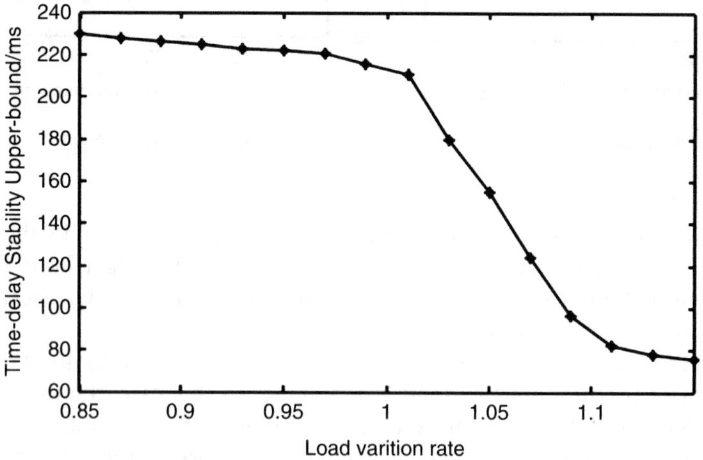

Figure 5.21 Time-delay stability upper bound of the 16-machine Markov jump system with different load variations.

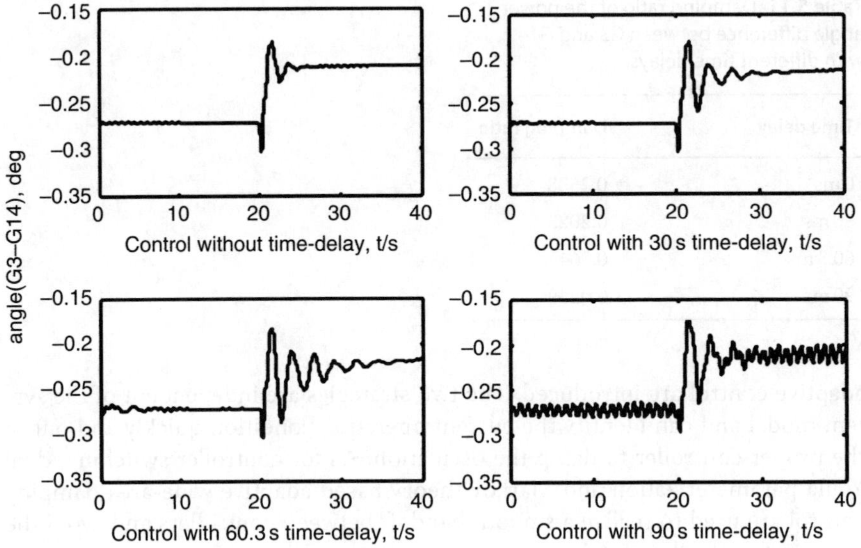

Figure 5.22 Dynamic response of the power angle difference between G3 and G14 with different time delays.

trend as the four-machine system. When the load is below the rated value, the influence of the load variation on the time-delay stability upper bound is small. However, when the load is above the rated value, the time-delay stability upper bound decreases significantly as the load increases and thus the time-delay stability of the power system will be affected.

Suppose that the system load jumps from 110% to 120% and the conservativeness of the proposed method is examined. In this case, the time-delay stability upper bound of the jump system is $h = 60.3$ ms. Set the time delay to be 0 ms, 30 ms, 60.3 ms, and 90 ms, respectively, and observe the dynamic response of the power angle difference between G3 and G14. The result is shown in Figure 5.22. It can be seen that, when the signal time-delay is below the time-delay stability upper bound, the system is able to damp the inter-area oscillation within 10 s. However, as the time delay increases, the damping effect decreases. Then the prony example is used to analyze the damping ratio of the power angle difference between G3 and G14 after a system jump, with the result shown in Table 5.11. It can be seen that when the system jumps, if the time delay of control signals reaches 60.3 ms, the damping ratio of the power angle difference will be 6.47%, which is below 10% and thus the controller will not meet the control requirements and the system will fall in a marginally stable state. Therefore, the proposed method is proved effective in solving the time-delay stability upper bound of the power system when the system load jumps.

In order to achieve accurate and efficient adaptive control aiming at a power system operating at a wide range of changing conditions, the system state identification matching and controller switching are analyzed and discussed in this chapter. With regard to operating condition identification matching, the federated Kalman filter and classification and regression tree-based wide-area

Table 5.11 Damping ratio of the power angle difference between G3 and G14 with different time delays.

Time delay	Damping ratio
0 ms	0.2538
30 ms	0.2023
60.3 ms	0.0647
90 ms	0.0289

adaptive control are introduced. The two strategies are independent of the system model and can identify the current operating condition quickly and put in the proper controller to damp the oscillation. As for controller switching, dual Youla parameterization and Markov theory based adaptive wide-area damping control are used to realize a smooth handoff between controllers and avoid the impact of controller switching on a power system.

5.3.2 Time-Delay Stability Analysis of Power System Based on the Fault Chains and Markov Process

With the scale and complexity of an interconnected power system ever increasing, a fault in a local area may easily develop into cascading failure. Therefore, a time-delay stability of the power system when multiple disturbances occur is of great concern. Since the above methods are not applicable to cascading faults, a new time-delay power system stability analysis method based on the fault chains and Markov process is proposed in this section, which is not only able to reveal the time-delay stability of the fault system but is also able to solve the time-delay stability upper bound of the system.

5.3.2.1 Basic Theory of Fault Chains

According to the fault chain theory, a fault is seldom caused by a single factor, but a fault usually occurs when multiple factors are satisfied at the same time. A fault chain can be expressed as

$$L_j = L_{j1}L_{j2}L_{j3}\cdots L_{jn} \tag{5.100}$$

where L_{ji} is the ith link in the fault chain L_j and n is the number of intermediate links. Faults in a power system are mostly line faults. Thus take a line fault as the unit component of the fault chain. A fault chain is formed as follows. First, randomly select any transmission line fault as the initial link of the fault chain and then through the prediction of the intermediate links, the next line fault can be determined. Finally, when constant-amplitude or increasing-amplitude low-frequency oscillation occurs, the fault chain reaches its terminal link.

When a power system is in normal operation, the lines are with some initial load. If a certain line is cut and in outage, the load on this line will be transferred to the other lines, which may result in overloading or maloperation of protection on the other lines, even ending up in cascading faults. In this section, α, β, and γ are used as the evaluation index of the line power flow variation rate, overload margin, coupling degree between the fault line, and the predicted line, respectively, and φ is used as the intermediate links prediction index, which decides the next fault line in the fault chain.

Suppose a fault occurs on line i, where S_i is the power flow on line i. Then α, β, and γ are

$$\alpha_{ij} = \left| \frac{S_{j(t_b)} - S_{j(t_f)}}{S_{j(t_f)}} \right| \tag{5.101}$$

$$\beta_{ij} = \left| \frac{S_{j(t_b)}}{S_{j\max}} \right| \tag{5.102}$$

$$\gamma_{ij} = \left| \frac{S_{j(t_b)} - S_{j(t_f)}}{S_{i(t_f)}} \right| \tag{5.103}$$

where $S_{i(t_f)}$ is the pre-fault power flow on line i, $S_{j(t_f)}$ is the pre-fault power flow on line j, $S_{j(t_b)}$ is the power flow on line j after a fault occurs on line I, and $S_{j\max}$ is the maximum power flow allowed on line j.

Define the intermediate links prediction index φ as

$$\varphi_{ij} = \frac{\alpha_{ij}\beta_{ij}\gamma_{ij}}{\left| \frac{S_{i\max}}{S_i} \right|} = \left| \frac{S_i}{S_{i\max}} \frac{S_{j(t_b)} - S_{j(t_f)}}{S_{j(t_f)}} \frac{S_{j(t_b)}}{S_{j\max}} \frac{S_{j(t_b)} - S_{j(t_f)}}{S_{i(t_f)}} \right| \tag{5.104}$$

In (5.104), the bigger φ_{ij} is, the more the non-fault line j is affected by the fault line i and the more probable that line j is the next fault line. Therefore, the state transfer probability caused by a power transfer can be calculated as follows:

$$p_{ij} = \frac{\varphi_{ij}}{\sum_{k=1}^{n} \varphi_{ik} - \varphi_{ii}} \tag{5.105}$$

According to the above analysis, the development of a fault chain can be described as the evolution of a series of events with conditional probabilities. For the links in a fault chain, the generation of each link is only related to the former link, and has nothing to do with the previous faults in the system. The conditional probability distribution of the future state of each link is only dependent on the current state. Therefore, the development of the fault chain is a Markov process. With the Markov theory, the fault chain can be described as follows. Suppose $r_t = r(t)$ is the system modal, a homogeneous Markov process defined in the finite modal set $S = \{1,2,\ldots,s\}$. For the fault chain $L = L_1 L_2 L_3 \cdots L_m$, first the intermediate link L_i is taken as the modal of random Markov process, that is, $r_i = r(i)$ (the finite modal set S corresponds to the number of intermediate links in fault chain L). Then the Markov transfer probability matrix $\boldsymbol{\pi}$ is calculated according to (5.105) and the Markov process based on the fault chain is established. Then the time-delay stability of the power system in the case of cascading faults is analyzed using the Lyapunov–Krasovskii theory. Finally, the maximum time delay that the fault system could bear is solved quantitatively with GEVP.

5.3.2.2 Stability Analysis of Time-Delay Markov Jump System

Time-Delay Markov Power System Model The model of a time-delay Markov jump system can be expressed as

$$\begin{cases} \dot{x}_{r(t)}(t) = A_{r(t)}x(t) + A_{dr(t)}x(t-h_t) \\ z_{r(t)}(t) = B_{r(t)}x(t) + B_{dr(t)}x(t-h_t) \\ x_{r(t)}(t) = \delta_{r(t)}(t), t \in \left[-\bar{h}, 0\right] \end{cases} \tag{5.106}$$

In (5.106), the time delay h_t meets the following conditions:

$$0 \le h_t \le \bar{h},\ \dot{h}_t < \mu \tag{5.107}$$

where $x(t) \in \mathbf{R}^n$ and $z(t) \in \mathbf{R}^n$ are the system state vector and output vector, respectively, A_1, A_{di}, B_i, B_{di} are given matrices with appropriate dimensions, h_t is the system time delay, and $r_t = r(t)$ is the system modal, a homogeneous Markov process defined in the finite modal set $S = \{1, 2, \ldots, s\}$ (S corresponds to the number of intermediate links in the fault chain L). The state transfer probability of the homogeneous Markov process is

$$p_{ij} = P\left(r_{t+\Delta} = j \mid r_t = i\right) = \begin{cases} \pi_{ij}\Delta + o(\Delta), & i \ne j \\ 1 + \pi_{ii}\Delta + o(\Delta), & i = j \end{cases} \tag{5.108}$$

where

$$\lim_{\Delta \to \infty} \frac{o(\Delta)}{\Delta} = 0 \qquad (\Delta > 0) \tag{5.109}$$

In (5.108), the state transfer probability p_{ij} can be calculated according to (5.105); $\boldsymbol{\pi}$ is the Markov state transfer probability matrix and its element π_{ij} represents the transfer rate that renders the system modal at state i when time = t and renders the system modal at state j when time = $t + \Delta$; $\pi_{ii} = -\sum_{i \ne j} \pi_{ij} \quad \left(\pi_{ij} \ge 0, i \ne j\right)$.

Theorem 5.5 For the jump time-delay system described in (5.106), with given scalar quantities $\bar{h} > 0$ and μ, if there exist $P = P^T > 0$, $Q = Q^T \ge 0$, $R_i = R_i^T \ge 0$, $K = K^T > 0$, $Z_i = Z_i^T \ge 0$, $U_i = U_i^T \ge 0$, $V_i = V_i^T \ge 0$, and $W_i = W_i^T \ge 0$ ($i = 1, 2, \ldots, S$), $N = [N_1 \quad N_2 \quad N_3]^{\mathrm{T}}$, $L = [L_1 \quad L_2 \quad L_3]^{\mathrm{T}}$, $M = [M_1 \quad M_2 \quad M_3]^{\mathrm{T}}$ that satisfy (5.110), then the Markov jump time-delay system in (5.106) that meets the conditions in (5.107) is asymptotically stable:

$$\begin{bmatrix} \Omega & \bar{h}N & \bar{h}L & \bar{h}M & \bar{h} \cdot \bar{A}_i(Z_1 + Z_2) \\ * & -\bar{h}Z_1 & 0 & 0 & 0 \\ * & * & -\bar{h}Z_1 & 0 & 0 \\ * & * & * & -\bar{h}Z_2 & 0 \\ * & * & * & * & -\bar{h}(Z_1 + Z_2) \end{bmatrix} < 0 \tag{5.110}$$

where

$$\Omega = \Omega_1 + \Omega_2 + \Omega_2^T$$

$$\Omega_2 = \begin{bmatrix} N+M & L-N & -L-M \end{bmatrix}$$

$$\Omega_1 = \begin{bmatrix} \begin{matrix} P_i A_i + A_i^T P_i + Q + R_i + \bar{h}K + \\ \sum_{j=1}^{S} \pi_{ij}\left(R_j + P_j - W_j\right) \end{matrix} & P_i A_{di} & 0 \\ * & \begin{matrix} -(1-\mu)Q - \\ \sum_{j=1}^{S} \pi_{ij} U_j \end{matrix} & 0 \\ * & * & \begin{matrix} -R_i - \\ \sum_{j=1}^{S} \pi_{ij} V_j \end{matrix} \end{bmatrix}$$

$$\bar{A}_i = \begin{bmatrix} A_i & A_{di} & 0 \end{bmatrix}$$

$$i = \{1, 2, \ldots, N\}$$

Proof: First, form the following Lyapunov–Krasovskii functional:

$$\begin{aligned} V(x_t, t) = {} & x^T(t) P_i x(t) + \int_{t-h_t}^{t} x^T(s) Q x(s) ds \\ & + \int_{t-\bar{h}}^{t} x^T(s) R_i x(s) ds \\ & + \int_{-\bar{h}}^{0} \int_{t+\theta}^{t} x^T(s) K x(s) ds d\theta \\ & + \int_{-\bar{h}}^{0} \int_{t+\theta}^{t} \dot{x}^T(s) (Z_1 + Z_2) \dot{x}(s) ds d\theta \end{aligned} \tag{5.111}$$

Consider that the elements in the transfer rate matrix $\boldsymbol{\pi}$ of the system in (5.106) meets $\pi_{ii} + \sum_{i \neq j} \pi_{ij} = 0 (\pi_{ij} \geq 0, i \neq j)$. Thus for any matrix $W_i = W_i^T \geq 0$ with appropriate dimension:

$$-x^T(t) \sum_{j=1}^{S} \pi_{ij} W_j x(t) = 0 \tag{5.112}$$

$$-x^T(t - h_t) \sum_{j=1}^{S} \pi_{ij} U_j x(t - h_t) = 0 \tag{5.113}$$

$$-x^T(t - \bar{h}) \sum_{j=1}^{S} \pi_{ij} V_j x(t - \bar{h}) = 0 \tag{5.114}$$

Meanwhile, according to the Newton–Leibniz formula, for any matrices N, L, and M with appropriate dimensions, the following equations are true:

$$2\zeta^T(t)N\left[x(t)-x(t-h_t)-\int_{t-h_t}^{t}\dot{x}(s)ds\right]=0 \tag{5.115}$$

$$2\zeta^T(t)L\left[x(t-h_t)-x(t-\bar{h})-\int_{t-\bar{h}}^{t-h_t}\dot{x}(s)ds\right]=0 \tag{5.116}$$

$$2\zeta^T(t)M\left[x(t)-x(t-\bar{h})-\int_{t-\bar{h}}^{t}x(s)ds\right]=0 \tag{5.117}$$

where $\zeta(t)=\left[x^T \quad x^T(t-h_t) \quad x^T(t-\bar{h})\right]^T$. The weak infinitesimal operator of the random process $\{x_t, r_t, t\geq 0\}$ on $V(x_t, t)$ is

$$\Gamma\left[V(x_t,t)\right]=\frac{\delta V}{\delta t}+\dot{x}_t^T\frac{\delta V}{\delta x}|_{r_t=t}+\sum_{j=1}^{s}\pi_{ii}V(x_t,\, t,\, j,\, i)$$

that is,

$$\begin{aligned}\Gamma\left[V(x_t,t,i)\right]=&\,2x^T(t)P_i\dot{x}(t)\\&+x^T(t)Qx(t)-\left(1-\dot{h}_t\right)x^T(t-h_t)Qx(t-h_t)\\&+x^T(t)R_ix(t)-x^T\left(t-\bar{h}\right)R_ix\left(t-\bar{h}\right)\\&+\bar{h}x^T(t)Kx(t)-\int_{t-\bar{h}}^{t}x^T(s)Kx(s)ds\\&+\bar{h}\dot{x}^T(t)(Z_1+Z_2)\dot{x}(t)\\&-\int_{t-\bar{h}}^{t}\dot{x}^T(s)(Z_1+Z_2)\dot{x}(s)ds\\&+\sum_{j=1}^{s}\pi_{ij}x^T(t)P_jx(t)\\&+\int_{t-\bar{h}}^{t}x^T(s)\left(\sum_{j=1}^{s}\pi_{ij}R_j\right)x(s)ds\end{aligned} \tag{5.118}$$

In order to lower the conservativeness of the proposed method, free weighting items constructed with the transfer rate matrix $\boldsymbol{\pi}$ ((5.112) to (5.114)) and free weighting items constructed with the Newton–Leibniz formula ((5.115) to (5.117)) are introduced to the right side of (5.118). At the same time, the whole time-varying time-delay interval $\left[0,\, \bar{h}\right]$ is divided into two subintervals $\left[0,\, \underline{h}\right]$ and $\left[\underline{h},\, \bar{h}\right]\left(\underline{h}>0\right)$. Thus,

$$\begin{aligned}\dot{V}(x)\leq&\, x^T(t)(P_iA_i+A_i^TP_i+Q+R_i+\bar{h}K\\&+\sum_{j=1}^{S}\pi_{ij}\left(P_j-W_j\right)x(t)\\&+2x^T(t)P_iA_{di}x(t-h_t)\end{aligned}$$

$$
\begin{aligned}
&-x^T(t-h_t)[(1-\mu)Q+\sum_{j=1}^{S}\pi_{ij}U_j]x^T(t-h_t)\\
&-x^T(t-\bar{h})\left[R_i+\sum_{j=1}^{S}\pi_{ij}V_j\right]x(t-\bar{h})\\
&+\bar{h}\dot{x}^T(t)(Z_1+Z_2)\dot{x}(t)-\int_{t-\bar{h}}^{t}\dot{x}^T(s)Z_2\dot{x}(s)ds\\
&-\int_{t-h_t}^{t}\dot{x}^T(s)Z_1\dot{x}(s)ds-\int_{t-\bar{h}}^{t-h_t}\dot{x}^T(s)Z_1\dot{x}(s)ds\\
&+2\zeta^T(t)N\left[x(t)-x(t-h_t)-\int_{t-h_t}^{t}\dot{x}(s)ds\right]\\
&+2\zeta^T(t)L\left[x(t-h_t)-x(t-\bar{h})-\int_{t-h_t}^{t}\dot{x}(s)ds\right]\\
&+2\zeta^T(t)M\left[x(t)-x(t-\bar{h})-\int_{t-\bar{h}}^{t}\dot{x}(s)ds\right]\\
&\le\zeta^T(t)(t)[\Phi+\Phi_v]\zeta(t)-\int_{t-h_t}^{t}\Delta_1 ds\\
&-\int_{t-\bar{h}}^{t}\Delta_2 ds-\int_{t-\bar{h}}^{t-h_t}\Delta_3 ds\\
&<\zeta^T(t)(t)[\Phi+\Phi_v]\zeta(t)
\end{aligned}
\tag{5.119}
$$

where

$$
\begin{aligned}
\Phi_v=&\bar{h}\cdot\bar{A}^T(Z_1+Z_2)\bar{A}+\bar{h}\cdot NZ_1^{-1}N^T\\
&+\bar{h}\cdot LZ_1^{-1}L^T+\bar{h}\cdot MZ_2^{-1}M^T
\end{aligned}
\tag{5.120}
$$

$$
\Delta_1=\left[\zeta^T(t)N+\dot{x}(t)Z_1\right]Z_1^{-1}\left[N^T\zeta(t)+Z_1\dot{x}(t)\right]
\tag{5.121}
$$

$$
\Delta_2=\left[\zeta^T(t)M+\dot{x}(t)Z_2\right]Z_2^{-1}\left[M^T\zeta(t)+Z_2\dot{x}(t)\right]
\tag{5.122}
$$

$$
\Delta_3=\left[\zeta^T(t)L+\dot{x}(t)Z_1\right]Z_1^{-1}\left[L^T\zeta(t)+Z_1\dot{x}(t)\right]
\tag{5.123}
$$

When $\Phi+\Phi_v\le 0$, according to the Schur complement theorem [4], (5.119) is equivalent to (5.110). Thus according to the Lyapunov stability theory, the time-delay Markov system in (5.106) is asymptotically stable.

Time-Delay Stability Upper Bound The matrix inequality in (5.110) could only be used to identify whether the system is stable or not; the time-delay stability upper bound of the system is still unknown. Consider that the solving of the time-delay stability upper bound is a convex optimization problem with a linear inequality constraint and manifests itself in the form of the GEVP (generalized eigenvalue). In this section the GEVP method is used to calculate the time-delay stability upper bound of the system. Since (5.110) is not a standard GEVP form and cannot be solved with the GEVP method directly, the following transformation is made, that is, multiplying both the left side and right side of (5.110) with matrix $diag\left\{I,1/\bar{h}\cdot I,1/\bar{h}\cdot I,1/\bar{h}\cdot I,1/\bar{h}\cdot I\right\}$:

$$\begin{bmatrix} \Omega & N & L & M & \bar{A}_i(Z_1+Z_2) \\ * & -Z_1/\bar{h} & 0 & 0 & 0 \\ * & * & -Z_1/\bar{h} & 0 & 0 \\ * & * & * & -Z_2/\bar{h} & 0 \\ * & * & * & * & -(Z_1+Z_2)/\bar{h} \end{bmatrix} < 0 \tag{5.124}$$

Set $\Omega = \Omega' + \bar{h}K$, and according to the nature of Schur complement,

$$\begin{bmatrix} \Omega' & N & L & M & \bar{A}_i(Z_1+Z_2) & K \\ * & -Z_1/\bar{h} & 0 & 0 & 0 & 0 \\ * & * & -Z_1/\bar{h} & 0 & 0 & 0 \\ * & * & * & -Z_2/\bar{h} & 0 & 0 \\ * & * & * & * & -(Z_1+Z_2)/\bar{h} & 0 \\ * & * & * & * & * & -K/\bar{h} \end{bmatrix} < 0 \tag{5.125}$$

Set $-Z_1/\bar{h} < -T_1, -Z_2/\bar{h} < -T_2, -K/\bar{h} < -T_3$ and $d = 1/\bar{h}$, so that

$$\begin{bmatrix} T_1 & 0 & 0 \\ * & T_2 & 0 \\ * & * & T_3 \end{bmatrix} < d\cdot \begin{bmatrix} Z_1 & 0 & 0 \\ * & Z_2 & 0 \\ * & * & K \end{bmatrix} \tag{5.126}$$

In (5.125), substitute $-Z_1/\bar{h}, -Z_2/\bar{h}, -K/\bar{h}$ with $-T_1, -T_2, -T_3$ respectively, so that

$$\begin{bmatrix} \Omega' & N & L & M & \bar{A}_i(Z_1+Z_2) & K \\ * & -T_1 & 0 & 0 & 0 & 0 \\ * & * & -T_1 & 0 & 0 & 0 \\ * & * & * & -T_2 & 0 & 0 \\ * & * & * & * & -(T_1+T_2) & 0 \\ * & * & * & * & * & -T_3 \end{bmatrix} < 0 \tag{5.127}$$

According to (5.124) to (5.127), solving the time-delay stability upper bound $\bar{h}$ could be transformed to the following optimization problem:

$$\min\ d \qquad s.t. (5.126), (5.127) \tag{5.128}$$

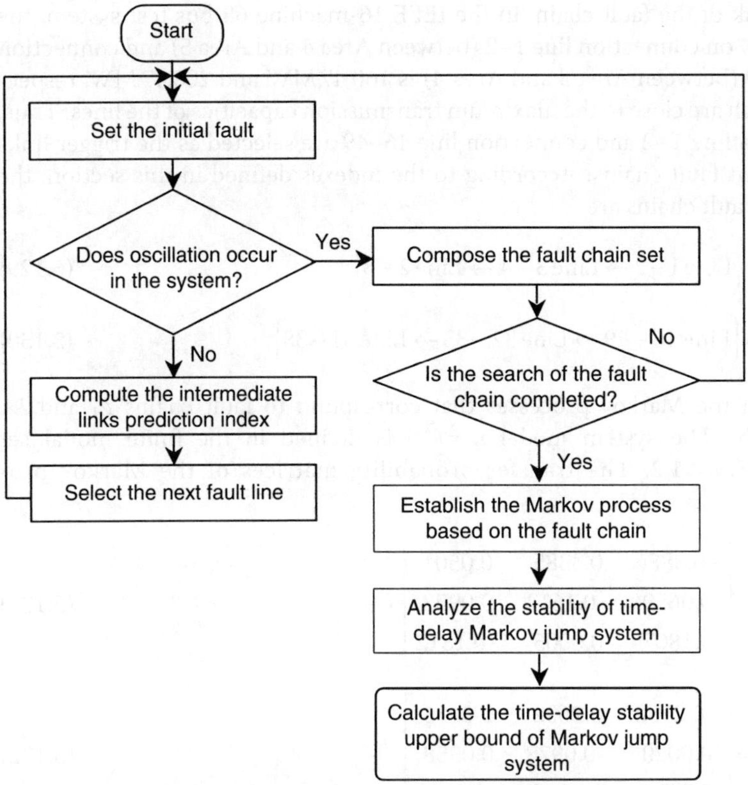

Figure 5.23 Flowchart of the time-delay stability analysis.

By solving (5.128), the minimum d could be gained. Then, according to $h = 1/d$, the time-delay stability upper bound of thre power system in the case of cascading failure could be obtained. A flowchart of the proposed algorithm is shown in Figure 5.23.

5.3.2.3 Simulation Analysis

Taking the IEEE 16-machine 68-bus system as the test system, the time-delay stability upper bound of power system based on the fault chains and Markov process is calculated and the effectiveness and low conservativeness of the proposed method is verified through time-domain simulation. The IEEE 16-machine 68-bus New England–New York interconnected system is shown in Figure 3.2. The test system can be divided into five areas. Areas 1, 2, and 3 are equivalent systems, Area 4 is the New York system, and Area 5 is the New England system. The generators apply the sixth-order detailed model, the excitation system applies the IEEE-DC1 excitation, and the load applies the WECC load model – 50% constant active power load, 50% constant reactive power impedance load, and 20% dynamic load.

Faults in the power system are mostly line faults. In view of the importance of inter-area connection lines, the connection line fault is taken as the trigger (initial) link of the fault chain. In the IEEE 16-machine 68-bus test system, the power flow on connection line 1–2 (between Area 4 and Area 5) and connection line 46–49 (between Area 3 and Area 4) is 160.47 MW and 208.47 MW, respectively, which are close to the maximum transmission capacities of the lines. Thus, connection line 1–2 and connection line 46–49 are selected as the trigger links for different fault chains. According to the indexes defined in this section, the predicted fault chains are

$$L_1 = \left\{\text{Line}\,1-2 \rightarrow \text{Line}\,3-4 \rightarrow \text{Line}\,2-3\right\} \tag{5.129}$$

$$L_2 = \left\{\text{Line}\,46-49 \rightarrow \text{Line}\,32-33 \rightarrow \text{Line}\,31-38\right\} \tag{5.130}$$

Establish the Markov processes that correspond to fault chains L_1 and L_2, respectively. The system modal $r_t = r(t)$ is defined in the finite modal set $S_m = \{1,2,3\}, m = 1,2$. The transfer probability matrices of the Markov processes are

$$\pi_{L_1} = \begin{bmatrix} -0.3886 & 0.3385 & 0.0501 \\ 0.0659 & -0.1612 & 0.0953 \\ 0.1807 & 0.0005 & -0.1812 \end{bmatrix} \tag{5.131}$$

$$\pi_{L_2} = \begin{bmatrix} -0.2788 & 0.2602 & 0.0186 \\ 0.0020 & -0.0978 & 0.0958 \\ 0.0022 & 0.0016 & -0.0038 \end{bmatrix} \tag{5.132}$$

Fault Chain L1 Concerning the fault chain L_1, apply the state matrix A_i and time-delay matrix A_{di} $(i = 1, 2, 3)$ of the test system (after a Schur-order reduction) to (5.126) and (5.127), and $\bar{v}$ the time-delay stability upper bound can be calculated using the LMI method: $\bar{h}$ = 82.9 ms.

In order to verify the feasibility and effectiveness of the proposed method, the damping controller is designed in the test system based on the H_2/H_∞ control method. Set the time delay to be 0 ms, 40 ms, 82.9 ms, and 120 ms respectively, and observe the dynamic responses of the power angle difference between G8 and G15 and the power angle difference between G1 and G16. The results are shown in Figure 5.24. It can be seen that when the time delay is below the time-delay stability upper bound, the system remains stable in the development of the fault chain and is able to damp the inter-area oscillation quickly. When the time delay reaches the time-delay stability upper bound, the system is in a weak damping state at the initial link of the fault chain, and gradually recovers to the stable state with the development of the fault chain. When the time delay rises above the time-delay stability upper bound, the system gradually turns from a weak damping state to an unstable state with the generation of the fault chain.

The prony algorithm [6] is used to calculate the damping ratios of the G8–G15 relative power angle curve and the G1–G16 relative power angle curve at the

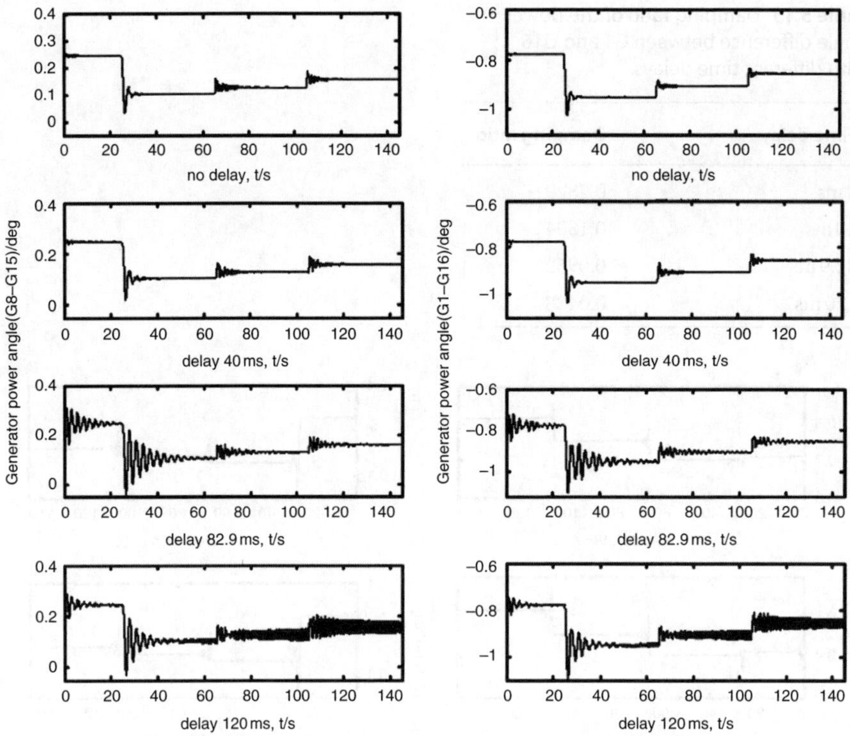

Figure 5.24 Dynamic responses of the generator relative power angle with different time delays.

Table 5.12 Damping ratio of the power angle difference between G8 and G15 with different time delays.

Time delay	Damping ratio
0 ms	0.1974
40 ms	0.1359
82.9 ms	0.0849
120 ms	0.0155

terminal link of the fault chain, giving the results shown in Tables 5.12 and 5.13, respectively. It can be seen that, when a cascading fault occurs in the system and the time delay is 82.9 ms, the damping ratios of the relative power angle curves are 8.49% and 9.62%, respectively, which are below 10% and thus the controller will not meet the control requirements. Therefore, by combining the fault chain and Markov process, the proposed method is proved effective in time-delay stability analysis of the fault power system. The time-delay stability upper bound calculated with the proposed method is close to the actual value.

Table 5.13 Damping ratio of the power angle difference between G1 and G16 with different time delays.

Time delay	Damping ratio
0 ms	0.2829
40 ms	0.1804
82.9 ms	0.0962
120 ms	0.0488

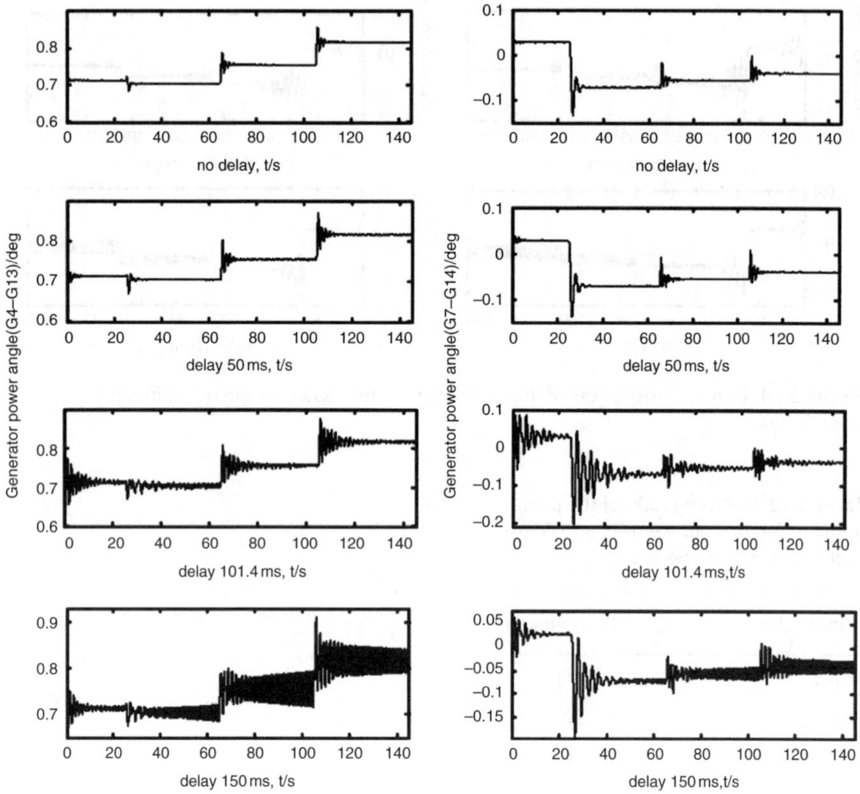

Figure 5.25 Dynamic responses of the generator relative power angle with different time delays.

Fault Chain L2 Concerning the fault chain L_2 according to the transfer probability matrix π_{L_2} and the given coefficient matrices A_i and A_{di} $(i = 1, 2, 3)$, the time-delay stability upper bound can be calculated using the GEVP method: $\bar{h} = 101.4$ ms. To verify the effectiveness of the proposed method, set the time delay to be 0 ms, 50 ms, 101.4 ms, and 150 ms, respectively, and observe the dynamic responses of the power angle difference between G4 and G13 and the power angle difference between G7 and G14. The results are shown in Figure 5.25.

It can be seen that, when the time delay is below the time-delay stability upper bound, the links in the fault chain are all able to damp the inter-area oscillation within 20 s. However, as the time delay increases from 0 ms to 101.4 ms, the damping effect decreases and the generator relative power angle curve fluctuates considerably at each link of the fault chain. When the time delay increases to 150 ms, the generator relative power angle curve experiences constant-amplitude or increasing-amplitude oscillation, which means the system is unstable.

Again, the prony algorithm is used to calculate the damping ratios of the G4–G13 relative power angle curve and the G7–G14 relative power angle curve at the terminal link of the fault chain, with the results shown in Tables 5.14 and 5.15, respectively. It can be seen that when the time delay is 0 ms, the damping ratios are 17.59% and 15.75%, respectively. When the time delay is 40 ms, the damping ratios are 11.81% and 11.54%, respectively. In these two cases, the damping ratios are relatively large and the system is stable. However, when the time delay reaches 101.4 ms, the damping ratios become 5.16% and 5.92%, which are below 10% and thus the system will be only marginally stable. Therefore, by combining the fault chain and Markov process, the proposed method is proved effective in analyzing the time-delay stability of the fault power system and calculating the time-delay stability upper bound.

Table 5.14 Damping ratio of the power angle difference between G4 and G13 with different time delays.

Time delay	Damping ratio
0 ms	0.1759
40 ms	0.1181
82.9 ms	0.0516
120 ms	0.0034

Table 5.15 Damping ratio of the power angle difference between G7 and G14 with different time delays.

Time delay	Damping ratio
0 ms	0.1575
40 ms	0.1154
82.9 ms	0.0592
120 ms	0.0041

Table 5.16 Time-delay stability upper bound $\bar{h}$ obtained using different methods.

μ	0.6	0.8	1.6
Theorem 5.5 in this section	0.7356	0.6150	0.5124
Reference [14]	0.4927	0.4261	0.3860
Theorem 1 in reference [15]	0.4428	0.3795	0.3469
Theorem 2 in reference [15]	0.4492	0.4341	0.4314
Reference [16]	0.5159	0.4814	0.4789

Conservativeness Verification Under the premise of fairness, the conservativeness of the proposed method is verified. Consider a classical two-modal time-delay Markov system, with the system parameters shown below:

$$\begin{aligned} A_1 &= \begin{bmatrix} -3.4888 & 0.8057 \\ -0.6451 & -3.2684 \end{bmatrix}, A_{d1} = \begin{bmatrix} -0.8620 & 1.2919 \\ -0.6841 & -2.0729 \end{bmatrix} \\ A_2 &= \begin{bmatrix} -2.4898 & 0.2895 \\ 1.3396 & -0.0211 \end{bmatrix}, A_{d2} = \begin{bmatrix} -2.8306 & 0.4978 \\ -0.8436 & -1.0115 \end{bmatrix} \\ \pi &= \begin{bmatrix} -0.1 & 0.1 \\ 0.8 & -0.8 \end{bmatrix} \end{aligned} \tag{5.133}$$

For different μ, the time-delay stability upper bounds $\bar{h}$ corresponding to different methods are calculated with the LMI method, shown in Table 5.16. Comparison between the results shows that the proposed method has relatively low conservativeness.

5.4 Summary

According to whether the system state and parameter jumps or not, this chapter gives a detailed introduction to the stability analysis methods for a non-jump time-delay system and a jump time-delay system.

For the non-jump system stability analysis, a stochastic time-delay stability analysis method based on the *Itô* differential is introduced. This method constructs an objective functional that is suitable for the time-delay stability analysis of a power system under stochastic disturbance. By adding free weighting items formed with the Newton–Leibniz formula to the weak infinitesimal generating operators of the objective functional, the conservativeness of this method is greatly lowered. Finally, the GEVP method is used to analyze the time-delay stability upper bound of a power system under different operating conditions quantitatively.

For the jump system stability analysis, a time-delay power system stability analysis method based on the Markov process is introduced. This method establishes a class of objective functionals considering the Markov transfer probability matrix to reveal the time-delay stability of a power system under multiple disturbances. On this basis, free weighting items formed according to the Newton–Leibniz formula and the transfer probability matrix, respectively, are introduced, so that the conservativeness of the system is lowered and full use is made of the information of the Markov transfer probability matrix. Finally, the maximum time delay that a power system could allow for as a system parameter or structure varies is calculated quantitatively.

Appendix A

$$\varepsilon = 0.0097 \tag{A1}$$

$$\begin{aligned} P = [&8.8496\ \ -0.3754\ \ 2.1098\ \ 0\ \ 0\ \ 0\ \ 0;\\ &-0.3754\ \ 0.0521\ \ 0.0008\ \ 0\ \ 0\ \ 0\ \ 0;\\ &2.1098\ \ 0.0008\ \ 3.0010\ \ 0\ \ 0\ \ 0\ \ 0;\\ &0\ \ 0\ \ 0\ \ 0.0047\ \ 0.0510\ \ 0.0060\ \ 0.0000;\\ &0\ \ 0\ \ 0\ \ 0.0510\ \ 24.6979\ \ 0.4548\ \ 0.0042;\\ &0\ \ 0\ \ 0\ \ 0.0060\ \ 0.4548\ \ 0.5345\ \ 0.0018;\\ &0\ \ 0\ \ 0\ \ 0.0000\ \ 0.0042\ \ 0.0018\ \ 0.0000] \end{aligned} \tag{A2}$$

$$\begin{aligned} Q = [&1.0468\ \ 0.0073\ \ 0.0000\ \ 0\ \ 0\ \ 0\ \ 0;\\ &0.0073\ \ 0.2825\ \ 0.0000\ \ 0\ \ 0\ \ 0\ \ 0;\\ &0.0000\ \ 0.0000\ \ 0.0003\ \ 0\ \ 0\ \ 0\ \ 0;\\ &0\ \ 0\ \ 0\ \ 0.0010\ \ -0.0028\ \ -0.0029\ \ -0.0000;\\ &0\ \ 0\ \ 0\ \ -0.0028\ \ 0.9341\ \ 0.1227\ \ 0.0008;\\ &0\ \ 0\ \ 0\ \ -0.0029\ \ 0.1227\ \ 0.2075\ \ 0.0014;\\ &0\ \ 0\ \ 0\ \ -0.0000\ \ 0.0008\ \ 0.0014\ \ 0.0000] \end{aligned} \tag{A3}$$

$$\begin{aligned} K = [&0.9941\ \ 0.0062\ \ 0.0000\ \ 0\ \ 0\ \ 0\ \ 0;\\ &0.0062\ \ 0.3576\ \ 0.0000\ \ 0\ \ 0\ \ 0\ \ 0;\\ &0.0000\ \ 0.0000\ \ 0.0106\ \ 0\ \ 0\ \ 0\ \ 0;\\ &0\ \ 0\ \ 0\ \ 0.0181\ \ -0.0024\ \ -0.0040\ \ -0.0002;\\ &0\ \ 0\ \ 0\ \ -0.0024\ \ 0.8867\ \ 0.1005\ \ 0.0007;\\ &0\ \ 0\ \ 0\ \ -0.0040\ \ 0.1005\ \ 0.2927\ \ 0.0019;\\ &0\ \ 0\ \ 0\ \ -0.0002\ \ 0.0007\ \ 0.0019\ \ 0.0001] \end{aligned} \tag{A4}$$

$$\begin{aligned} R = 1.0\mathrm{e}-004 * [&0.0214\ \ 0.0001\ \ 0.0000\ \ 0\ \ 0\ \ 0\ \ 0;\\ &0.0001\ \ 0.0141\ \ -0.0000\ \ 0\ \ 0\ \ 0\ \ 0;\\ &0.0000\ \ -0.0000\ \ 0.1140\ \ 0\ \ 0\ \ 0\ \ 0;\\ &0\ \ 0\ \ 0\ \ 0.0152\ \ -0.0003\ \ -0.0029\ \ 0.0003;\\ &0\ \ 0\ \ 0\ \ -0.0003\ \ 0.0213\ \ 0.0003\ \ 0.0001;\\ &0\ \ 0\ \ 0\ \ -0.0029\ \ 0.0003\ \ 0.0174\ \ 0.0000;\\ &0\ \ 0\ \ 0\ \ 0.0003\ \ 0.0001\ \ 0.0000\ \ 0.0040] \end{aligned} \tag{A5}$$

$$Z = 1.0e-005*[\begin{matrix} 0.2159 & -0.0005 & 0.0000 & 0 & 0 & 0 & 0; \\ -0.0005 & 0.2222 & 0.0003 & 0 & 0 & 0 & 0; \\ 0.0000 & 0.0003 & 0.2357 & 0 & 0 & 0 & 0; \\ 0 & 0 & 0 & 0.1591 & 0.0004 & 0.0015 & -0.0034; \\ 0 & 0 & 0 & 0.0004 & 0.2161 & -0.0002 & 0.0003; \\ 0 & 0 & 0 & 0.0015 & -0.0002 & 0.2190 & 0.0002; \\ 0 & 0 & 0 & -0.0034 & 0.0003 & 0.0002 & 0.0127 \end{matrix}] \tag{A6}$$

References

1 Mao, X.R. (1997) *Stochastic Differential Equations and Applications*, Horwood, Chichester.

2 Xu, S.Y. and Chen, T.W. (2002) Robust H^∞ control for uncertain stochastic systems with state delay. *IEEE Transactions on Automatic Control*, **47**(12), 2089–2094.

3 Gu, K. (2000) An integral inequality in the stability problem of time-delay systems. *Proceedings of the IEEE Decision and Control Conference on Decision and Control*, vol. **3**, pp. 2805–2810.

4 Boyd, S., El-Ghaoui, L., Feron, E., *et al.* (1994) *Linear Matrix Inequalities in System and Control Theory*, SIAM, Philadelphia.

5 Pal, B. and Chaudhuri, B. (2005) *Robust Control in Power Systems*, Springer, New York.

6 Rudnowski, D.J., Johnson, T., and Hauer, J.F. (1999) Making prony analysis more accurate using multiple signals. *IEEE Transactions on Power Systems*, **14**(1), 226–231.

7 Yue, D. (2005) Delay-dependent exponential stability of stochastic systems with time-varying delay, nonlinearity and Markovian switching. *IEEE Transactions on Automatic Control*, **50** (2), 217–222.

8 Yun, C., Anke, X., Renquan, L., *et al.* (2008) On robustly exponential stability of uncertain neutral systems with time-varying delays and nonlinear perturbations. *Nonlinear Analysis*, **68**(8), 2464–2470.

9 Pena, R., Clare, J., and Asher, G. (1996) Doubly-fed induction generators using back-to-back PWM converters and its applications to variable-speed wind-energy generation. *Proceedings of the Institution of Electrical Engineers*, **153**(3), 231–241.

10 Zhang, B.Q., Zhang, B.M., and Wu, W.C. (2009) Time constant model analysis of double fed induction wind generator. *Proceedings of International Conference on Electrical Engineering.*

11 Zhang, B.Q., Zhang, B.M., and Wu, W.C. (2010) Small signal based reduced electromagnetic model analysis of double fed induction generator. *Automation of Electric Power Systems*, **34**(10), 51–55.

12 Akhmatov, V. (2003) On mechanical excitation of electricity producing wind turbinesat grid faults. *Wind Engineering*, **27**(4), 257–272.

13 Ni, Y., Chen, S., and Zhang, B. (2001) *Power System Dynamics Theory and Analysis*, Tsinghua University Press, Beijing.

14 Xu, S., Lam, J., and Mao, X. (2007) Delay-dependent control and filtering for uncertain Markovian jump systems with time-varying delays. *IEEE Transactions on Circuits and Systems I: Regular Papers*, **54**(9), 2070–2077.

15 Wang, J. and Luo, Y. (2008) Further improvement of delay-dependent stability for Markov jump systems with time-varying delay. *Proceedings of the 7th World Congress on Intelligent Control and Automation*, pp. 6319–6324.

16 Zhao, X.D. and Zeng, Q.S. (2010) Delay-dependent stability analysis for Markovian jump systems with interval time-varying delays. *International Journal of Automation and Computing*, **7**(2), 224–229.

6

Wide-Area Robust Control

6.1 Introduction

Due to the uncertainty of the grid structure, parameter, and external disturbance, it is difficult to describe the model of a power system accurately, which greatly affects the effects of power system stability control. Conventional control methods are mostly based on traditional control theory, and when the accurate models of system and disturbance can be acquired, the damping characteristics of such methods are relatively good. However, once uncertainty exists in the models, it will be difficult to guarantee the damping effects. Comparatively, robust control methods have an incomparable advantage in dealing with uncertainties in a power system parameter and external disturbance, etc. The mixed H_2/H_∞ control theory has also won favor with many experts and scholars due to its comprehensive consideration of factors such as power system stability and robustness.

The H_2 control theory and H_∞ control theory both use norm bounded conditions to describe the uncertainty in external disturbance on a power system. Optimal control theory with the H_2 norm of a system as the performance index has good dynamic and steady-state performance, but the robust stability is poor. Comparatively, the H_∞ control can obtain good robust stability, but the dynamic and steady-state regulating performance is relatively poor. Therefore, if H_2 control and H_∞ control are combined, the resulting H_2/H_∞ mixed control can guarantee both good regulating performance and good robust stability for the system. However, many problems in H_2/H_∞ mixed control still need to be resolved, such as an internal uncertainty description, control objective optimization, and error tracking control.

Concerning different links in the H_2/H_∞ mixed controller, this chapter gives a detailed introduction from the following aspects:

1) For the internal uncertainty description problem, a multiobjective robust H_2/H_∞ control considering uncertainties for damping oscillation is proposed, the convex polytope theory is used to model the uncertainty at the operating point, and the polyhedron theory is applied to H_2 control and H_∞ control to obtain the feedback control input vector that satisfies the polyhedral uncertainty.

2) For the control objective optimization problem, the eigenvalue transfer factor is used to expand the stability region and the non-convex stability region, which could render the system in stable state rapidly, is defined according to the non-convex region control theory. Also the complete regulation control theory is used to transfer the system eigenvalues to the target region.
3) For error tracking control, the Youla parameterization control theory is used to reflect the error between the estimated output value and the actual input value in the control action, thus realizing tracking control of the error.

6.2 Robust Control for Internal Uncertainties

6.2.1 Multiobjective Robust H_2/H_∞ Control Considering Uncertainties for Damping Oscillation

This section deals with the robust H_2/H_∞ control problem for damping power system low-frequency oscillations considering the uncertainties. In the process of designing the damp controller, uncertain parameters are introduced into the perturbation matrix and both the H_∞ disturbance depression and the optimal H_2 performances of the closed-loop control system are satisfied simultaneously for all allowable uncertainties. To solve the proposed multiobjective optimization problem, linear matrix inequalities (LMIs) are deduced based on the LMI optimization technology to obtain the output feedback controller. Simulation results of a four-machine 11-bus IEEE test system show that the proposed method is effective and robust.

6.2.1.1 Multiobject Robust H_2/H_∞ Control

The linear time invariable model of multiobject robust H_2/H_∞ control considering uncertainties is described as [1]

$$\begin{cases} \dot{\mathbf{x}}(t)=(\mathbf{A}+\Delta\mathbf{A})\mathbf{x}(t)+\mathbf{B}_2 w(t)+(\mathbf{B}_1+\Delta\mathbf{B}_1)\mathbf{u}(t) \\ \mathbf{z}_\infty=\mathbf{C}_\infty\mathbf{x}(t)+\mathbf{E}_\infty w(t)+\mathbf{D}_\infty\mathbf{u}(t) \\ \mathbf{z}_2=\mathbf{C}_2\mathbf{x}(t)+\mathbf{E}_2 w(t)+\mathbf{D}_2\mathbf{u}(t) \end{cases} \tag{6.1}$$

where $\mathbf{x}\in R^n$ is a state vector, $\mathbf{u}\in R^m$ is a control input vector, $w\in R^m$ is disturbance input, $\mathbf{z}_\infty$ and $\mathbf{z}_2$ are H_∞ and H_2 performance variables, $A\in R^{n\times n}$ is a state matrix, $B_2\in R^{n\times m}$ is a control input matrix, and $\Delta\mathbf{A}$ and $\Delta\mathbf{B}_1$ are perturbation matrices. Assume that the perturbation matrices are norm-bounded as follows:

$$\begin{bmatrix}\Delta\mathbf{A} & \Delta\mathbf{B}_1\end{bmatrix}=\mathbf{H}\mathbf{F}\begin{bmatrix}\mathbf{E}_1 & \mathbf{E}_2\end{bmatrix} \tag{6.2}$$

where $\mathbf{F}^\mathrm{T}\mathbf{F}\le\mathbf{I}$, $\mathbf{E}_1$ and $\mathbf{E}_2$ are known matrices representing the structure of uncertainties, and $\mathbf{B}_1$, $\mathbf{B}_2$, $\mathbf{C}_\infty$, $\mathbf{D}_\infty$, $\mathbf{E}_\infty$, $\mathbf{C}_2$, $\mathbf{D}_2$, and $\mathbf{E}_2$ are known constant matrices that describe a nominal system. The multiobjective robust H_2/H_∞ control for damping oscillation is to design a controller $\mathbf{U}=\mathbf{K}(\mathbf{s})\mathbf{Y}_\mathbf{m}$, with the closed system satisfying the following constraints:

1) The closed system is asymptotic stability and the poles of the closed system are located in the left-half plane.

2) The H_∞-norm of the closed transfer function $\mathbf{T}_{WZ\infty}$ does not exceed the given maximum bound γ_∞, that is, $\|\mathbf{T}_{WZ\infty}\|_\infty < \gamma_\infty$, to guarantee that the closed system is robust stable.
3) The H_2-norm of the closed transfer function $\mathbf{T}_{WZ2}$ does not exceed the given maximum bound γ_2, that is, $\|\mathbf{T}_{WZ2}\|_2 < \gamma_2$, to guarantee that the closed system gives a good performance.

The problem is to look for a controller with minimum γ_2 satisfying the above constraints. It can be seen that the performance index $[\gamma_\infty \quad \gamma_2]$ appears together; therefore, the controller designers can adjust the two performance indexes to acquire appropriate parameters of controllers.

6.2.1.2 Design of LMI-Based H_2/H_∞ Controller

The uncertainty matrices need to be dealt with in the deducing process. Therefore, two theorems are quoted.

Theorem 6.1 [2]. $\mathbf{D_0}$, $\mathbf{F_0}$, and $\mathbf{E_0}$ are given matrices and $\mathbf{Y_0}$ is a symmetric matrix. Then

$$\mathbf{Y_0} + \mathbf{D_0 F_0 E_0} + \mathbf{E_0^T F_0^T D_0^T} < 0 \tag{6.3}$$

where $\mathbf{Y_0}$ is tenable for those matrices $\mathbf{F}$ satisfying $\mathbf{F^T F} < \mathbf{I}$, if and only if there exists a positive constant μ and satisfies

$$\mathbf{Y_0} + \mu \mathbf{D_0 D_0^T} + \mu^{-1} \mathbf{E_0^T E_0} < 0 \tag{6.4}$$

The output $\mathbf{Y_m}$ is described as follows:

$$\mathbf{Y_m} = \mathbf{C_m Z} + \mathbf{D_m U} + \mathbf{E_m W} \tag{6.5}$$

Consider the state space of the output-feedback controller:

$$\mathbf{Z_c} = \mathbf{A_c Z_c} + \mathbf{B_c Y_m} \quad \mathbf{U} = \mathbf{C_c Z_c} \tag{6.6}$$

where $\mathbf{Z_c}$ is the state vector of controller, $\mathbf{A_c}$, $\mathbf{B_c}$, and $\mathbf{C_c}$ are matrices to be determined. The closed system can be obtained by substituting (6.5) and (6.6) into system (6.1):

$$\begin{cases} \bar{\mathbf{Z}} = \left(\mathbf{A_{cl}} + \bar{\mathbf{H}}\mathbf{F}\bar{\mathbf{E}}\right)\bar{\mathbf{Z}} + \mathbf{B_{cl} W} \\ \mathbf{Z_\infty} = \bar{\mathbf{C}}_\infty \bar{\mathbf{Z}} + \mathbf{E_\infty W} \\ \mathbf{Z_2} = \bar{\mathbf{C}}_2 \bar{\mathbf{Z}} \end{cases} \tag{6.7}$$

where

$$\bar{\mathbf{Z}} = [\mathbf{Z} \quad \mathbf{Z_c}]^{\mathrm{T}}, \quad \mathbf{A_{cl}} = \begin{bmatrix} \mathbf{A} & \mathbf{B_1 C_c} \\ \mathbf{B_c C_m} & \mathbf{A_c} + \mathbf{B_c D_m C_c} \end{bmatrix}, \quad \bar{\mathbf{H}} = [\mathbf{H} \quad \mathbf{0}]^{\mathrm{T}}, \quad \bar{\mathbf{E}} = [\mathbf{E_1} \quad \mathbf{E_2 C_c}],$$

$$\mathbf{B_{cl}} = [\mathbf{B_2} \quad \mathbf{B_c E_m}]^{\mathrm{T}}, \bar{\mathbf{C}}_\infty = [\mathbf{C_\infty} \quad \mathbf{D_\infty C_c}], \bar{\mathbf{C}}_2 = [\mathbf{C_2} \quad \mathbf{D_2 C_c}]$$

The necessary and sufficient condition satisfying aspects (1) and (2) in Section 6.1 is that there exists a positive definite matrix **P** such that

$$\begin{bmatrix} \left(\mathbf{A}_{cl}+\bar{\mathbf{H}}\mathbf{F}\bar{\mathbf{E}}\right)^{T} & \mathbf{B}_{cl} & \mathbf{P}\bar{\mathbf{C}}_{\infty}^{T} \\ \mathbf{B}_{cl}^{T} & -\mathbf{I} & \mathbf{E}_{\infty}^{T} \\ \bar{\mathbf{C}}_{\infty}\mathbf{P} & \mathbf{E}_{\infty} & -\gamma_{\infty}^{2}\mathbf{I} \end{bmatrix} < \mathbf{0} \tag{6.8}$$

Uncertainty matrices are included in formula (6.8) and need to be removed, although (6.8) should be transformed to the format of LMI. Inequality (6.8) satisfies matrix **F** when $\mathbf{F}^{T}\mathbf{F} < \mathbf{I}$, if and only if there exists a constant $\mu > 0$ such that

$$\begin{bmatrix} \mathbf{A}_{cl}\mathbf{P}+\mathbf{P}\mathbf{A}_{cl}^{T}+\mu\bar{\mathbf{H}}\bar{\mathbf{H}}^{T}+\mu^{-1}\mathbf{P}\bar{\mathbf{E}}^{T}\bar{\mathbf{E}}\mathbf{P} & \mathbf{B}_{cl} & \mathbf{P}\bar{\mathbf{C}}_{\infty}^{T} \\ \mathbf{B}_{cl}^{T} & -\mathbf{I} & \mathbf{E}_{\infty}^{T} \\ \bar{\mathbf{C}}_{\infty}\mathbf{P} & \mathbf{E}_{\infty} & -\gamma_{\infty}^{2}\mathbf{I} \end{bmatrix} < \mathbf{0} \tag{6.9}$$

The formula above is equivalent to

$$\begin{bmatrix} \mathbf{A}_{cl}\mathbf{P}+\mathbf{P}\mathbf{A}_{cl}^{T} & \mathbf{B}_{cl} & \mathbf{P}\bar{\mathbf{C}}_{\infty}^{T} & \bar{\mathbf{H}} & \mathbf{P}\bar{\mathbf{E}}^{T} \\ \mathbf{B}_{cl}^{T} & -\mathbf{I} & \mathbf{E}_{\infty}^{T} & \mathbf{0} & \mathbf{0} \\ \bar{\mathbf{C}}_{\infty}\mathbf{P} & \mathbf{E}_{\infty} & -\gamma_{\infty}^{2}\mathbf{I} & \mathbf{0} & \mathbf{0} \\ \bar{\mathbf{H}}^{T} & \mathbf{0} & \mathbf{0} & -\mu^{-1}\mathbf{I} & \mathbf{0} \\ \bar{\mathbf{E}}\mathbf{P} & \mathbf{0} & \mathbf{0} & \mathbf{0} & -\mu\mathbf{I} \end{bmatrix} < \mathbf{0} \tag{6.10}$$

Partitioned matrices of **P** and its inverse are shown as follows:

$$\mathbf{P} = \begin{bmatrix} \mathbf{P}_{11} & \mathbf{P}_{12} \\ \mathbf{P}_{12}^{T} & \mathbf{P}_{22} \end{bmatrix}, \mathbf{P}^{-1} = \begin{bmatrix} \mathbf{P}_{i11} & \mathbf{P}_{i12} \\ \mathbf{P}_{i12}^{T} & \mathbf{P}_{i22} \end{bmatrix} \tag{6.11}$$

where $\mathbf{P}_{11}$ and $\mathbf{P}_{i11}$ are positive symmetric matrices. $\mathbf{P}\mathbf{P}^{-1} = \mathbf{I}$ results in $\mathbf{P}_{12}\mathbf{P}_{i12}^{T} = \mathbf{I} - \mathbf{P}_{11}\mathbf{P}_{i11}$.

Define

$$\mathbf{F}_1 = \begin{bmatrix} \mathbf{P}_{11} & \mathbf{I} \\ \mathbf{P}_{12}^{T} & \mathbf{0} \end{bmatrix} \mathbf{F}_2 = \begin{bmatrix} \mathbf{I} & \mathbf{P}_{i11} \\ \mathbf{0} & \mathbf{P}_{i12}^{T} \end{bmatrix} \tag{6.12}$$

Then the relative relations can be deduced from (6.12) as

$$\mathbf{P}\mathbf{F}_2 = \mathbf{F}_1\ \mathbf{F}_2^{T}\mathbf{P}\mathbf{F}_2 = \mathbf{F}_2^{T}\mathbf{F}_1 = \begin{bmatrix} \mathbf{P}_{11} & \mathbf{I} \\ \mathbf{I} & \mathbf{P}_{i11} \end{bmatrix} \tag{6.13}$$

It can be seen that (6.10) is a nonlinear inequality, so suitable variables should be found to transform (6.10) to a linear matrix inequality. Therefore, these variables can be shown as below:

$$\begin{cases} \hat{\mathbf{A}} = \mathbf{P_{i11}AP_{11}} + \mathbf{P_{i11}B_1C_cP_{12}^T} + \mathbf{P_{i12}B_cC_mP_{11}} \\ \quad + \mathbf{P_{i12}A_cP_{12}^T} + \mathbf{P_{i12}B_cD_mC_cP_{12}^T} \\ \hat{\mathbf{B}} = \mathbf{P_{i12}B_c} \\ \hat{\mathbf{C}} = \mathbf{C_cP_{12}^T} \end{cases} \tag{6.14}$$

The left multiplication diag($[\mathbf{F_2^T\ I\ I\ I}]$) and the right multiplication diag($[\mathbf{F_2^T\ I\ I\ I}]$) in (6.10) with a combination of (6.14) result in

$$\begin{bmatrix} \mathbf{L_{11}} & \mathbf{L_{12}} \\ \mathbf{L_{21}} & \mathbf{L_{22}} \end{bmatrix} < 0 \tag{6.15}$$

where

$$\mathbf{Z_2} = \bar{\mathbf{C}}_2\bar{\mathbf{Z}}, \mathbf{L_{11}} = \begin{bmatrix} \mathbf{AP_{11}} + \mathbf{B_1C} + \mathbf{P_{11}A^T} + (\mathbf{B_1C})^\mathbf{T} & \mathbf{A} + \mathbf{A^T} \\ \mathbf{A} + \mathbf{A^T} & \mathbf{P_{i11}^TA} + \mathbf{A^TP_{i11}} + \mathbf{BC_m} + (\mathbf{BC_m})^\mathbf{T} \end{bmatrix},$$

$$\mathbf{L_{21}} = \begin{bmatrix} \left[\mathbf{B_2^T}\ \ \mathbf{B_2^TP_{i11}} + (\mathbf{BE_m})^\mathbf{T}\right] \\ \left[\mathbf{C_\infty P_{11}} + \mathbf{D_\infty C}\ \ \mathbf{C_\infty}\right] \\ \left[\mathbf{H^T}\ \ \mathbf{H^TP_{i11}}\right] \\ \left[\mathbf{E_1P_{11}} + \mathbf{E_2C}\ \ \mathbf{E_1}\right] \end{bmatrix}, \mathbf{L_{22}} = \begin{bmatrix} -\mathbf{I} & \mathbf{E}_\infty^T & \mathbf{0} & \mathbf{0} \\ \mathbf{E_\infty} & -\gamma_\infty^2\mathbf{I} & \mathbf{0} & \mathbf{0} \\ \mathbf{0} & \mathbf{0} & -\mu^{-1}\mathbf{I} & \mathbf{0} \\ \mathbf{0} & \mathbf{0} & \mathbf{0} & -\mu\mathbf{I} \end{bmatrix}$$

The necessary and sufficient condition satisfying aspect (3) in Section 6.1 is that if and only if there exists symmetric positive matrix **P** and **Q** for any given scalar $\gamma_2 > 0$ such that

$$(\mathbf{A_{cl}} + \bar{\mathbf{H}}\mathbf{F}\bar{\mathbf{E}})\mathbf{P} + \mathbf{P}(\mathbf{A_{cl}} + \bar{\mathbf{H}}\mathbf{F}\bar{\mathbf{E}})^\mathbf{T} + \mathbf{B_{cl}B_{cl}^T} < 0 \tag{6.16}$$

$$\begin{bmatrix} -\mathbf{Q} & \bar{\mathbf{C}}_2\mathbf{P} \\ \mathbf{P}\bar{\mathbf{C}}_2^\mathbf{T} & -\mathbf{P} \end{bmatrix} < 0 \tag{6.17}$$

$$Trace(\mathbf{Q}) < \gamma_2^2 \tag{6.18}$$

The left multiplication diag($[\mathbf{F_2^T\ I}]$) and the right multiplication diag($[\mathbf{F_2^T\ I}]$) on (6.17) with a combination of (6.14) result in

$$\begin{bmatrix} -\mathbf{Q} & \mathbf{C_2P_{11}} + \mathbf{D_2C} & \mathbf{C_2} \\ (\mathbf{C_2P_{11}} + \mathbf{D_2C})^\mathbf{T} & -\mathbf{P_{11}} & -\mathbf{I} \\ \mathbf{C_2^T} & -\mathbf{I} & -\mathbf{P_{i11}} \end{bmatrix} \tag{6.19}$$

Then the corollary can be obtained from the above deduced steps. There exist solutions $\mathbf{P}_{11}$, $\mathbf{P}_{i11}$, $\hat{\mathbf{A}}$, $\hat{\mathrm{B}}$, $\hat{\mathbf{C}}$, and $\mathbf{Q}$ for the closed system (6.7) and, given scalar $\gamma_\infty > 0$ and $\mu > 0$, by solving (6.15), (6.18), and (6.19). The optimum problem can be written in this form as follows:

$$\min_{\mathbf{P}_{11},\mathbf{P}_{i11},\hat{\mathbf{A}},\hat{\mathbf{B}},\hat{\mathbf{C}},\mathbf{Q}} \gamma_2^2 \tag{6.20}$$

$\mathbf{P}_{12}$ and $\mathbf{P}_{i12}$ can be obtained by singular value decomposition of (6.11), then $\mathbf{A}_c$, $\mathbf{B}_c$, and $\mathbf{C}_c$ can be computed, and the controller (6.6) is the robust H_2/H_∞ controller after considering uncertainties.

6.2.1.3 Test System

A four-generator system with 11 buses, as shown in Figure 3.1, is considered for the damping control design. All the generators use a detailed model and the high-speed excitation is installed in each generator. The active loads are 50% constant current and 50% constant impedance; reactive loads all have constant impedance. The supplementary control for a high-voltage direct circuit (HVDC) is used to form a robust controller when considering uncertainties. The control structure for an HVDC small-signal modulator is shown in reference [2]. The oscillation modes can be obtained from the eigenvalue analysis and are displayed in Table 6.1. There are two inner-area modes and one inter-area mode.

6.2.1.4 Testing Results and Analysis

Frequency Domain Results and Analysis The performance of the controllers was evaluated with different operating cases, as shown in Table 6.2.

The damping action is quite robust with respect to the outage of each of the tie lines between buses 7–8 and buses 8–9, which are tie lines. In addition, the robustness is shown of the damping action in the case of various tie line flows. Cases 4–6 correspond to 600 MW, 400 MW, and −230 MW through tie line 7–8, respectively. Constant impedance (CI), constant current (CC), and constant power (CP) are also demonstrated to test the robustness of controllers. The first case is the base case. The open- and closed-loop eigenvalues of these test cases are shown Table 6.3. This table shows the numerical results with an H_2/H_∞ robust controller. The controller 1 represents the H_2/H_∞ robust controller considering uncertainty and the controller 2 represents the H_2/H_∞ robust controller without considering uncertainties. From this table, it can be seen that the controller considering uncertainty improves the damping of the three critical modes to a more adequate level and is more effective.

Table 6.1 Eigenvalue of electromechanical modes in IEEE four-machine systems without controllers.

Mode	Eigenvalue	Frequency	Damping ratio	Mode type
1	$-0.6307 \pm 7.2639_i$	1.1561	0.0865	Inner-area
2	$-0.6309 \pm 7.3466_i$	1.1692	0.0856	Inner-area
3	$-0.0053 \pm 3.9285_i$	0.6252	0.0014	Inter-area

Table 6.2 Case bank.

No.	Tie lie flow	Outage of tie line	Load characteristic
1	400 MW	No outage	50% CC + 50% CI
2	400 MW	Line 7–8 outage	50% CC + 50% CI
3	400 MW	Line 8–9 outage	50% CC + 50% CI
4	600 MW	No outage	50% CC + 50% CI
5	200 MW	No outage	50% CC + 50% CI
6	−340 MW	No outage	50% CC + 50% CI
7	400 MW	No outage	50% CC + 50% CP

Table 6.3 Electromechanical modes in four-machine systems with different controllers under different operating conditions.

		Mode 1		Mode 2		Mode 3	
Cases		Frequency	Damp	Frequency	Damp	Frequency	Damp
1	Non	1.1561	0.0865	1.1692	0.0856	0.6252	0.0014
	Con1	1.1601	0.1327	1.1548	0.1085	0.6260	0.3025
	Con2	1.1546	0.1056	1.1674	0.1018	0.6042	0.2184
2	Non	1.1514	0.0857	1.1614	0.0847	0.5054	0.0037
	Con1	1.1521	0.1278	1.1527	0.1082	0.4884	0.4495
	Con2	1.1562	0.1017	1.1563	0.1042	0.4605	0.3319
3	Non	1.1519	0.0893	1.1640	0.0841	0.5094	−0.0016
	Con1	1.1571	0.1319	1.1485	0.1082	0.4948	0.4338
	Con2	1.1626	0.0991	1.1504	0.1066	0.4665	0.3201
4	Non	1.1634	0.0677	1.1468	0.0923	0.5557	−0.0057
	Con1	1.1593	0.1060	1.1426	0.1085	0.5555	0.3549
	Con2	1.1615	0.0802	1.1458	0.1070	0.5273	0.2532
5	Non	1.1572	0.0999	1.1688	0.0844	0.6448	0.0094
	Con1	1.1533	0.1529	1.1609	0.1080	0.6431	0.2954
	Con2	1.1536	0.1169	1.1697	0.1068	0.6237	0.2173
6	Non	1.1536	0.1088	1.1711	0.0816	0.6491	0.0129
	Con1	1.1489	0.1622	1.1637	0.1077	0.6458	0.2956
	Con2	1.1513	0.1258	1.1713	0.1056	0.6272	0.2192
7	Non	1.1634	0.0678	1.1467	0.0924	0.5554	−0.0061
	Con1	1.1593	0.1064	1.1425	0.1087	0.5555	0.3548
	Con2	1.1615	0.0804	1.1456	0.1072	0.5270	0.2528

Time Domain Results and Analysis To demonstrate the effectiveness of the H_2/H_∞ robust control-based HVDC supplementary controller over a range of operating conditions, four time-domain simulations have been carried out for 10 s to further investigate the performance of the H_2/H_∞ robust controller. The four simulations are on different operating conditions and a temporary three-phase fault was applied to the tie line between buses 7 and 8. The fault is cleared by the removal of the faulted circuit after 50 ms (3 cycles). This contingency can excite inter-area oscillations, which can then be examined for the performance of the damping controllers. Figure 6.1 exhibits relative angular separations between

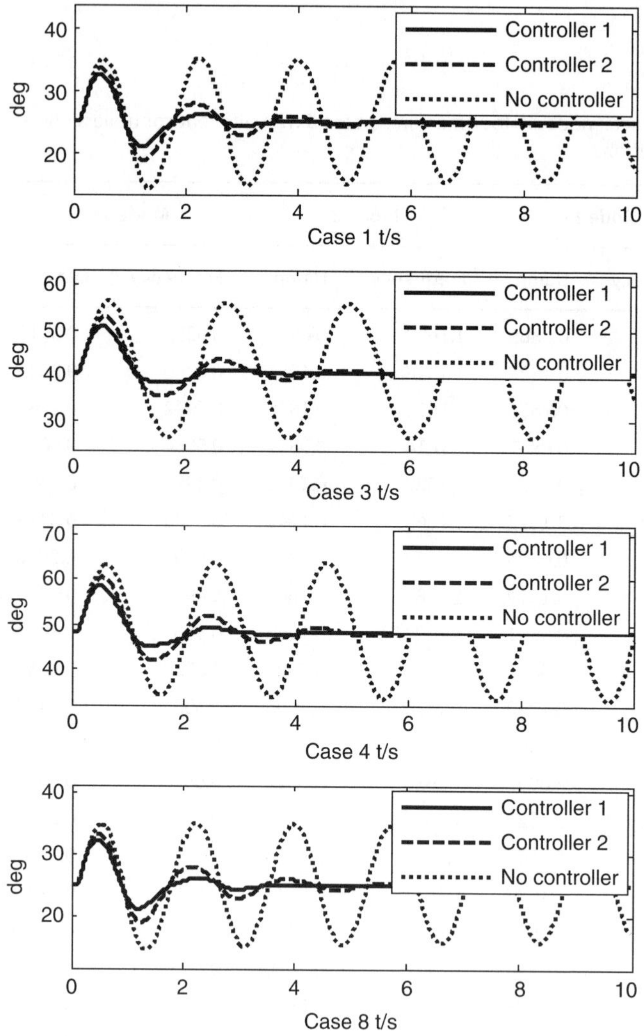

Figure 6.1 The δ_{13} simulation curves in cases with different controllers under different operating conditions.

machines G1 and G3. It can be seen that oscillations settle down within the desired performances specification of 2 s with a controller considering uncertainty compared with that of a controller without considering uncertainty.

6.2.2 Robust H_2/H_∞ Control Strategy Based on Polytope Uncertainty

Polytope uncertainty is suitable for time-varying uncertainty analysis in power systems. The addition of an uncertain factor increases the robustness of a closed-loop system. First, wide-area information measured by PMUs is used as controller input to control the power transferred by HVDC. Then an extended LMI approach is used in the controller's design considering the uncertainty. In the paper the state feedback control and design of the controller are chosen based on a reduced system. The two-area four-machine system is studied to validate the adaptability of the controller. Different types of fault and operating conditions are considered and the simulation result shows that the controller with uncertainty can deal with these situations well.

6.2.2.1 Mathematical Background

The power system can be described by the state space model as follows [3]:

$$\begin{aligned} \dot{\boldsymbol{x}} &= A\boldsymbol{x} + \boldsymbol{B}_1\boldsymbol{w} + \boldsymbol{B}_2\boldsymbol{u} \\ \boldsymbol{z}_\infty &= \boldsymbol{C}_1\boldsymbol{x} + \boldsymbol{D}_{11}\boldsymbol{w} + \boldsymbol{D}_{12}\boldsymbol{u} \\ \boldsymbol{z}_2 &= \boldsymbol{C}_2\boldsymbol{x} + \boldsymbol{D}_{22}\boldsymbol{u} \end{aligned} \tag{6.21}$$

where $\boldsymbol{x} \in \boldsymbol{R}^{\mathrm{n}}$ is the system state vector, $\boldsymbol{u} \in \boldsymbol{R}^{\mathrm{m}}$ is the control input vector, $\boldsymbol{w} \in \boldsymbol{R}^{\mathrm{q}}$ is the outside disturbance input vector, and $\boldsymbol{z}_\infty \in \boldsymbol{R}^{r_1}$ and $\boldsymbol{z}_2 \in \boldsymbol{R}^{r_2}$ are the controlled system output vectors. $\boldsymbol{A}$, $\boldsymbol{B}_1$, $\boldsymbol{B}_2$, $\boldsymbol{C}_1$, $\boldsymbol{D}_{11}$, $\boldsymbol{D}_{12}$, $\boldsymbol{C}_2$, and $\boldsymbol{D}_{22}$ are constant matrixes with proper dimensions.

The feedback controller is given as

$$\boldsymbol{u} = \boldsymbol{K}\boldsymbol{x} \tag{6.22}$$

The corresponding closed-loop system is

$$\begin{aligned} \dot{\boldsymbol{x}} &= \left(\boldsymbol{A} + \boldsymbol{B}_2\boldsymbol{K}\right)\boldsymbol{x} + \boldsymbol{B}_1\boldsymbol{w} \\ \boldsymbol{z}_\infty &= \left(\boldsymbol{C}_1 + \boldsymbol{D}_{12}\boldsymbol{K}\right)\boldsymbol{x} + \boldsymbol{D}_{11}\boldsymbol{w} \\ \boldsymbol{z}_2 &= \left(\boldsymbol{C}_2 + \boldsymbol{D}_{22}\boldsymbol{K}\right)\boldsymbol{x} \end{aligned} \tag{6.23}$$

The designed controller should achieve the following goals:

- Maintain the H_∞-norm of the closed-loop transfer functions from w to $\boldsymbol{z}_\infty$, $\boldsymbol{T}_{wz_\infty}(\mathrm{s})$, below some prescribed value $\gamma_1 > 0$.
- Maintain the H_2-norm of the closed-loop transfer functions from w to z_2, $\boldsymbol{T}_{wz_2}(\mathrm{s})$, below some prescribed value $\gamma_2 > 0$.

The closed-loop system (6.22) satisfies $\|\boldsymbol{T}_{wz_\infty}(s)\|_\infty < \gamma_1$, making $\mathbf{min}\|\boldsymbol{T}_{wz_2}(s)\|_2$, if and only if

$$\mathbf{min}\|\boldsymbol{T}_{wz_2}(s)\|_2 \tag{6.24}$$

s.t.

$$\begin{bmatrix} -(V+V^T) & M^T & N^T & V^T \\ M & -X_2 & 0 & 0 \\ N & 0 & -I & 0 \\ V & 0 & 0 & -X_2 \end{bmatrix} < 0 \tag{6.25}$$

$$\begin{bmatrix} -(V+V^T) & M_1^T & N_1^T & V^T & 0 \\ M_1 & -X_\infty & 0 & 0 & B_1 \\ N_1 & 0 & -\gamma_1 I & 0 & D_{11} \\ V & 0 & 0 & -X_\infty & 0 \\ 0 & B_1^T & D_{11}^T & 0 & -\gamma_1 I \end{bmatrix} < 0 \tag{6.26}$$

$$\begin{bmatrix} X_2 & B_1 \\ B_1^T & Q \end{bmatrix} > 0 \tag{6.27}$$

where, X_2 and X_∞ are symmetric positive definite matrixes V, H are real matrixes, and $M = AV + B_2H + X_2$, $N = C_2V + D_{22}H$, and $M_1 = AV + B_2H + X_\infty$, $N_1 = C_1V + D_{12}H$. If X_2^*, X_∞^*, V^*, H^* is the optimal solution of this optimization problem, then the feedback control law is $K = HV^{-1}$. In order to make the controller adapt the effect of uncertainty, consider the state space models as follows:

$$\begin{aligned} \dot{x} &= (A+\Delta A)x + B_1 w + (B_2 + \Delta B_2)u \\ z_\infty &= C_1 x + D_{11} w + D_{12} u \\ z_2 &= C_2 x + D_{22} u \end{aligned} \tag{6.28}$$

The closed-loop system of (6.28) satisfies $\|T_{wz_\infty}(s)\|_\infty < \gamma_1$, making $\mathbf{min}\|T_{wz_2}(s)\|_2$ if and only if symmetric positive definite matrices X_2, X_∞ real matrices V, and H exist, which make the following $3N$ LMIs have a solution at the uncertainty polytope $(A_i, B_{1,i}, B_{2,i}, C_{1,i}, D_{11,i}, D_{12,i}, C_{2,i}, D_{22,i})$, $i = 1, \ldots, N$, then

$$\mathbf{min}\, Trace(Q) \tag{6.29}$$

s.t.

$$\begin{bmatrix} -(V+V^T) & M_i^T & N_i^T & V^T \\ M_i & -X_2 & 0 & 0 \\ N_i & 0 & -I & 0 \\ V & 0 & 0 & -X_2 \end{bmatrix} < 0 \tag{6.30}$$

$$\begin{bmatrix} -\left(V+V^T\right) & M_{1i}^T & N_{1i}^T & V^T & 0 \\ M_{1i} & -X_\infty & 0 & 0 & B_{1,i} \\ N_{1i} & 0 & -\gamma_1 I & 0 & D_{11,i} \\ V & 0 & 0 & -X_\infty & 0 \\ 0 & B_{1,i}^T & D_{11,i}^T & 0 & -\gamma_1 I \end{bmatrix} < 0 \tag{6.31}$$

$$\begin{bmatrix} X_2 & B_{1,i} \\ B_{1,i}^T & Q \end{bmatrix} > 0 \tag{6.32}$$

where $M_i = A_i V + B_{2,i} H + X_2$, $N_i = C_{2,i} V + D_{22,i} H$, $M_{1i} = A_i V + B_{2,i} H + X_\infty$, and $N_{1i} = C_{1,i} V + D_{12,i} H$. Then $u = HV^{-1}x$. According to the polytope theory [4], if the controller designed suits the vertex of the polytope, then it also suits the inner part of the polytope.

6.2.2.2 Test System

A two-area four-machine system with parallel DC and AC ties are studied to test the validity of the method given in the paper. Different operating conditions are considered during the simulations. The generator adopts subtransient models. We suppose $H=6.5$; then the system is uniform damp. An SMA method is used to eliminate extra state variables, so relative rotor angle and relative machine speed are used for the following analysis.

Table 6.4 is the small-signal analysis result under normal operating conditions, where the system consists of one negative damped inter-area mode and two local-area modes. Under normal operating situations, the power transferred from area 1 to area 2 is 400 MW, of which the DC ties account for 200 MW. A simplified excitation system is considered on machine 2 and the DC exciter is considered on machines 1, 3, and 4. All machines are without PSS.

The control method of the DC ties adapts schemes of a rectifier converter with constant current control and an inverter converter with constant extinction angle control. From the paper showing that the power transferred by DC ties is controlled to damp the oscillation, we select

$$C_1 = \begin{bmatrix} C_{1a} \\ C_{1b} \end{bmatrix}, C_2 = \begin{bmatrix} C_{2a} \\ C_{2b} \end{bmatrix}, D_{12} = \begin{bmatrix} D_{12a} \\ D_{12b} \end{bmatrix} = D_{22} \tag{6.33}$$

Table 6.4 Eigenvalues of electromechanical modes in the two-area four-machine system.

Mode	Frequency (Hz)	Damping ratio	Type
1	0.6081	−0.0183	Inter-area mode
2	1.0795	0.1138	Local-area mode
3	1.0919	0.0795	Local-area mode

where

$$C_{1a} = \begin{bmatrix} 1 & & & \\ & 1 & & \\ & & \ddots & \\ & & & 1 \end{bmatrix}_{6\times 6}, C_{1b} = [0]_{1\times 6} \tag{6.34}$$

$$C_{2a} = \begin{bmatrix} 1 & & & \\ & 1 & & \\ & & \ddots & \\ & & & 1 \end{bmatrix}_{6\times 6}, C_{2b} = [0]_{1\times 6} \tag{6.35}$$

$$D_{12a} = [0]_{6\times 1}, D_{12b} = [1]_{1\times 1} \tag{6.36}$$

The considered disturbances existing on a relative rotor angle and at a relative machine speed selects

$$B_1 = \begin{bmatrix} 1 & & & & & \\ & 1 & & & & \\ & & 1 & & & \\ & & & 1 & & \\ & & & & 1 & \\ & & & & & 1 \end{bmatrix}_{6\times 6}, D_{11} = [0]_{7\times 6} \tag{6.37}$$

while A and B_2 change inside the following polytope:

$$\Omega_{(A,B_2)} \overset{\Delta}{=} \left\{ (A, B_2) \middle| (A, B_2) = \sum_{i=1}^{N} \alpha_i (A, B_2); \alpha_i \geq 0, \sum_{i=1}^{N} \alpha_i = 1 \right\} \tag{6.38}$$

Consider A and B_2 with 30% uncertainty; the vertex of the polytope is $[0.7A \quad 0.7B_2]$, $[0.7A \quad 1.3B_2]$, $[1.3A \quad 0.7B_2]$, $[1.3A \quad 1.3B_2]$. Using Matlab to solve (6.29) to (6.32), we can obtain the parameters of the state feedback controller.

Three kinds of faults are considered. Fault type 1 is that machine 1 input mechanical power decreases 0.3 p.u. at 1 s and machine 2 input mechanical power increases 0.3 p.u. at the same time. Fault type 2 is to apply a three-phase to ground fault at machine 1 and recover after 0.08 s. Fault type 3 is to apply a three-phase to ground fault on one of the ties between 13 and 101 and recover after 0.08 s and the 3 s after the fault the voltage on machine 1 decreases to 0.75 p.u. and recovers after 0.5 s.

6.2.2.3 Testing Results and Analysis

The frequency domain result is given in Figure 6.2. The controller is designed under normal operating conditions. Figure 6.2 shows the location of both open-loop and closed-loop eigenvalues while power transferred by the ties is 400 MW,

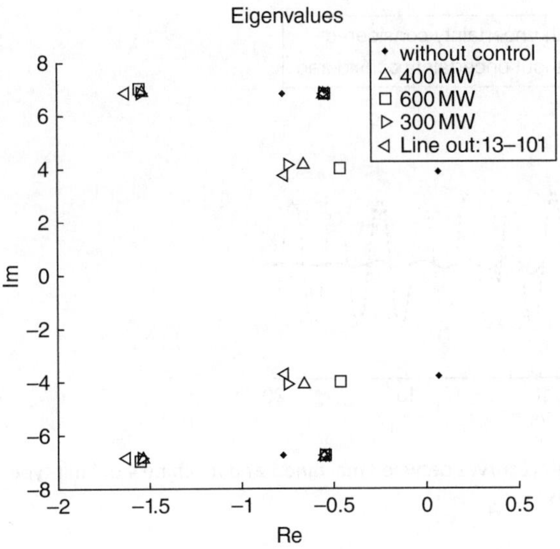

Figure 6.2 Open-loop and closed-loop eigenvalues of the system.

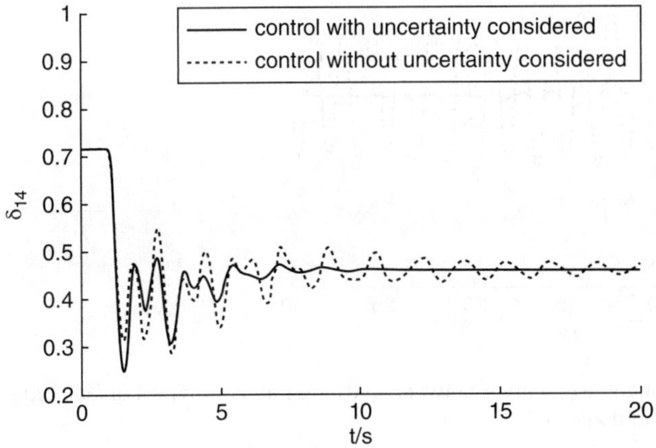

Figure 6.3 The relative rotor angle δ_{14} curves between machine 1 and machine 4 of fault type 1 under given operating conditions.

600 MW, 300 MW and lines out from there; one of the tie lines is between 13 and 101. As the closed-loop eigenvalues are pushed to the stable left plane, it shows that the controller has strong robustness for different operating conditions. Simulation results of different fault types under different operating conditions are given below. The simulation results from Figures 6.3 to 6.5 indicate that taking uncertainty into account during the design of the controller makes it fit different operating conditions well. The closed-loop system can still satisfy the robust need under severe operating conditions. In the simulation three different types of fault are discussed; the system with a feedback controller is suitable for different operating conditions and the oscillations are well damped.

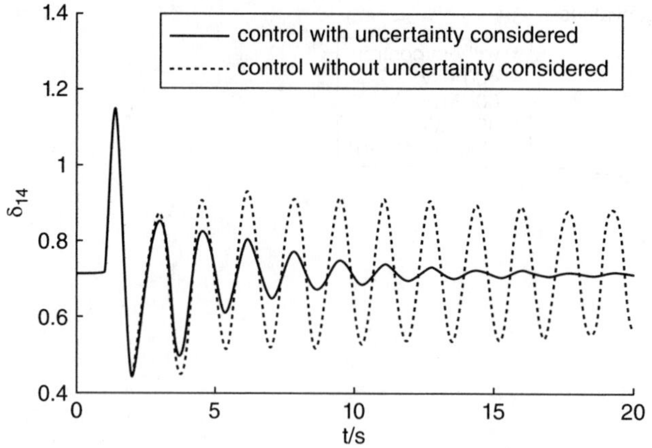

Figure 6.4 The relative rotor angle δ_{14} curves between machine 1 and machine 4 of fault type 2 under given operating conditions.

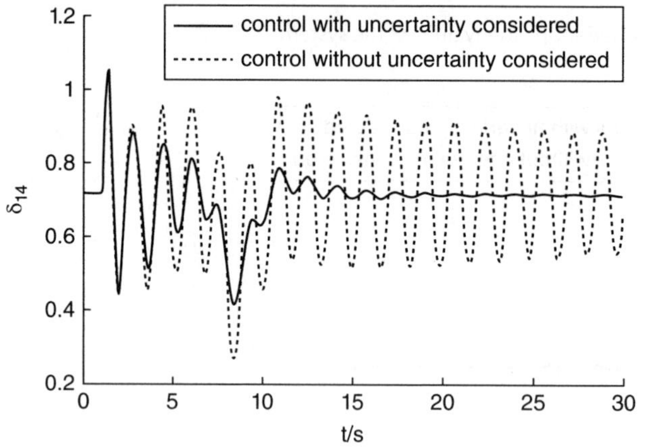

Figure 6.5 The relative rotor angle δ_{14} curves between machine 1 and machine 4 of fault type 3 under given operating conditions.

6.3 Optimal Robust Control

6.3.1 Wide-Area Damping Robust Control Based on Nonconvex Stable Region

In view of the low frequency, low damping ratio and long duration characteristics of the inter-area oscillation mode, the stable region is first extended by the eigenvalue shifted factor in this section. Furthermore, a nonconvex stable region is designed that can stabilize the system rapidly. Then a mixed H_2/H_∞ multiobjective robust control strategy based on the nonconvex stable region is proposed considering the perturbation and system uncertainty. Compared with the traditional H_2/H_∞ control strategy, the proposed strategy is more effective and robust.

6.3.1.1 H_2/H_∞ Robust Control Strategy Based on the Nonconvex Stable Region

Nonconvex Stable Region

Lemma 6.1 [5]. The Lyapunov first method. For the plant $\sum:(\boldsymbol{A}, \boldsymbol{B}, \boldsymbol{C}, \boldsymbol{D})$,

$$\begin{aligned} \dot{\boldsymbol{x}} &= \boldsymbol{A}\boldsymbol{x} + \boldsymbol{B}\boldsymbol{u} \\ \boldsymbol{y} &= \boldsymbol{C}\boldsymbol{x} + \boldsymbol{D}\boldsymbol{u} \end{aligned} \tag{6.39}$$

where $\boldsymbol{x}$ is the system state vector, $\boldsymbol{u}$ is the control input vector, $\boldsymbol{y}$ is the output vector, $\boldsymbol{A}$ is the system state matrix, $\boldsymbol{B}$ is the control input matrix, $\boldsymbol{C}$ is the output matrix, and $\boldsymbol{D}$ is the feedforward matrix.

In the equilibrium state where $\boldsymbol{x}_e = 0$, the necessary and sufficient condition for the system to be asymptotically stable is that all eigenvalues of matrix **A** have negative real parts.

Lemma 6.2 [6]. The fact that all the eigenvalues of matrix $\boldsymbol{A} \in \boldsymbol{R}^{n \times n}$ have negative real parts equals the fact that there is a symmetric matrix $\boldsymbol{X} > \boldsymbol{0}$ that satisfies $\boldsymbol{A}\boldsymbol{X} + \boldsymbol{X}\boldsymbol{A}^T < \boldsymbol{0}$.

The oscillation modes of the system can be shifted to the stable region in the left half-plane by Lemmas 6.1 and 6.2. However, this method cannot guarantee enough damping for all the oscillation modes. Thus, according to the Gutman theorem, as long as there is a positive definite symmetric matrix $\boldsymbol{X}$ that meets equation (6.40), all eigenvalues of matrix $\boldsymbol{A}$ should be in the stable region $\boldsymbol{M}$ described by equation (6.41):

$$\sum_{k,l} c_{kl} \boldsymbol{A}^k \boldsymbol{X} \left(\boldsymbol{A}^T\right)^l < 0 \tag{6.40}$$

$$\boldsymbol{M} = \left\{ z \in \mathbf{C}: \sum_{0 \le k, l \le m} c_{kl} z^k \bar{z}^l < 0 \right\} \tag{6.41}$$

where z is any point on the complex plane, $\bar{z}$ is the conjugate of z, **M** is the symmetrical matrix, c_{kl} is the kth row and ith column element, and m is the order of **M**.

By the Gutman theorem, the stable region can be extended from the left half-plane to a previously set region, which not only improves the damping performance of the system but also makes choosing of the stable region more flexible. However, it should be noted that the regions solved using this theorem are limited to convex regions. As for nonconvex regions, the stability of the regions cannot be determined since there is no positive symmetric matrix **X** that meets equation (6.40). To solve this problem, the eigenvalue shifted factor is used in this paper to extend the stable region. Then a non-convex region R_1 in plane λ is designed, shown in the shaded area in Figure 6.6(a). The corresponding set of this area $\boldsymbol{M}_{R_1}$ is as follows:

$$\boldsymbol{M}_{R_1} = \left\{ \boldsymbol{z} \in \mathbf{C}: \left|\boldsymbol{z} - \boldsymbol{c}_0\right| - \boldsymbol{r} > 0, \mathrm{Re}(\boldsymbol{z}) < 0 \right\} \tag{6.42}$$

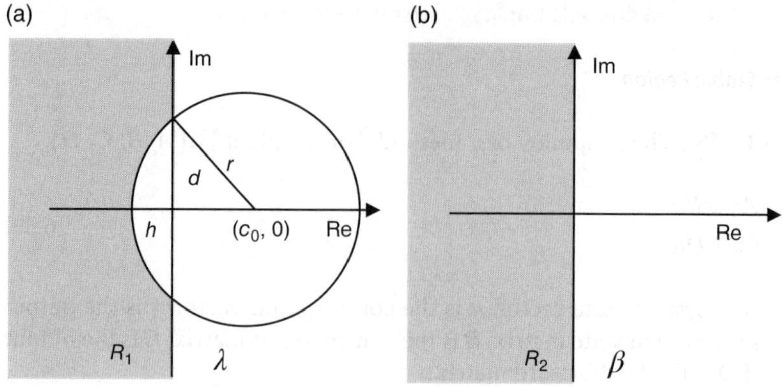

Figure 6.6 The mapping of complex planes.

where r is the radius, $(c_0, 0)$ are the coordinates of the center, and $\boldsymbol{d} = \sqrt{\boldsymbol{r}^2 - \boldsymbol{c}_0^2}$ is the intersection of the nonconvex region and the imaginary axis.

The system regulation time is inversely proportional to the real part of the closed-loop eigenvalue. The farther the eigenvalue is from the imaginary axis, the shorter is the system regulation time [7]. Meanwhile, the amplitude will be larger as the inter-area oscillation frequency becomes lower, since the number of generators involved increases and more energy is focused on the tie line. Therefore the non-convex region R_1 is selected in this paper. On the one hand, the inter-area modes with lower frequencies can be more easily stabilized since they are farther from the imaginary axis and have larger damping ratios and shorter regulation times. On the other hand, the inter-area modes with higher frequencies can still meet the requirement of damping control even with a slight drop in their damping ratios. Normally, the interval of an inter-area oscillation mode is 0.1–0.8 Hz. To ensure that oscillation modes with the frequency of 0.8 Hz are able to meet the requirement of damping control, d should be 7.6 and h should be 0.9, according to $d = 2\pi f$ and $h = r - c_0$. Then the corresponding boundary damping is 0.1, which meets the requirement that damping ratios of inter-area oscillation modes should be no less than 0.1–0.3 [8].

H_2/H_∞ Robust Control Strategy Based on the Nonconvex Stable Region The linear time-invariant system of the mixed H_2/H_∞ multiobjective robust control is given by [9]

$$\begin{cases} \dot{\boldsymbol{x}}(t) = \boldsymbol{A}\boldsymbol{x}(t) + \boldsymbol{B}_1\boldsymbol{w}(t) + \boldsymbol{B}\boldsymbol{u}(t) \\ \boldsymbol{z}_\infty = \boldsymbol{C}_1\boldsymbol{x}(t) + \boldsymbol{D}_{11}\boldsymbol{w}(t) + \boldsymbol{D}_{12}\boldsymbol{u}(t) \\ \boldsymbol{z}_2 = \boldsymbol{C}_2\boldsymbol{x}(t) + \boldsymbol{D}_{21}\boldsymbol{w}(t) + \boldsymbol{D}_{22}\boldsymbol{u}(t) \end{cases} \tag{6.43}$$

where $\boldsymbol{x}$ is the system state vector, $\boldsymbol{u}$ is the control input vector, $\boldsymbol{w}$ is the external perturbation input vector, $\boldsymbol{z}_\infty$ and $\boldsymbol{z}_2$ are the controlled vectors related to the performance indexes of H_∞ and H_2 respectively, $\boldsymbol{A}$ is the system state matrix, $\boldsymbol{B}_1$ is the gain matrix of input $\boldsymbol{w}$, $\boldsymbol{B}$ is the control input matrix, $\boldsymbol{C}_1$, $\boldsymbol{D}_{11}$, and $\boldsymbol{D}_{12}$ are the weight matrixes of state variables, perturbation input, and control input related to the performance indicator of H$_\infty$, respectively, and C_2, D_{21}, and D_{22} are the weight matrixes of state variables, perturbation input, and control input related to the performance index of H$_2$, respectively.

The feedback control law based on the eigenvalue transfer factor is as follows:

$$u(t) = FB^T x(t) \tag{6.44}$$

where F denotes the eigenvalue shifted factor since it acts as a weight coefficient and is related to the mapping.

Apply equation (6.44) to equation (6.43), so that the closed-loop system equation can be gained:

$$\begin{cases} \dot{x}(t) = \left(A + BFB^T\right)x(t) + B_1 w(t) \\ z_\infty = \left(C_1 + D_{12}FB^T\right)x(t) + D_{11}w(t) \\ z_2 = \left(C_2 + D_{22}FB^T\right)x(t) + D_{21}w(t) \end{cases} \tag{6.45}$$

The eigenvalue equation is given by

$$\left|\lambda I - \left(A + BFB^T\right)\right| = 0 \tag{6.46}$$

The constraints guaranteeing that the closed-loop eigenvalues are stable in the nonconvex region R_1 are difficult to obtain in the plane λ. Therefore, region R_1 is mapped to region R_2, that is, the left half-plane of β, shown in Figure 6.6(b). The mapping relationship between region R_1 and R_2 is as follows:

$$\beta = \frac{r - c_0 + \lambda}{r + c_0 - \lambda}, \ \lambda = \frac{r(\beta - 1)}{\beta + 1} + c_0 \tag{6.47}$$

Based on (6.47), the eigenvalue equation of the closed-loop system on plane β is given by

$$\left|U(F)\beta + V(F)\right| = 0 \tag{6.48}$$

where $U(F)$ and $V(F)$ are both coefficient matrices expressed in terms of variable F and $U(F) = (r + c_0)I - (A + BFB^T)$ and $V(F) = -(r - c_0)I - (A + BFB^T)$.

It can be deduced from equation (6.47) that the sufficient and necessary condition for the eigenvalues in plane λ to fall in region R_1 is that the corresponding eigenvalues in plane β fall in the left half-plane R_2. It can be deduced from Lemma 6.1 that the sufficient and necessary condition for the system to be stable is that the real parts of all the eigenvalues are negative, that is, there exists a nonnegative definite eigenvalue shifted factor matrix F that satisfies

$$U^{-1}(F)V(F) > 0 \tag{6.49}$$

Thus, the condition for the closed-loop system to be asymptotically stable is

$$G = \left\{F \in R^{n \times n} : V(F) > 0 \| U(F) < 0\right\} \tag{6.50}$$

From Lemma 6.2, it can be deduced that the equivalent condition of equation (6.50) is that there is a symmetrical matrix $X > 0$ that meets the following requirement:

$$G = \left\{ F \in R^{n \times n} : G_1 < 0 \| G_2 > 0 \right\} \tag{6.51}$$

where $G_1 = VX + XV^T, G_2 = UX + XU^T$

It should be noted that the points on the imaginary axis outside the arc in Figure 6.6(a) correspond to the undamped inner-area oscillation modes. A detailed model is utilized in this paper so that all of the inner-area oscillation modes can be damped slightly and stabilized. In addition, this paper mainly focuses on the wide-area damping control of inter-area oscillation. The inner-area oscillation modes with higher frequencies can be damped with the help of local controllers (e.g., the PSS). Therefore, the points on the imaginary axis outside the arc in Figure 6.6(a) are not considered, which means that the stable region in Figure 6.6(b) only corresponds to the arc cut by the imaginary axis in Figure 6.6(a).

In order to apply multiobjective robust control to the system, the following conditions need to be satisfied apart from the ones that enable the closed-loop system to be asymptotically stable:

1) H_∞ performance: when w is regarded as a disturbance signal with finite energy, the H_2 norm of T_{wz_∞}, which is the closed-loop transfer function from w to z_∞, should not exceed the given upper bound γ, that is, $\|T_{wz_\infty}(s)\| < \gamma$, in order to ensure that the closed-loop system is robust and stable.
2) H_2 performance: when w is regarded as a white noise signal with unit spectral density, the H_2 norm of $T_{wz_2}(s)$, which is the closed-loop transfer function from w to z_2, should not exceed the given upper bound η, that is, $\|T_{wz_2}(s)\| < \eta$, in order to ensure that the system performance measured by the H_2 norm is steady and good.

The sufficient and necessary condition that meets both conditions 1 and 2 is as follows:

$$\begin{bmatrix} A_{cl}X + XA_{cl}^T & B_1 & X\left(C_1 + D_{12}FB^T\right)^T \\ B_1^T & -\gamma I & D_{11}^T \\ \left(C_1 + D_{12}FB^T\right)X & D_{11} & -\gamma I \end{bmatrix} < 0 \tag{6.52}$$

$$\begin{bmatrix} Q & \left(C_2 + D_{22}FB^T\right)X \\ X\left(C_2 + D_{22}FB^T\right)^T & Q^T \end{bmatrix} > 0 \tag{6.53}$$

$$Trace(Q) < \eta^2 \tag{6.54}$$

where $A_{cl} = A + BFB^T$.

The H_2/H_∞ robust control based on the nonconvex stable region can be transformed to optimization of the linear inequalities in (6.55). The performance indicator $\alpha\gamma + \beta Trace(\boldsymbol{Q})$ is the weighted combination of H_2 and H_∞ performances, where α and β are the weights of the H_2 and H_∞ performances:

$$
\begin{aligned}
&\min_{\gamma,X,N,Q} \alpha\gamma + \beta Trace(\boldsymbol{Q}) \\
&G_1 < 0 \| G_2 > 0 \\
&\begin{bmatrix} \boldsymbol{Q} & C_2\boldsymbol{X} + D_{22}\boldsymbol{NB}^T \\ \boldsymbol{X}C_2 + D_{22}\boldsymbol{N}^T\boldsymbol{B}^T & \boldsymbol{Q}^T \end{bmatrix} > \boldsymbol{0} \\
&\begin{bmatrix} \boldsymbol{AX} + \boldsymbol{XA}^T + \boldsymbol{BNB}^T + \boldsymbol{BN}^T\boldsymbol{B}^T & \boldsymbol{B_1} & \boldsymbol{X}C_1^T + \boldsymbol{BN}^T D_{12}^T \\ \boldsymbol{B}_1^T & -\gamma\boldsymbol{I} & \boldsymbol{D}_{11}^T \\ C_1\boldsymbol{X} + \boldsymbol{D}_{12}\boldsymbol{NB}^T & \boldsymbol{D}_{11} & -\gamma\boldsymbol{I} \end{bmatrix} < \boldsymbol{0} \\
&Trace(\boldsymbol{Q}) < \eta^2
\end{aligned}
\tag{6.55}
$$

where $G_1 = -2(r - c_0)\boldsymbol{X} - \boldsymbol{AX} - \boldsymbol{XA}^T - \boldsymbol{BNB}^T - \boldsymbol{BN}^T\boldsymbol{B}^T$ and $\mathbf{G}_2 = 2(r + c_0)\boldsymbol{X} - \boldsymbol{AX} - \boldsymbol{XA}^T - \boldsymbol{BNB}^T - \boldsymbol{BN}^T\boldsymbol{B}^T$.

Set

$$
\begin{cases} \boldsymbol{FB}^T\boldsymbol{X} = \boldsymbol{NB}^T \\ \boldsymbol{MB}^T = \boldsymbol{B}^T\boldsymbol{X} \end{cases}
$$

where $\boldsymbol{N}$ and $\boldsymbol{X}$ can be gained by calculating the constraint conditions above; then $\boldsymbol{M} = \boldsymbol{B}^T\boldsymbol{XB}(\boldsymbol{B}^T\boldsymbol{B})^{-1}$ and the eigenvalue transfer factor $\boldsymbol{F} = \boldsymbol{NM}^{-1}$. Apply $\boldsymbol{F}$ to equation (6.44) so that the feedback control vector is $\boldsymbol{u}(t) = \boldsymbol{NM}^{-1}\boldsymbol{B}^T\boldsymbol{x}(t)$.

6.3.1.2 Practical Examples

Four-Machine Test System The design of the robust H_2/H_∞ damping controller based on the nonconvex stable region is conducted in a two-area four-machine system using the thyristor controlled series capacitor (TCSC), and the damping performance and robust performance are analyzed. The generator applies the detailed model. The excitation system applies fast excitation and the load under the benchmark load model is a mixed model of 50% constant impedance and 50% constant current. The TCSC is between tie lines 8 and 9 with the compensation degree of 40%. The structure of the supplementary TCSC controller is shown in Figure 6.7, where $\boldsymbol{T} = 0.05$ is the first-order inertia constant, $\boldsymbol{T}_a = 0.4$ is the lead time constant, $\boldsymbol{T}_b = 0.02$ is the lag time constant, and $X_{\max} = 3$ and $X_{\min} = -3$ are the maximum and minimum outputs. This paper focuses on the damping of

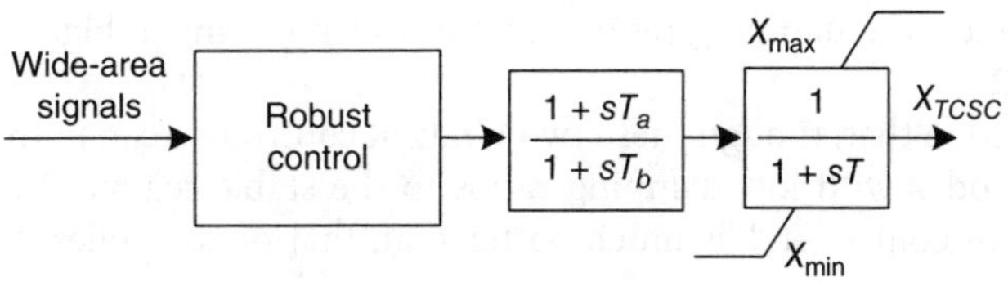

Figure 6.7 Structure of the TCSC supplementary controller.

Table 6.5 Dominant modes in a four-machine system without controllers.

Mode	Frequency (Hz)	Damping ratio	Mode type
1	0.4926	0.0086	Inter-area
2	1.0854	0.0773	Inner-area
3	1.0948	0.0724	Inner-area

inter-area low-frequency oscillation by the robust H_2/H_∞ control based on the nonconvex stable region using TCSC. Generator 4 is regarded as the reference generator and the wide-area feedback signals of TCSC are the speed differences of the other generators from generator 4.

When the system operates in open-loop, the dominant oscillation modes of the system and its damping ratios can be obtained by using the mode analysis method, shown in Table 6.5. It can be seen that there are two inner-area modes and one inter-area mode with weak damping. When the system is disturbed, stability cannot be regained rapidly. If the oscillation is not effectively damped, it will cause severe damage to the system. Simulation tests are carried out using the DSA tools.

In comparison with the traditional H_2/H_∞ controller, the effectiveness and superiority of the proposed method is examined in the following cases:

1) The transmission power on the tie line is 421 MW in the steady state. At $t = 0.1$ s, a temporary three-phase short-circuit fault occurs on line 4–13. The fault is cleared at $t = 0.2$ s without changing the topology.
2) The transmission power on the tie line is 421 MW in the steady state. At $t = 0.1$ s, a permanent three-phase short-circuit fault occurs on line 4–13. The fault line is tripped out at $t = 0.2$ s.
3) The load consists of 60% constant impedance and 40% constant current. At $t = 0.1$ s, a temporary three-phase short-circuit fault occurs on line 9–10. The fault is cleared at $t = 0.15$ s without changing the topology.
4) The load consists of 40% constant impedance and 60% constant current. At $t = 0.1$ s, a temporary three-phase short-circuit fault occurs on line 9–10. The fault is cleared at $t = 0.15$ s without changing the topology.

When the system operates in closed-loop, with the H_2/H_∞ robust damping controller based on the nonconvex stable region and the traditional H_2/H_∞ robust damping controller applied, respectively, the oscillation modes of the system are obtained, shown in Table 6.6. It can be seen that for mode 1, the damping ratio with controller 1 is 27.83%, which is more than two times that of controller 2. For the other three modes, the damping ratio with controller 1 is much bigger than that of controller 2.

It is indicated in Table 6.6 that, though these two kinds of controllers can both damp the oscillation modes with low damping ratios to the stable region, the damping performance of controller 1 is much better than that of controller 2, that is the robust H_2/H_∞ control based on the nonconvex stable region is more effective in damping inter-area low-frequency oscillations. In the four cases of

Table 6.6 Dominant inter-area modes of the test system operating in open-loop and closed-loop systems.

Cases		Without controller	Controller 1	Controller 2
1	Frequency (Hz)	0.4926	0.4450	0.4664
	Damp ratio	0.0086	0.2783	0.1211
2	Frequency (Hz)	0.5274	0.5321	0.5352
	Damp ratio	0.0089	0.2251	0.1158
3	Frequency (Hz)	0.4937	0.4796	0.4985
	Damp ratio	0.0093	0.2152	0.1268
4	Frequency (Hz)	0.4922	0.4865	0.4769
	Damp ratio	0.0091	0.2825	0.1376

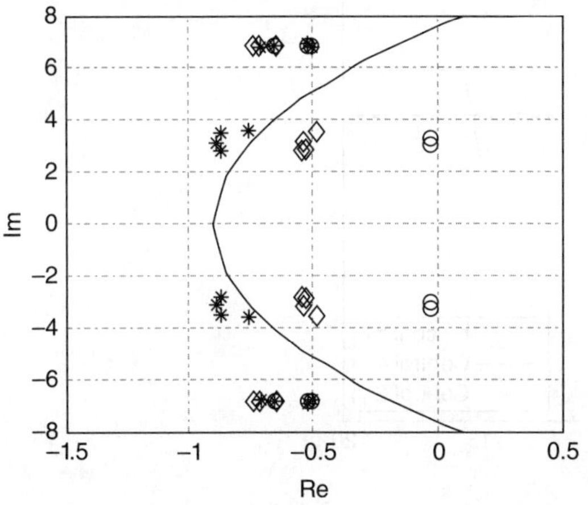

Figure 6.8 Eigenvalues of open-loop and closed-loop systems.

perturbations mentioned above, the eigenvalues of the closed-loop system all lay in the nonconvex stable region, as shown in Figure 6.8, where the circles represent eigenvalues of the open-loop system and the stars and diamonds represent eigenvalues of the closed-loop system applying controller 1 and controller 2, respectively.

Time-domain simulations of the relative angular difference between the generators of the two areas and the tie line power in the four cases are shown in Figures 6.9 to 6.12.

The dotted curves represent the system dynamic responses without controllers. The dot–dash curves represent dynamic responses with a traditional H_2/H_∞ damping controller and the solid curves represent the dynamic responses with the H_2/H_∞ damping controller based in the nonconvex stable region. The oscillations are able to be damped within 15–17 s with controller 2. Compared with controller 2, controller 1 is able to damp the oscillations within 5–7 s. It can be seen that the H_2/H_∞ robust damping controller based on the nonconvex stable

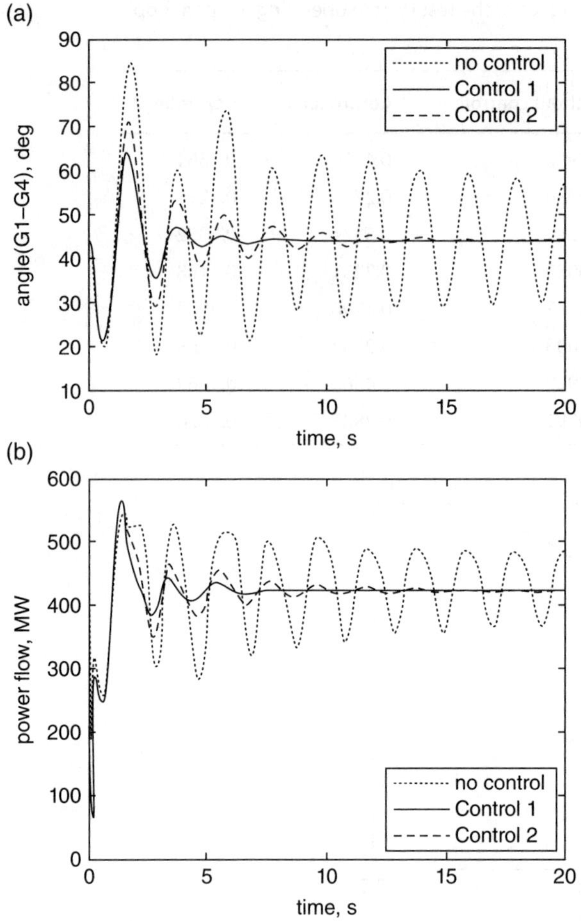

Figure 6.9 System dynamic responses in the first case.

region is able to damp inter-area low-frequency oscillations quickly and effectively under different perturbations, with better robustness than the traditional H_2/H_∞ robust controller.

Sixteen-Machine Test System In order to further verify the validity and generality of the proposed control method, the 16-machine 68-bus system is taken for simulation analysis. There is an HVDC line between bus 41 and 42, an SVC at bus 17, and an energy storage device at bus 52.

Modal analysis of the open-loop system is conducted and the dominant modes of low-frequency oscillation are gained, as well as the corresponding damping ratios of open-loop, shown in Table 6.7. It can be seen that there are four oscillation modes with low damping ratios in the system. The damping effects of the closed-loop system applying controller 1 and controller 2, respectively, are shown in Table 6.7. It can be seen that controller 1 displays a better performance than controller 2.

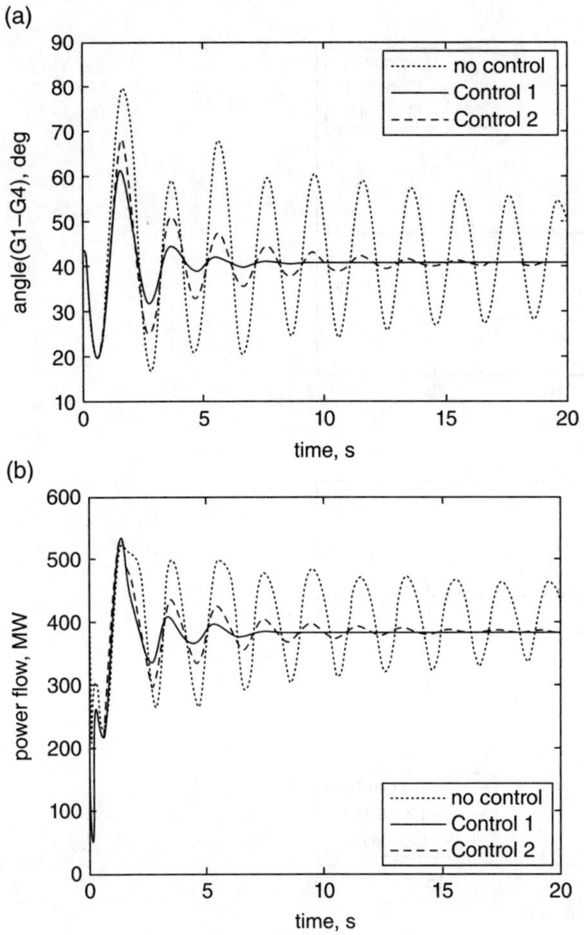

Figure 6.10 System dynamic responses in the second case.

Suppose the system is in a normal operation condition (which is named as case 5 in this paper) and at $t = 0.1$ s, a three-phase fault occurs on line 8–9. The fault lasts for 50 ms and is cleared at $t = 0.15$ s. The dynamic responses of the system are shown in Figure 6.13. The dynamic response without control is represented by the dashed line. The dynamic response with controller 1 as the damping control is represented by the solid line and the dynamic response with controller 2 is represented by the dash–dot line. Figure 6.13(a) shows the angular difference between generators 1 and 14 and Figure 6.13(b) shows the power flow on the tie line 1–2. It can be seen that controller 1 is more effective than controller 2.

The robustness of controller 1 against generation rescheduling is then investigated. Suppose the output of generator 16 drops to 80% of the rated capacity and a three-phase fault occurs on line 8–9 (which is named as case 6); the dynamic responses of the system are shown in Figure 6.14. It can be seen that the mixed

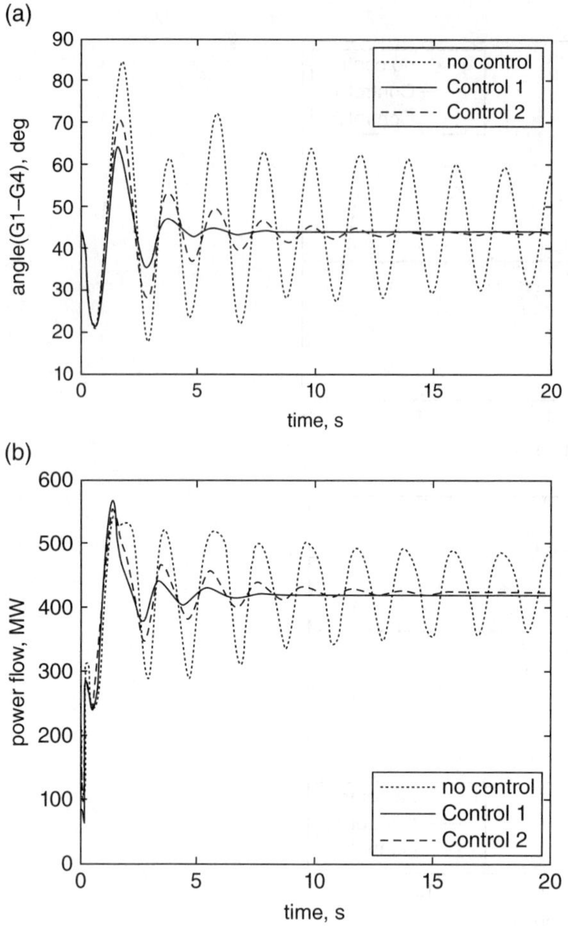

Figure 6.11 System dynamic responses in the third case.

H_2/H_∞ multiobjective robust control based on the nonconvex stable region can still quickly restore the system to a new stable state in the case of generation rescheduling.

6.3.2 Wide-Area Damping Robust H_2/H_∞ Control Strategy Based on Perfect Regulation

The application of PMU measurements and perfect regulation with robust control to damp inter-area oscillations is investigated. Combined with the robust control theory, a perfect regulation control with robust performance is obtained. The closed-loop eigenvalues gradually increase from the underdamping or negative damping to positive damping. A four-machine 13-bus study system is considered. The damping performance of the controllers is examined in the frequency and time domains. The simulation results show that the control scheme is able to robustly damp out inter-area oscillations following possible disturbances.

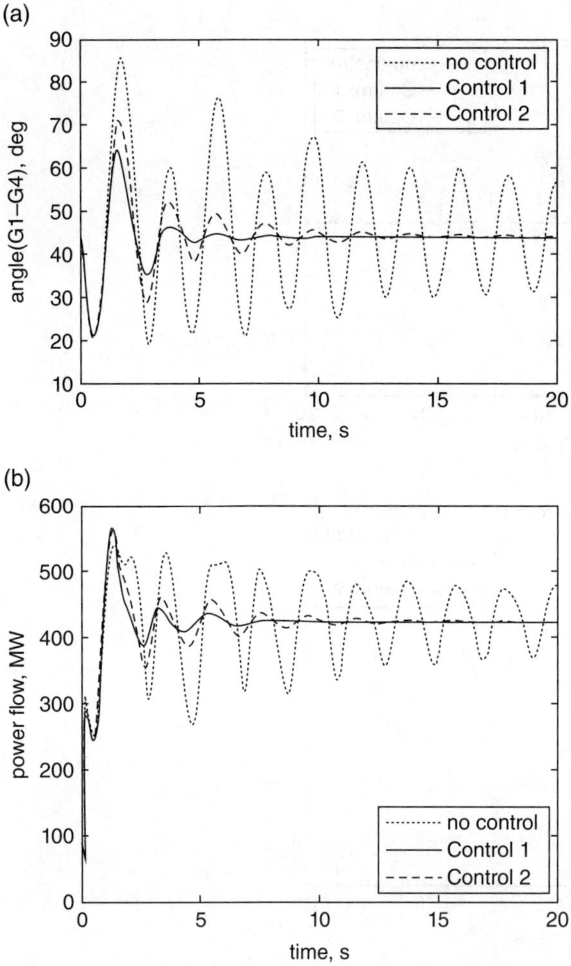

Figure 6.12 System dynamic responses in the fourth case.

Table 6.7 Dominant oscillation modes of the test system operating in open-loop and closed-loop systems.

Mode		Open-loop	Controller 1	Controller 2
1	Frequency (Hz)	0.4286	0.4218	0.4239
	Damping ratio	0.0038	0.5325	0.1496
2	Frequency (Hz)	0.5373	0.5219	0.5371
	Damping ratio	0.0159	0.2269	0.1458
3	Frequency (Hz)	0.6447	0.6653	0.6541
	Damping ratio	0.0428	0.1762	0.0938
4	Frequency (Hz)	0.7957	0.7834	0.7746
	Damping ratio	0.0421	0.1746	0.0732

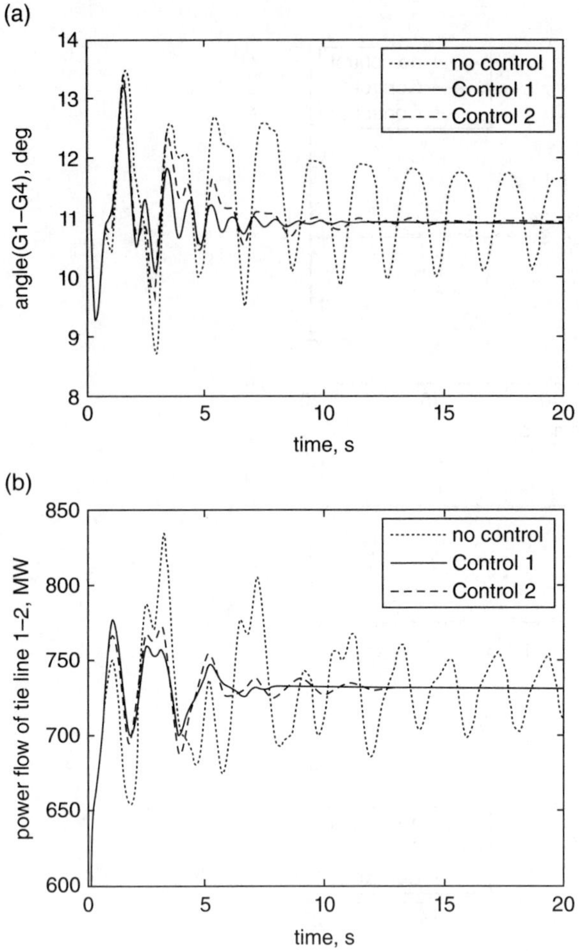

Figure 6.13 System dynamic responses in case 5.

6.3.2.1 Mathematical Background

The nominal state space form of a multimachine power system [10] is as shown below:

$$\begin{aligned}\dot{\mathbf{X}} &= \mathbf{AX} + \mathbf{Bu} \\ \mathbf{Y} &= \mathbf{CX}\end{aligned} \tag{6.56}$$

where

$$\mathbf{A} = \begin{bmatrix} \mathbf{0} & \omega_0\mathbf{I} & \mathbf{0} \\ \mathbf{A}_{21} & \mathbf{A}_{22} & \mathbf{A}_{23} \\ \mathbf{A}_{31} & \mathbf{0} & \mathbf{A}_{33} \end{bmatrix}, \mathbf{B} = \begin{bmatrix} \mathbf{0} \\ \mathbf{0} \\ \mathbf{B}_3 \end{bmatrix}, \mathbf{x} = \begin{bmatrix} \delta \\ \omega \\ \mathbf{z} \end{bmatrix} \tag{6.57}$$

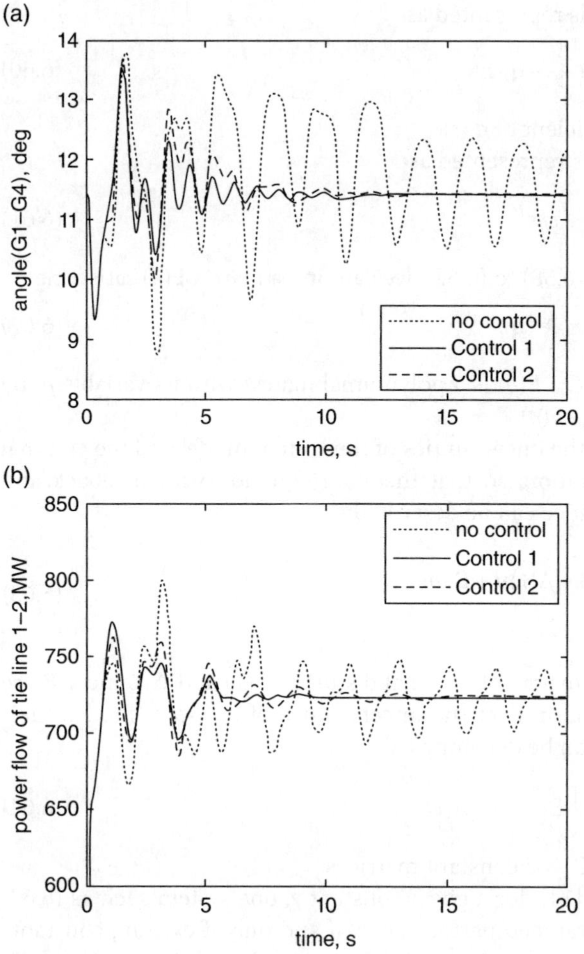

Figure 6.14 System dynamic responses in the case 6.

where $\boldsymbol{\delta}$, $\boldsymbol{\omega}$ are differences from values at the reference machine – a large external equivalent – and z is the state associated with the controls chosen, Exciters, PSSs, DC lines, SVCs, load modulation, FACTS, etc. Matrices are formed numerically with perturbations from a load flow equilibrium using Matlab suite PST. For some controls, $\mathbf{A}_{31} = \mathbf{0}$ and $\mathbf{A}_{33}$ is diagonal. Most controls have $\mathbf{A}_{32} = \mathbf{0}$ [11].

According to reference [12], for a system (**A, B, C**) there exists a unique polynomial matrix $\mathbf{Q}(s)$ [13] such that

$$\lim_{s\to\infty} \mathbf{C}(s\mathbf{I}-\mathbf{A})^{-1}\mathbf{B}*\mathbf{Q}(s) = \mathbf{L} \tag{6.58}$$

where $\mathbf{L}$ is a nonsingular constant matrix. Let

$$\mathbf{BQ}(s) = (s\mathbf{I}-\mathbf{A})\mathbf{T}(s) + \mathbf{W} \tag{6.59}$$

where $\mathbf{T}(s)$ is a polynomial matrix and $\mathbf{W}$ is constant matrix.

The ith column of $\mathbf{Q}(s)$ is represented as

$$\mathbf{q}_i(s) = \mathbf{q}_{i1}s + \mathbf{q}_{i2}s^2 + \cdots + \mathbf{q}_{ij}s^j \tag{6.60}$$

where j is the largest coefficient of $\mathbf{q}_i(s)$.

The ith column of $\mathbf{T}(s)$ is represented as

$$\mathbf{t}_i(s) = \mathbf{t}_{i1} + \mathbf{t}_{i2}s + \cdots + \mathbf{t}_{ij}s^{j-1} \tag{6.61}$$

Based on the equations (6.58) to (6.62), we can find a matrix $\mathbf{F}(\rho)$ satisfying

$$\mathbf{F}(\rho)\mathbf{t}_{ij} = \mathbf{q}_{ij-1} - a_{ij}\rho^{j-k+1}\mathbf{q}_{ij} \tag{6.62}$$

where $k \subseteq j$, $\mathbf{q}_{i0} = \mathbf{0}$, $\alpha_{ij} \in C^-$. $\mathbf{F}(\rho)$ is a polynomial matrix with its variable ρ. By regulating ρ we can change $\mathbf{F}(\rho)$.

In reality, we must take the uncertainties of the system model and the external disturbance into consideration, so that the state equation with feedback and uncertainties and disturbance can be described:

$$\begin{aligned} \dot{\mathbf{X}} &= (\mathbf{A} + \Delta\mathbf{A})\mathbf{X} + (\mathbf{B} + \Delta\mathbf{B})\mathbf{u} + \mathbf{B_d}\omega \\ \mathbf{Y} &= \mathbf{CX} \end{aligned} \tag{6.63}$$

where $\mathbf{B_d}$ is the weighted matrix of external disturbance ω and $\mathbf{\Delta A}$ and $\mathbf{\Delta B}$ are uncertain matrices that can reflect the uncertainties of a system model; their norms are bounded and can be decomposed as

$$\begin{bmatrix} \Delta\mathbf{A} & \Delta\mathbf{B} \end{bmatrix} = \mathbf{HF}\begin{bmatrix} \mathbf{E}_1 & \mathbf{E}_2 \end{bmatrix} \tag{6.64}$$

where $\mathbf{F}^{\mathrm{T}}\mathbf{F} \le \mathbf{I}$ and $\mathbf{E}_1$ and $\mathbf{E}_2$ are constant matrices.

According to reference [14], for a given constant γ, one system such as (6.63) possesses the $\mathrm{H}_2/\mathrm{H}_\infty$ guaranteed performance, if and only if existing constants $\alpha > 0, \beta > 0$, and symmetric positive definite matrix $\mathbf{P}$ and matrix $\mathbf{V}$ satisfy the inequality (6.65). If the inequality (6.65) has the feasible solution $\alpha, \beta, \mathbf{P}, \mathbf{V}$, $\mathbf{u} = \mathbf{VP}^{-1}\mathbf{x}$ is a guaranteed performance controller, the closed loop H_∞ norm is smaller than γ, and the H_2 norm has an up-limitation $J = Trace(\mathbf{B_d}^{\mathrm{T}}\mathbf{P}^{-1}\mathbf{B_d})$, then

$$\begin{bmatrix} -\mathbf{X} & \mathbf{0} & (\mathbf{AP}+\mathbf{BV})^{\mathrm{T}} & (\mathbf{E}_1\mathbf{P}+\mathbf{E}_2\mathbf{V})^{\mathrm{T}} & \mathbf{CP} & \mathbf{CP} \\ \mathbf{0} & -\alpha\gamma^2\mathbf{I} & \alpha\mathbf{B_d} & \mathbf{0} & \mathbf{0} & \mathbf{0} \\ \mathbf{AP}+\mathbf{BV} & \alpha\mathbf{B_d} & -\mathbf{P}+\beta\mathbf{HH}^{\mathrm{T}} & \mathbf{0} & \mathbf{0} & \mathbf{0} \\ \mathbf{E}_1\mathbf{P}+\mathbf{E}_2\mathbf{V} & \mathbf{0} & \mathbf{0} & -\beta\mathbf{I} & \mathbf{0} & \mathbf{0} \\ \mathbf{CP} & \mathbf{0} & \mathbf{0} & \mathbf{0} & -\alpha\mathbf{I} & \mathbf{0} \\ \mathbf{CP} & \mathbf{0} & \mathbf{0} & \mathbf{0} & \mathbf{0} & -\mathbf{I} \end{bmatrix} < 0 \tag{6.65}$$

Under the premise that we have the feedback matrix $\mathbf{F}(\rho)$, using the $\mathbf{F}(\rho)\mathbf{P}$ substitute $\mathbf{V}$ in inequality (6.65), we can then solve this inequality using the Matlab/LMI toolbox to reduce the extent of variable ρ, and thus the controller will always fulfill the H_2/H_∞ guaranteed performance, allowing us to adjust ρ so that the control energy becomes smaller.

6.3.2.2 Testing Results and Analysis

A four-machine 13-bus test system of IEEE is employed for this study. The generator adopts the second-order model with damping and bus 4 and bus 14 and adopt the constant impedance load. The dominant modes of the system are shown in Table 6.8.

Frequency-Domain Results and Analysis The open- and closed-loop eigenvalues of this modified test system are shown in Figure 6.15, which shows, that by increasing in the parameter ρ, the real part of the eigenvalues in the open loop will also increase. This means that the closed-loop system has enough damping to suppress the oscillation.

Through the simulation test analysis, when the parameter $\rho \in (0,40)$, the frequency of the closed-loop dominant modes have little distortion but the damping ratio increases with ρ. When ρ changes to the extent of (0, 15), the feedback matrix $\mathbf{F}(\rho)$ will always enable the inequality (6.65) to have a solution. When $\rho = 10$, the dominant modes of the closed-loop system are as shown in Table 6.9.

Table 6.8 Dominant modes of the open-loop system.

Modes	Frequency	Damping ratio	Mode type
1	1.1989	0.0306	Inner-area
2	0.8902	0.0433	Inner-area
3	0.2710	0.1344	Inter-area

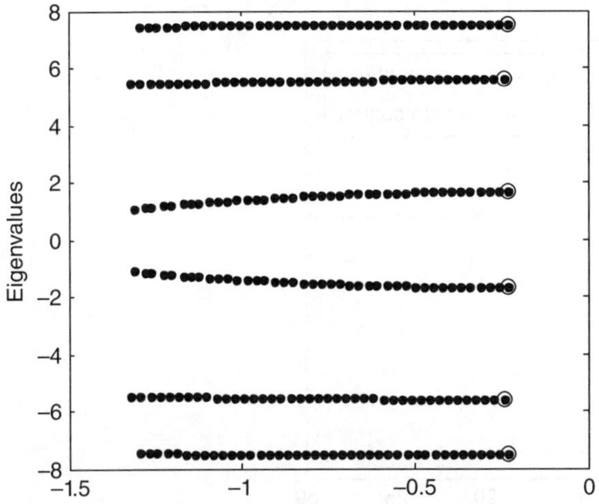

Figure 6.15 The root loci with parameter ρ increasing.

Table 6.9 Dominant modes of the closed-loop system.

Modes	Frequency	Damping ratio	Mode type
1	1.1996	0.0450	Inner-area
2	0.8888	0.0804	Inner-area
3	0.2545	0.3694	Inter-area

Time-Domain Results and Analysis

1) Case 1. In the time domain, after step changes in the all of the controlled variables are conducted, the speed of machine 1 relative to 4 without and with the control are obtained as shown in Figure 6.16.
2) Case 2. When three faults in one tie line from bus 3 to bus 101 happen at the time $t = 0.1$ s, the fault clearing time is $t = 0.2$ s, the speed of machine 1 relative to 4 without and with the control are obtained as shown in Figure 6.17, and the real power of tie line from bus 3 to bus 101 without and with the control are obtained as shown in Figure 6.18.
3) Case 3 When the structure of the system has cut one line from bus 101 to bus 13, and the system has the same fault as in Case 2, the speeds of machine 1 relative to 4 without and with the control are obtained as shown in Figure 6.19 and the real power of the tie line from bus 3 to bus 101 without and with the control are obtained as shown in Figure 6.20.

From Figures 6.16 to 6.20, it is evident that considering the external disturbance or uncertainties of the system, with the control design based on PR and H_2/H_∞ control theory, the oscillations of the system can all be damped. This example demonstrates the significant potential for using this control for stability improvement. With this control, the transmission system can operate reliably at a higher loading level without having stability problems.

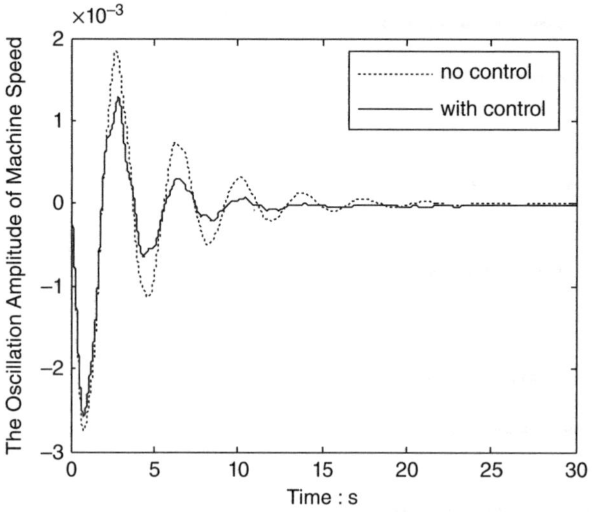

Figure 6.16 Speed of G1–G4 without and with the control.

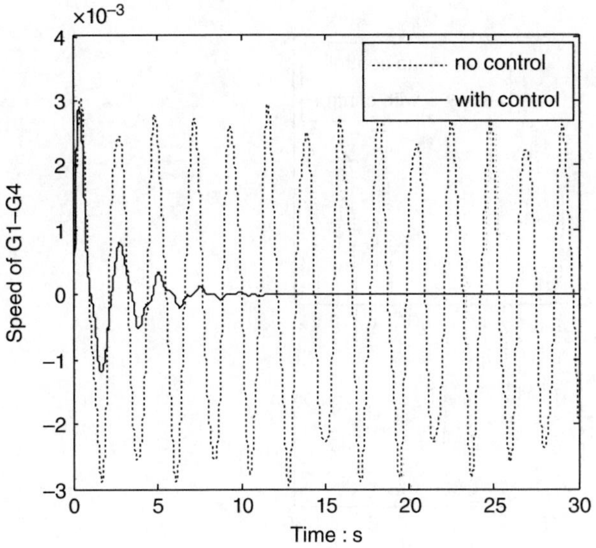

Figure 6.17 Speeds of G1–G4 without and with control.

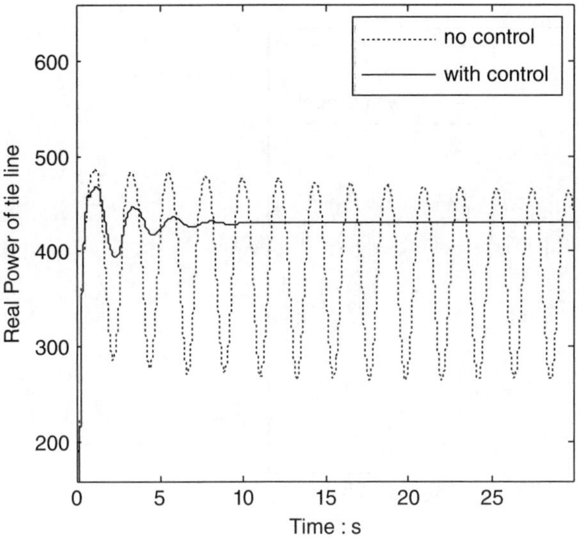

Figure 6.18 The real power of the tie line without and with control.

6.4 Error Tracking Robust Control

Traditional control theories such as linear optimal control and pole assignment are often reliant on accurate system parameters, and are incapable of achieving a good control effect when there are uncertainties in the system parameter. The H_2/H_∞ robust control utilizes the parameter perturbation method and the convex polytopic model to describe system uncertainties, and with H_2/H_∞ as the performance index, an optimal control strategy can be gained to guarantee

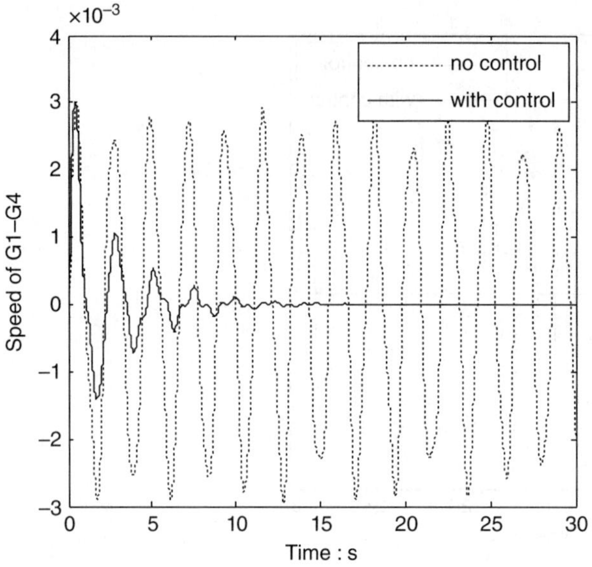

Figure 6.19 Speed of G1–G4 without and with control.

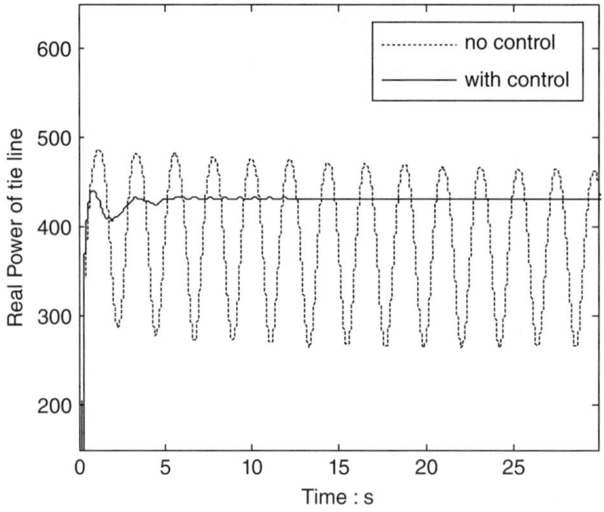

Figure 6.20 The real power of tie line without and with control.

stability of the control system under uncertain conditions. However, the robust limitation will be exhibited when a large-scale power system is operating during a wide range of changing conditions. Especially under some serious conditions, the predesigned damping controller might lose its effectiveness. The Youla parameterization control can produce tracking control of the error by applying the difference between the estimated output value and the actual input value to the control link. Thus tracking control of the output error can be realized as well as H_2/H_∞ robust control of the system. Simulation results show that the proposed method has a good performance for the damping effect and robust control.

6.4.1 Control Algorithm

Considering uncertainties caused by the parameter measurement error, estimation error, and external disturbance, the power system state equation can be written as [15]

$$\begin{cases} \dot{\boldsymbol{x}} = \boldsymbol{A}\boldsymbol{x} + \boldsymbol{B}\boldsymbol{u} + \boldsymbol{B}1\boldsymbol{\omega} \\ \boldsymbol{y} = \boldsymbol{C}\boldsymbol{x} \end{cases} \tag{6.66}$$

where $\boldsymbol{A} = \boldsymbol{A}_0 + \boldsymbol{\Delta A}$, $\boldsymbol{B} = \boldsymbol{B}_0 + \boldsymbol{\Delta B}$, $\boldsymbol{\omega}$ is the external disturbance and $\boldsymbol{B}_1$ is its weighting matrix, and $\boldsymbol{\Delta A}$ and $\boldsymbol{\Delta B}$ are unknown perturbation real matrices that reflect the uncertainties of the parameter measurement error and estimation error.

Suppose the norms of $\boldsymbol{\Delta A}$ and $\boldsymbol{\Delta B}$ are bounded and have the following form:

$$\begin{bmatrix} \Delta \boldsymbol{A} & \Delta \boldsymbol{B} \end{bmatrix} = \boldsymbol{E}\boldsymbol{F}\begin{bmatrix} \boldsymbol{H}_a & \boldsymbol{H}_b \end{bmatrix} \tag{6.67}$$

where $\boldsymbol{F}^{\mathrm{T}}\boldsymbol{F} \leq \boldsymbol{I}$, and $\boldsymbol{E}$, $\boldsymbol{H}_a$, and $\boldsymbol{H}_b$ are all known constant matrices and reflect the structure information of the uncertain parameters.

Taking the Youla parameterization matrix and the state observer into account, a proper feedback gain $\boldsymbol{F}$ and estimation gain $\boldsymbol{L}$ can be chosen so that $\boldsymbol{A}+\boldsymbol{BF}$ and $\boldsymbol{A}+\boldsymbol{LC}$ are stable. Therefore, the Youla controller based on state observation is obtained, with the structure shown in Figure 6.21.

In Figure 6.21, $\boldsymbol{Q}$ is the Youla parameterization matrix, which works to apply the error between the estimated output value and the actual input value to the control link, thus guaranteeing stability of the closed-loop control system, as well as meeting the H_2/H_∞ control performance demand [16]. Suppose the parameterization matrix $\boldsymbol{Q}$ is

$$\boldsymbol{Q} = \begin{bmatrix} \boldsymbol{A}_q & \boldsymbol{B}_q \\ \boldsymbol{C}_q & 0 \end{bmatrix} \tag{6.68}$$

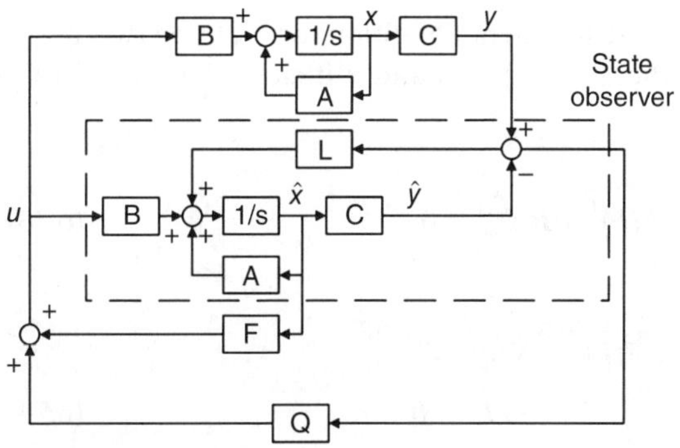

Figure 6.21 System structure

Then the system state equation can be

$$\begin{cases} \dot{\boldsymbol{x}} = \boldsymbol{Ax} + \boldsymbol{BF\hat{x}} + \boldsymbol{BC}_q\boldsymbol{x}_q + \boldsymbol{B}_1\boldsymbol{\omega} \\ \dot{\hat{\boldsymbol{x}}} = (\boldsymbol{A} + \boldsymbol{BF} + \boldsymbol{LC})\hat{\boldsymbol{x}} - \boldsymbol{LCx} + \boldsymbol{BC}_q\boldsymbol{x}_q \\ \dot{\boldsymbol{x}}q = \boldsymbol{A}_q\boldsymbol{x}_q - \boldsymbol{B}_q\boldsymbol{Cx} + \boldsymbol{B}_q\boldsymbol{C\hat{x}} \\ \boldsymbol{y} = \boldsymbol{Cx} \\ \hat{\boldsymbol{y}} = \boldsymbol{C\hat{x}} \\ \boldsymbol{y}_q = \boldsymbol{C}_q\boldsymbol{x}_q \end{cases} \tag{6.69}$$

where $\boldsymbol{x}$ is the original system state vector, $\hat{\boldsymbol{x}}$ is the state vector of the state observer, $\boldsymbol{x}_q$ is the state vector of the Youla parameterization matrix, $\boldsymbol{y}$, $\hat{\boldsymbol{y}}$, and $\boldsymbol{y}_q$ are the output vectors of the original system, the state observer, and Youla parameterization matrix, respectively, and $\boldsymbol{A}_q$, $\boldsymbol{B}_q$, and $\boldsymbol{C}_q$ are the corresponding state matrices of the parameterization matrix $\boldsymbol{Q}$.

Based on (6.69), when taking the Youla parameterization matrix into account, the system matrix becomes

$$\boldsymbol{G}_q = \begin{bmatrix} \boldsymbol{A} & \boldsymbol{BF} & \boldsymbol{BC}_q & 0 & \boldsymbol{B}_1 \\ -\boldsymbol{LC} & \boldsymbol{A}+\boldsymbol{BF}+\boldsymbol{LC} & \boldsymbol{BC}_q & 0 & 0 \\ -\boldsymbol{B}_q\boldsymbol{C} & \boldsymbol{B}_q\boldsymbol{C} & \boldsymbol{A}_q & 0 & 0 \\ \boldsymbol{C} & 0 & 0 & 0 & 0 \\ 0 & \boldsymbol{C} & 0 & 0 & 0 \\ 0 & 0 & \boldsymbol{C}_q & 0 & 0 \end{bmatrix} \tag{6.70}$$

In order to guarantee that (6.70) meets the demand of H_2/H_∞ norms being bounded, the following two inequalities should be guaranteed [17]:

$$(\boldsymbol{A}_c + \Delta\boldsymbol{A}_c)\boldsymbol{X} + \boldsymbol{X}(\boldsymbol{A}_c + \Delta\boldsymbol{A}_c)^{\mathrm{T}} + \boldsymbol{B}_c\boldsymbol{B}_c^{\mathrm{T}} < 0 \tag{6.71}$$

$$\begin{bmatrix} (\boldsymbol{A}_c + \Delta\boldsymbol{A}_c)\boldsymbol{X} + \boldsymbol{X}(\boldsymbol{A}_c + \Delta\boldsymbol{A}_c)^{\mathrm{T}} & \boldsymbol{XB}_c & \boldsymbol{C}^{\mathrm{T}} \\ \boldsymbol{B}_c^{\mathrm{T}}\boldsymbol{X} & -\gamma\boldsymbol{I} & \boldsymbol{0} \\ \boldsymbol{C} & \boldsymbol{0} & -\gamma\boldsymbol{I} \end{bmatrix} < 0 \tag{6.72}$$

Equation (6.71) is the necessary and sufficient condition for $\boldsymbol{G}_q$ to be two-norm bounded and equation (6.72) is the the necessary and sufficient condition for $\boldsymbol{G}_q$ to be ∞-norm bounded.

Applying (6.67) to (6.71) yields

$$(\boldsymbol{A}_c + \boldsymbol{EFH})\boldsymbol{X} + \boldsymbol{X}(\boldsymbol{A}_c + \boldsymbol{EFH})^{\mathrm{T}} + \boldsymbol{B}_c\boldsymbol{B}_c^{\mathrm{T}} < 0 \tag{6.73}$$

Applying (6.67) to (6.72) yields

$$\begin{bmatrix} (\boldsymbol{A}_c + \boldsymbol{EFH})\boldsymbol{X} = \boldsymbol{X}(\boldsymbol{A}_c + \boldsymbol{EFH})^{\mathrm{T}} & \boldsymbol{XB}_c & \boldsymbol{C}^{\mathrm{T}} \\ \boldsymbol{B}_c^{\mathrm{T}}\boldsymbol{X} & -\gamma\boldsymbol{I} & \boldsymbol{0} \\ \boldsymbol{C} & \boldsymbol{0} & -\gamma\boldsymbol{I} \end{bmatrix} < 0 \tag{6.74}$$

where

$$A_c = \begin{bmatrix} A & BF & BC_q \\ -LC & A_o + B_oF & B_oC_q \\ -BC_q & B_qC_o & A_q \end{bmatrix} \text{ and } B_c = \begin{bmatrix} B_1 & 0 & 0 \end{bmatrix}^{\mathrm{T}}$$

The two constraints in (6.73) and (6.74) both contain the uncertainty matrix ***EFH***, which needs to be separated first so that the constraints can be transformed into linear inequality constraints for solution. The following lemma is needed during this process.

Lemma 6.3 Given the appropriately dimensioned matrixes $\boldsymbol{Y}_0$, $\boldsymbol{D}_0$, and $\boldsymbol{E}_0$, where $\boldsymbol{Y}_0$ is symmetrical, if and only if there exists a positive constant $\mu > 0$, which meets $\boldsymbol{Y}_0 + \mu \boldsymbol{D}\boldsymbol{D}_0^{\mathrm{T}} + \mu^{-1}\boldsymbol{E}_0\boldsymbol{E}_0^{\mathrm{T}} < 0$, is

$$\boldsymbol{Y}_0 + \boldsymbol{D}_0\boldsymbol{F}_0\boldsymbol{E}_0 + \boldsymbol{E}_0^T\boldsymbol{F}_0^T\boldsymbol{D}_0^T < 0 \tag{6.75}$$

where $\boldsymbol{F}$ can be any matrix that meets $\boldsymbol{F}^T\boldsymbol{F} < \boldsymbol{I}$.

According to Lemma 6.1, for any $\boldsymbol{F}$ that meets $\boldsymbol{F}^T\boldsymbol{F} < \boldsymbol{I}$, (6.73) is tenable if and only if there exists a constant $\mu > 0$, which meets

$$A_c X + X A_c^{\mathrm{T}} + B_c B_c^{\mathrm{T}} + \mu^{-1}(HX)^{\mathrm{T}} HX < 0 \tag{6.76}$$

According to the property of the Schur complement [18], the quadratic term of the position matrix in (6.76) can be transformed to

$$\begin{bmatrix} A_c X + X A_c^{\mathrm{T}} + B_c B_c^{\mathrm{T}} + EE^{\mathrm{T}} & HX \\ (HX)^{\mathrm{T}} & -\mu^{-1} I \end{bmatrix} < 0 \tag{6.77}$$

Similarly, by separating the uncertainty matrix and transforming the quadratic term in (6.74),

$$\begin{bmatrix} A_c X + X A_c^{\mathrm{T}} & XB_c & C^{\mathrm{T}} & E & XH^{\mathrm{T}} \\ B_c^{\mathrm{T}} X & -\gamma I & 0 & 0 & 0 \\ C & 0 & -\gamma I & 0 & 0 \\ E^{\mathrm{T}} & 0 & 0 & -\mu I & 0 \\ HX & 0 & 0 & 0 & -\mu^{-1} I \end{bmatrix} < 0 \tag{6.78}$$

The uncertainties caused by the parameter measurement error and estimation error are fully considered in (6.77) and (6.78). However, uncertainties caused by variations of the system operating point have not yet been considered. Therefore, the convex polytopic model is integrated into inequalities containing system uncertainties, with different operating points as the vertexes of the convex polytopic model. The *i*th topic in the convex polytopic model is

$$\begin{bmatrix} A_{ci} X + X A_{ci}^{\mathrm{T}} + B_{ci} B_{ci}^{\mathrm{T}} + EE^{\mathrm{T}} & HX \\ (HX)^{\mathrm{T}} & -\mu^{-1} I \end{bmatrix} < 0 \tag{6.79}$$

$$\begin{bmatrix} A_{ci}X + XA_{ci}^{\mathrm{T}} & XB_{ci} & C^{\mathrm{T}} & E & HX \\ B_{ci}^{\mathrm{T}}X & -\gamma I & 0 & 0 & 0 \\ C & 0 & -\gamma I & 0 & 0 \\ E^{\mathrm{T}} & 0 & 0 & -\mu I & 0 \\ HX & 0 & 0 & 0 & -\mu^{-1}I \end{bmatrix} < 0 \tag{6.80}$$

Equations (6.79) and (6.80) could be transformed to the minimization problem of an objective function with a BMI (bilinear matrix inequality) constraint. Then. by applying the BMI solver in the Matlab BMI toolbox, the Q matrix that meets H_2/H_∞ control performance could be gained.

The state matrices of different operating points can be gained and applied to (6.79) and (6.80). Then, by simultaneous solution, the Youla parameterization matrix that meets both the parameter perturbation uncertainties and operating point uncertainties can be gained. Finally, the output error of the actual system to the nominal system can be introduced to the output feedback control link via the Youla parameterization matrix, thus achieving tracking control of the output error, as well as H_2/H_∞ robust control of the system.

6.4.2 Simulation Verification

6.4.2.1 Test System

The IEEE four-machine 11-bus system is taken as the test system to design a Youla parameterization controller based on DC modulation, as well as to analyze the damping effect and robust performance of the controller. The generators utilize the sixth-order transient model, the detailed parameters of which are listed in Table 6.10. The excitation system applies fast excitation and the load uses the hybrid model of 50% constant impedance and 50% constant current. The HVDC line is between the connection line 7–9 and is modulated by the HVDC supplementary controller. The control structure of the HVDC supplementary controller is shown in Figure 6.22, where $\Delta P_{\max}$, $\Delta P_{\min}$, $I_{\max}$, $I_{\min}$, $\alpha_{\max}$, and $\alpha_{\min}$ are amplitude limiting links. The wide-area input signals are the relative rotating speed of generators 1, 2, and 3 to Generator 4.

Meanwhile, the selective mode analysis (SMA) is applied to reduce the system orde and the state variables that are relevant to low-frequency oscillation and are easy to measure are reserved during the order reduction.

Table 6.10 Generator parameter of the test system.

Generator parameter	**Base MVA**	**x_l (p.u.)**	**r_a (p.u.)**	**x_d (p.u.)**	**x'_d (p.u.)**	**x''_d (p.u.)**	**T'_{do} (s)**
Generator 1–4	900	0.2	0	1.8	0.3	0.25	8.0
Generator parameter	**T''_{do} (s)**	**x_q (p.u.)**	**x'_q (p.u.)**	**x''_q (p.u.)**	**T'_{qo} (s)**	**T''_{qo} (s)**	**H (s)**
	0.03	1.7	0.55	0.25	0.4	0.05	6.5

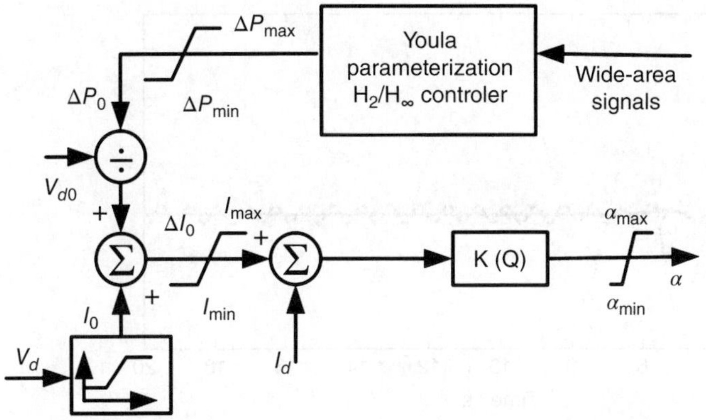

Figure 6.22 Control structure of the HVDC supplementary controller.

In the simulation process, five operating points are chosen to compose the convex polytopic model: normal operating point, 80% load, 90% load, 110% load, and 120% load. The parameter perturbation range is ±10%. In the following four cases, the proposed control strategy is verified concerning its effectiveness and excellence compared with the Youla parameterization control without considering system uncertainties.

1) Case 1. The system is 80% loaded. An instantaneous three-phase grounding fault occurs to a single line between bus 7 and bus 8 at $t = 0.1$ s and is cleared 0.05 s later.
2) Case 2. The system is 90% loaded. A permanent three-phase grounding fault occurs to a single line between bus 7 and bus 8 at $t = 0.1$ s and the fault line is cut 0.05 s later.
3) Case 3. The system is 110% loaded. A single AC line between bus 7 and bus 8 breaks at $t = 0.1$ s.
4) Case 4. The system is 120% loaded. The voltage of generator 1 drops to 0.85 p.u. at $t = 0.1$ s and is restored 0.05 s later.

6.4.2.2 Simulation Results and Analysis

Figures.6.23 to 6.26 show the time-domain waveforms of the relative speed difference for generators 1 to 4 corresponding to the above four cases, respectively. The dotted lines represent the dynamic responses of the open-loop system. The double-crossed lines represent the dynamic responses of the Youla parameterization controller closed-loop system without considering system uncertainties. The solid lines are the dynamic responses of the Youla parameterization controller closed-loop system considering system uncertainties.

It can be seen from Figures 6.24 to 6.26 that, when different types of fault occur at the operating points of the convex polytopic model, the Youla parameterization controller considering system uncertainties is capable of damping low-frequency oscillation faster and more effectively, with a better robust performance than the Youla parameterization controller without considering system uncertainties.

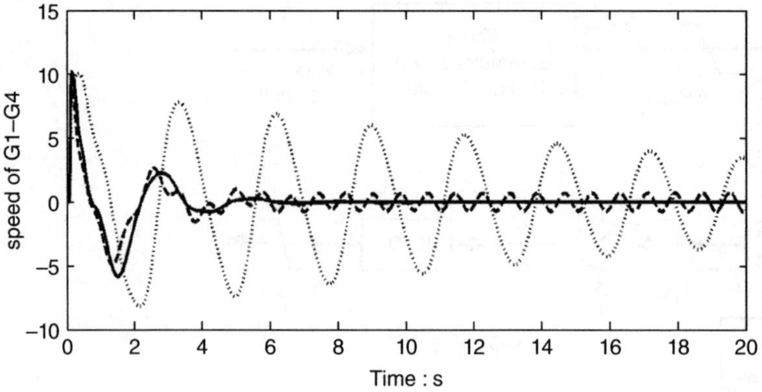

Figure 6.23 System dynamic responses in case 1.

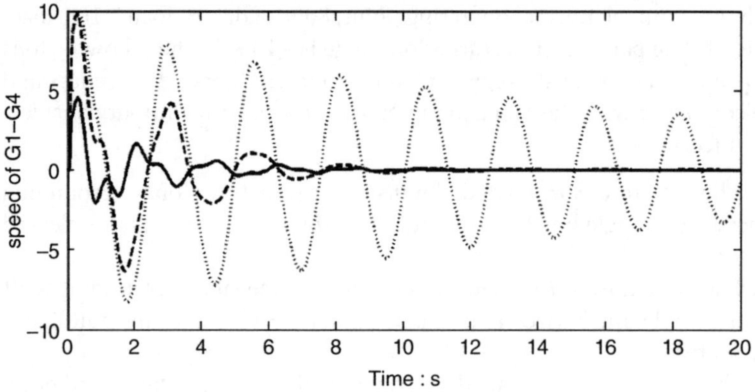

Figure 6.24 System dynamic responses in case 2.

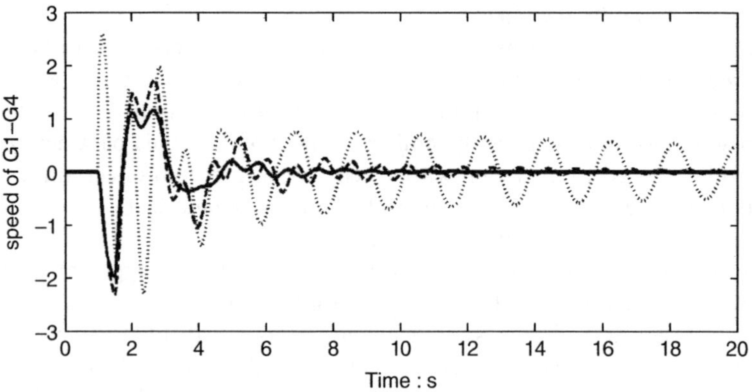

Figure 6.25 System dynamic responses in case 3.

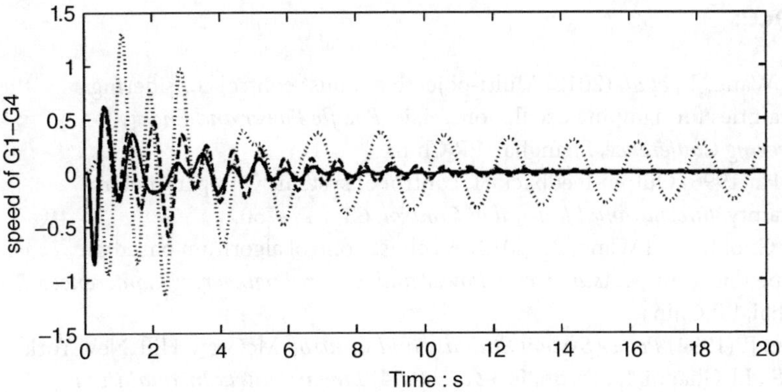

Figure 6.26 System dynamic responses in case 4.

6.5 Summary

This chapter discusses corresponding solutions to the problems that occur when designing the mixed H_2/H_∞ robust controller.

In robust control considering uncertainties, a multiobjective robust H_2/H_∞ control considering uncertainties for damping oscillation is introduced. The perturbation matrix is used to describe the uncertainty of parameter estimation and the uncertainty of measurement error, and the output feedback controller is obtained based on the LMI optimization technology. A robust H_2/H_∞ control strategy based on polytope uncertainty is proposed and the H_2/H_∞ inter-area oscillation control method based on the method lowers the conservativeness of the traditional H_2/H_∞ control method caused by using only one common Lyapunov matrix, by taking into account multiple internal uncertainties.

In optimal robust control, a nonconvex region control theory is introduced, which uses the eigenvalue transfer factor to expand the stability region and defines the nonconvex stability region that could rapidly render the system in a stable state; thus the H_2/H_∞ damping control strategy is more robust. A wide-area robust H_2/H_∞ damping control method unlimited by the stability region is introduced based on the complete regulation control theory. In addition, by using the complete regulation method to obtain the state feedback matrix, which can adjust the system damping value, a mixed H_2/H_∞ multiobjective robust control strategy is designed and optimization of the control objective is achieved.

In error tracking robust control, a Youla parameterization robust control strategy is proposed. Through the Youla parameterization matrix, the output error between the real system and the nominal system is introduced to the output feedback control, so that the system is able to meet the performance requirement of H_2/H_∞ robust control and realize error tracking. The proposed method is unaffected by internal or external uncertainties and is thus more accurate compared with traditional robust control methods.

References

1 Ma, J., Wang, T., *et al.* (2012) Multi-objective robust control considering uncertainties for damping oscillations. *Asia-Pacific Power and Energy Engineering Conference*, Shanghai, PR China.

2 Xie, L.H. (1996) Output feedback H_∞control of systems with parameters uncertainty. *International Journal of Control*, **63**, 741–750.

3 Ma, J., Guo, R., and Wang, Z. (2012) A robust control algorithm based on polytope uncertainty. *Asia-Pacific Power and Energy Engineering Conference*, Shanghai, PR China.

4 Kundur, P. (1994) *Power System Stability and Control*, McGraw-Hill, New York.

5 Boyd, S., El-Ghaoui, L., Feron, E., *et al.* (1994) *Linear Matrix Inequalities in System and Control Theory*, SIAM, Philadelphia.

6 Skogestad, S. and Postlethwaite, I. (2001) *Multivariable Feedback Control*, John Wiley & Sons, Inc., New York.

7 Pal, B. and Chaudhuri, B. (2005) *Robust Control in Power Systems*, Springer, New York.

8 Rogers, G. (2000) *Power System Oscillations*, Kluwer Academic Publishers, Dordrecht.

9 Ma, J., Wang, T., Yan, W., *et al.* (2014) Design of wide-area robust damping controller based on the non-convex stable region for inter-area oscillations. *International Journal of Electrical Power and Energy Systems*, **55**(2), 473–480.

10 Ma, J., Wang, Y., and Li, J. (2012) The perfect regulation based on robust control to damp inter-area oscillations, *Asia-Pacific Power and Energy Engineering Conference*, Shanghai, PR China, pp. 1–4.

11 Kamwa, I., Trudel, G., and Gérin-Lajoie, L. Robust design and coordination of multiple damping controllers using nonlinear constrained optimization. *IEEE Transactions on Power Systems*, **15**(3), 1084–1092.

12 Fang, D.Z., Yang, X., Chung, T.S., and Wong, K.P. (2004) Adaptive fuzzy-logic SVC damping controller using strategy of oscillation energy descent. *IEEE Transactions on Power Systems*, **19**(3), 1414–1421.

13 Mei, S., Shen, T., and Liu, K. (2008) *Modern Robust Control Theory and Application*, Tsinghua University Press, Beijing.

14 Majumder, R., Chaudhuri, B., and Pal, B.C. (2005) A probabilistic approach to model-based adaptive control for damping of interarea oscillations. *IEEE Transactions on Power Systems*, **20**(1), 367–374.

15 Ma, J., Wang, S., Gao, X., *et al.* (2014) Youla parameterization robust control strategy considering power system uncertainties, *Electric Power Components and Systems*, **42**, 1152–1157.

16 Wu, M., He, Y., and She, J. (2010) *Robust Control Theory*, Higher Education Press, Beijing.

17 Yu, L. (2002) *Robust Control/Linear Matrix Inequality Approach*, Tsinghua University Press, Beijing.

18 Kothare, M., Balakrishnan, V., and Morsri M. (1996) Robust constrained model predictive control using linear matrix inequalities. *Automatica*, **32**(10), 1361–1379.

7
Wide-Area Adaptive Control

7.1 Introduction

The rapid growth of the power system and complicated operating conditions has resulted in the severe problem of oscillatory instability. Furthermore, the oscillatory stability can be influenced by changes of generator, network, and load parameters, etc. However, the effect of stability control can be seriously affected by the uncertainty of network structure, parameter, and external disturbance.

Robust control has obvious advantages in copying with uncertain parameters and disturbances; for example, the disturbance is molded as an additional form of H_∞ robust control and a multiple form of μ robust control. However, the robust limitation will be exhibited when a large-scale power system is operating at a wide range of changing conditions. Especially, under some serious conditions, the predesigned damping controller might lose its effectiveness. An adaptive control method is thus a promising solution for accommodating a wider range of operating conditions for a large-scale power system.

In view of the above problems, the two important aspects of wide-area adaptive control are introduced in this chapter, which are operating condition identification matching and controller switching. With regard to operating condition identification, the federated Kalman filter and classification and regression tree theory are used to achieve optimum matching between the operating condition and the controller. As for controller matching, the dual Youla parameterization and Markov theory based on adaptive wide-area damping control are used to realize a smooth handoff between controllers, avoiding the impact of controller switching on the system.

Power System Wide-Area Stability Analysis and Control, First Edition. Jing Ma.

7.2 Adaptive Control Considering Operating Condition Identification

7.2.1 Federated Kalman Filter Based Adaptive Damping Control of Inter-Area Oscillations

7.2.1.1 Kalman Application to Power System

Design of Robust Control A damping controller is designed using H_2/H_∞ robust control to guarantee the good performance for each case. The H_2/H_∞ formulation with a pole-placement region is given with state space models of the form

$$\mathbf{P}(s):\begin{cases}\dot{\mathbf{x}} = \mathbf{A}\mathbf{x} + \mathbf{B}_1\zeta + \mathbf{B}_2\mathbf{u} \\ \mathbf{z}_\infty = \mathbf{C}_1\mathbf{x} + \mathbf{D}_{11}\zeta + \mathbf{D}_{12}\mathbf{u} \\ \mathbf{z}_2 = \mathbf{C}_2\mathbf{x} + \mathbf{D}_{21}\zeta + \mathbf{D}_{22}\mathbf{u}\end{cases} \tag{7.1}$$

where $\mathbf{x}$ is the state of the system, $\mathbf{u}$ is the control input, ζ is a disturbance, and $\mathbf{z}_\infty$ and $\mathbf{z}_2$ are for the H_2/H_∞ problems. In equation (7.1), these matrixes can be obtained from the power system toolbox (PST). The multiobjective robust H_2/H_∞ control for damping oscillation is to design a controller $\mathbf{u} = \mathbf{K}\mathbf{x}$. The state feedback controller results in a closed-loop state matrix $\mathbf{A}_{cl} = \mathbf{A} + \mathbf{B}_2\mathbf{K}$. Applying the result for the matrix $\mathbf{A}_{cl}$ gives the closed-loop state models as follows:

$$\mathbf{P}_{cl}(s):\begin{cases}\dot{\mathbf{x}} = \mathbf{A}_{cl}\mathbf{x} + \mathbf{B}_1\zeta \\ \mathbf{z}_\infty = \mathbf{C}_{1cl}\mathbf{x} + \mathbf{D}_{11}\zeta \\ \mathbf{z}_2 = \mathbf{C}_{2cl}\mathbf{x} + \mathbf{D}_{21}\zeta\end{cases} \tag{7.2}$$

where $\mathbf{A}_{cl} = \mathbf{A} + \mathbf{B}_2\mathbf{K}, \mathbf{C}_{1cl} = \mathbf{C}_1 + \mathbf{D}_{12}\mathbf{K}, \mathbf{C}_{2cl} = \mathbf{C}_2 + \mathbf{D}_{22}\mathbf{K}$, ζ is a limited energy disturbance signal, and $\mathbf{T}_{WZ\infty}$ and $\mathbf{T}_{WZ2}$ are the closed-loop transfer functions from ζ to output channels $\mathbf{z}_\infty$ and $\mathbf{z}_2$.The closed system should be satisfying the following constraints:

1) The H_∞-norm of the closed transfer function $\mathbf{T}_{WZ\infty}$ does not exceed the given maximum bound γ_∞, that is, $\|\mathbf{T}_{WZ\infty}\|_\infty < \gamma_\infty$, to guarantee that the closed system is robust stable, where

$$\mathbf{T}_{WZ\infty} = \mathbf{C}_{1cl}\left(s\mathbf{I} - \mathbf{A}_{cl}\right)^{-1}\mathbf{B}_1 + \mathbf{D}_{11} \tag{7.3}$$

2) The H_2-norm of the closed transfer function $\mathbf{T}_{WZ2}$ does not exceed the given maximum bound γ_2, that is, $\|\mathbf{T}_{WZ2}\|_2 < \gamma_2$, to guarantee good performance of the closed system:

$$\mathbf{T}_{WZ2} = \mathbf{C}_{2cl}\left(s\mathbf{I} - \mathbf{A}_{cl}\right)^{-1}\mathbf{B}_1 + \mathbf{D}_{21} \tag{7.4}$$

3) **D**-stability is also an additional constraint, as shown in Figure 7.1, where l_o and ξ_o are the maximum real part of eigenvalues and the minimum damping ratio, respectively. In the controller design process, the **D**-stability region has the values $l_0 = -0.5$ and $\xi_0 = 7.5\%$.

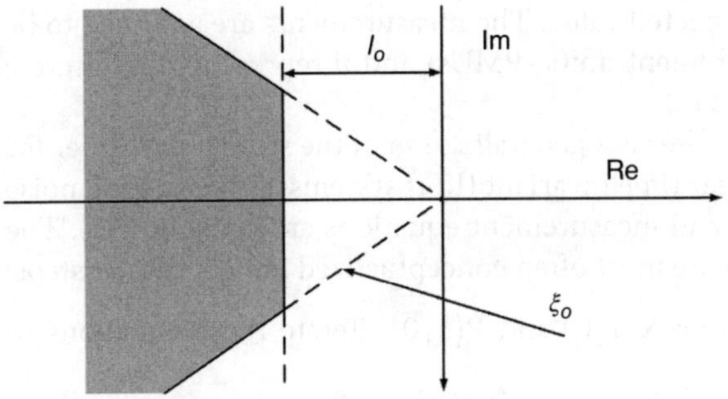

Figure 7.1 D-stability region.

Time-Varying Kalman Filter and χ^2 Variables The application of the Kalman filtering techniques to adaptive damping control generally requires a stochastic forcing function in the state equation. The difficulty is that the exact operating condition may be unknown. This problem can be handled by making the time-varying Kalman filter (TVKF) and controller for each operating condition and the most suitable controller will be chosen for the operators using TVKF [1].

To model the problem so that a Kalman filter may be used, it is necessary to write a discrete state equation for the parameters to be estimated in the following form, where the discrete state equation can be obtained by using the Matlab function c2d:

$$\begin{cases} \mathbf{x}(n) = \mathbf{\Phi}(n,n-1)\mathbf{x}(n-1) + \mathbf{\Gamma}(n,n-1)\boldsymbol{w}(n-1) \\ \mathbf{z}(n) = \mathbf{H}(n)\mathbf{x}(n) + \boldsymbol{\varepsilon}(n) \end{cases} \tag{7.5}$$

where $\boldsymbol{x}(n)$ is the state vector, $\boldsymbol{z}(n)$ is the measurements for observation, and $\mathbf{\Phi}$, $\mathbf{\Gamma}$, and $\mathbf{H}$ are matrices that are from discrete-time plant, respectively, $\mathbf{\Phi}(n,n-1)$ is the discrete time version of $\mathbf{A}$, $\boldsymbol{w}(n)$ is either the process (random state) noise or represents inaccuracies in the system model [2], $\boldsymbol{\varepsilon}(n)$ represents the measurement noise, and $\boldsymbol{w}(n)$ and $\boldsymbol{\varepsilon}(n)$ are assumed to be white Gaussian noise sequences. From equations (7.6) to (7.9), it is shown that the means of $\boldsymbol{w}(n)$ and $\boldsymbol{\varepsilon}(n)$ are zero and the covariance matrices are $\mathbf{Q}(n)$ and $\mathbf{R}(n)$, respectively. The process and measurement noise are assumed to be independent of each other and uncorrelated from sample to sample satisfying (7.10):

$$E\{\boldsymbol{w}(n)\} = 0 \tag{7.6}$$

$$E\{\boldsymbol{w}(j)\boldsymbol{w}^{\mathrm{T}}(n)\} = \mathbf{Q}(n)\delta_{jn} \tag{7.7}$$

$$E\{\boldsymbol{\varepsilon}(n)\} = 0 \tag{7.8}$$

$$E\{\boldsymbol{\varepsilon}(n)\boldsymbol{\varepsilon}^{\mathrm{T}}(j)\} = \mathbf{R}(n)\delta_{jn} \tag{7.9}$$

$$E\{\boldsymbol{w}(n)\boldsymbol{\varepsilon}^{\mathrm{T}}(j)\} = 0 \tag{7.10}$$

where $E[\cdot]$ represents the expected value. The measurements are assumed to be made by the phasor measurement units (PMUs) and n represents the discrete sampling time instant (n=1,2,...).

The time-varying Kalman filter is a generalization of the steady-state filter for time-varying systems or linear time invariant (LTI) systems with changing noise covariance. The plant state and measurement equations are given in (7.2). The discrete-time Kalman filters are most often conceptualized in two distinct steps:

Step I. Given initial conditions $\mathbf{x}(1|0)$ and $\mathbf{P}(1|0)$, iterate these equations to perform the filtering.

Step II. Update both the state estimates and error covariance matrices at each time sample.

Typically, the updated phases alternate with the prediction, advancing the state until the next scheduled observation and the update incorporating the observation. In this way, the time-varying Kalman filters can be formed for each case and work in parallel.

Let a tilde denote measurement residuals, that is,

$$\tilde{\mathbf{z}}(k)=\mathbf{z}(k)-\hat{\mathbf{z}}(k|k-1) \tag{7.11}$$

The measurement covariance matrix $\mathbf{S}_k$ of the ith time-varying Kalman filter is defined as follows:

$$\mathbf{S}_k=\mathrm{cov}(\tilde{z}) \tag{7.12}$$

The quadratic form is the only term depending on the actual observations. Let

$$\mathbf{C}(k)=\tilde{\mathbf{z}}^T(k)\mathbf{S}_k\tilde{\mathbf{z}}(k) \tag{7.13}$$

where $\mathbf{C}$ denotes the χ^2 variables. Since the mean of the measurement noise $\boldsymbol{\varepsilon}(n)$ is assumed to be zero, the Gaussian white noise with covariance $\mathbf{R}(n)$, then $\mathbf{C}(k)$ is distributed according to the chi-squared distribution with m degrees of freedom:

$$\mathbf{C}(k)\sim\chi^2(m) \tag{7.14}$$

where m is also the number of observations.

The model estimate is compared to the observation and this difference is scaled by the Kalman gain, which is used as feedback into the model to improve subsequent predictions. When the observations are from the matched operating condition and are sent to the corresponding Kalman filter, the χ^2 variable $\mathbf{C}$ should tend to be the number of observations since the expected value of $\mathbf{C}(k)$ can be computed as follows:

$$E(\mathbf{C}(k))=m \tag{7.15}$$

When the measurements and the Kalman filter do not match each other well, then the $E(\mathbf{C}(k))$ will become larger than m. In a power system, the estimation at one point cannot be trusted due to various disturbances. To obtain more accurate results for identification, equation (7.11) could be extended to points beyond k in the same way, yielding

$$\tilde{\mathbf{z}}(k+l)=\mathbf{z}(k+l)-\hat{\mathbf{z}}(k+l|k-1+l) \tag{7.16}$$

$$\mathbf{C}(k+l)=\tilde{\mathbf{z}}^T(k+l)\mathbf{S}_{k+l}\tilde{\mathbf{z}}(k+l) \tag{7.17}$$

The sum of the χ^2 variables can be greatly simplified for multiple points by writing

$$\mathbf{L}(k+l)=\sum_{i=1}^{k+l}\mathbf{C}(k+l) \tag{7.18}$$

Then $\mathbf{L}$ is distributed according to the chi-squared distribution with $m\times(k+l)$ degrees of freedom:

$$\mathbf{L}(k+l)\sim\chi^2(m\times(k+l)) \tag{7.19}$$

The Confidence Interval of Chi-Square Variables In this section, the confidence interval and chi-square testing are used to estimate the state of the power system operating condition and to choose the most suitable controller for damping the inter-area oscillations.

Specifically, $\mathbf{L}(k+l)$ is distributed according to the chi-squared distribution with $m\times(k+l)$ degrees of freedom, where the known mean of $L(k+l)$ is $m\times(k+l)$:

$$\mu=E(L(k+l))=m\times(k+l) \tag{7.20}$$

The confidence level α of the one-sided confidence interval is defined as

$$\Pr(\mathbf{L}(k+l)<\underline{\mathbf{L}}(k+l))=\alpha \tag{7.21}$$

where

$$\underline{\mathbf{L}}(k+l)=\chi^2_{1-\alpha}(m\times(k+l)) \tag{7.22}$$

The chi-squared distribution confidence interval can be obtained using (7.21):

$$\mathbf{CI}=(-\infty,\underline{\mathbf{L}}(k+l)) \tag{7.23}$$

If the chi-squared variable of the ith case satisfies $\mathbf{L}_i(k+l)\in\mathbf{CI}$, then the ith controller has enough confidence to be chosen. If $\mathbf{L}_i(k+l)\notin\mathbf{CI}$, then the controller should be rejected.

By summing more points, the shape of the chi-squared distribution is sharper and the number of cases in the specific confidence interval is less, and it then becomes easier to choose a suitable controller.

Two situations occurred in the process of choosing the most suitable damping controller for the changing power system operating conditions. One situation occurs when the changed operating condition is a good fit for the pre-prepared cases. In this situation, the variable **L** will lie in the specific confidence interval no matter how many points will be summed due to the good match. The matched controller can then perform well enough to damp out the oscillations.

The other situation occurs when there is no equivalent operating condition in the pre-prepared cases. In this situation, some points will initially be in the confidence interval, but as more observations are made, the fraction of points in the confidence interval will decrease. Assuming s cases lie in the confidence interval, a combined control can be formed as

$$\mathbf{K}_{com} = \beta_1\mathbf{K}_1 + \beta_2\mathbf{K}_2 + \cdots + \beta_s\mathbf{K}_s \tag{7.24}$$

The coefficients, $\beta_1, \beta_2, \ldots, \beta_s$ are determined by

$$\beta_i = \mathrm{Pr}_{Li} / \sum_{j=1}^{s} \mathrm{Pr}_{Lj} \tag{7.25}$$

where Pr_{Li} is the chi-square probability density for a given chi-square variable L_i and ν degrees of freedom:

$$\mathrm{Pr}_{Li} = \frac{L_i^{(\nu-2)/2}\exp(-L_i/2)}{2^{\nu/2}\Gamma(\nu/2)} \tag{7.26}$$

where $\Gamma(\cdot)$ is the gamma function.

Then the coefficients $\beta_1, \beta_2, \ldots, \beta_s$ could be expressed using (7.26) in terms of the chi-square variable L_i and ν degrees of freedom:

$$\beta_i = \frac{L_i^{(\nu-2)/2}\exp(-L_i/2)}{\sum_{j=1}^{s} L_j^{(\nu-2)/2}\exp(-L_j/2)} \tag{7.27}$$

The coefficients $\beta_1, \beta_2, \ldots, \beta_s$ represent weights for different predesigned controllers. From (7.27), it can be seen that the small chi-square variables $\mathbf{L}_i$ will magnify exponentially the coefficients of the ith case, which indicates the high matching degree with the$i$th case.

7.2.1.2 An Example Test System

Test System A 16-machine 68-bus test system, which is a reduced order model of the New England–New York interconnected system, is employed for this study. Group 4 contains machines that are a simplified representation of the New England system. Group 5 represents the New York system. The lines 1–2 and 8–9 are double circuits. The thick lines indicate the tie lines, which are the root source of the low-frequency inter-area oscillations. The tie lines connecting NETS and NYPS carry 720 MW. Group 3 exports 1163 MW to NYPS and imports 73 MW from Group 1 while 266 MW flows from Group 2 to NYPS and 75.8 MW from Group 2 to Group 1. In order to damp inter-area modes and study the effectiveness of adaptive control under different operating conditions, the test

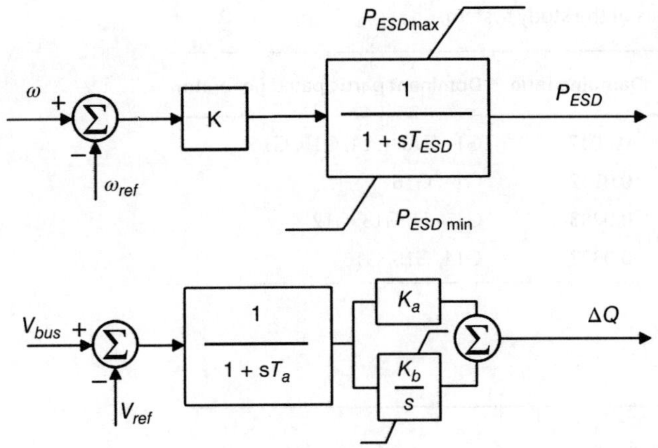

Figure 7.2 Small-signal model of the ESD unit.

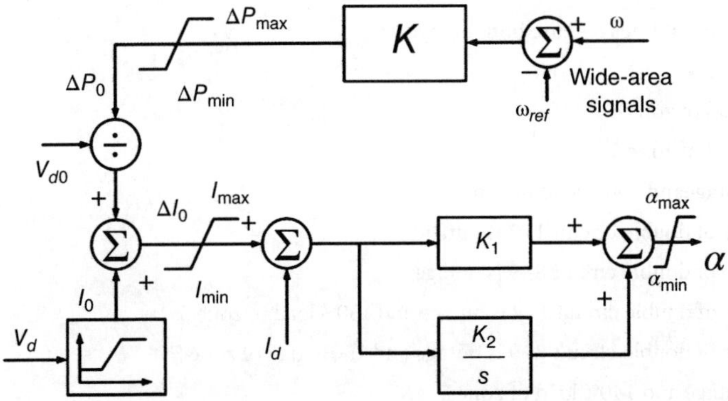

Figure 7.3 Control structure for the HVDC small-signal modulator.

system is modified by adding parallel HVDC (high-voltage direct current) lines and ESDs (energy storage devices). More controllers mean more feasibility for more cases.

Three DC lines are added in parallel with three tie lines (bus 41 to bus 42, bus 42 to bus 52, and bus 50 to bus 51) and three ESDs are installed at buses 37, 44, and 52, respectively. IEEE type DC1 excitation systems are installed in machines 1–9. The WECC load model is assumed at all load buses. The real power is 80% constant current, the reactive power is 80% constant admittance, and all loads support the remaining 20% as an induction motor load. The small-signal model of EDS and HVDC are shown in Figures 7.2 and 7.3, respectively. T_{ESD} and T_a are the responses times of the EDS involved with the measurement hardware.

The inter-area modes are obtained from the eigenvalue analysis and are displayed in Table 7.1. There are four critical inter-area modes. Generators 14, 15, and 16 are oscillating against generators 1 to 13 under mode 1, generator 14 is oscillating against generator 16 under mode 2, generators 12 and 13

Table 7.1 Inter-area modes of the study system.

Mode	Frequency(Hz)	Damping ratio	Dominant participated generator
1	0.2777	−0.0047	G1–G13, G14, G15, G16
2	0.4171	0.0142	G14, G16
3	0.6565	0.0283	G1–G9, G13, G12
4	0.7697	0.0323	G14, G15, G16

Table 7.2 Case bank.

No.	Cases
1	Base case
2	Double circuit lines 1–2 are both outage
3	Double circuit lines 8–9 are both outage
4	1–27 outage
5	140% load of zone 4
6	150% load of zone 5
7	1–27 outageand 150% load of zone 4
8	One line of double circuit 1–2 is outage
9	One line of double circuit 8–9 is outage
10	One line of double circuit 1–2 is outage and 150% load of zone 5
11	One line of double circuit 8–9 is outage and 150% load of zone 5
12	1–27 outage and 140% load of zone 5

are oscillating against generators 1 to 9 under mode 3, and generators 14 and 16 are oscillating against generator 15 under mode 4.

Twelve typical cases, which show great differences from each other, are shown in Table 7.2, and are used to investigate the performance of the adaptive damping controller based on the federated Kalman filters. The line 1–2 and line 8–9 are all double circuit lines. Cases 1 and 2 are double circuit line outages, which are more critical than only one of the double circuit line outages. The WECC load model is assumed at all load buses. The real power is 80% constant current, the reactive power is 80% constant admittance, and all loads support the remaining 20% as an induction motor load. The simulation environment with the WECC load model is stricter than that with a constant load model. The operating conditions show large differences from each other, as shown in Table 7.2, so it is impossible to find a single optimization feedback gain matrix for all the cases. Furthermore, for serious contingences, it is impossible to find a feasible feedback gain just using only one or two FACTS devices or HVDC links.This has been verified using the Matlab LMI toolkit, which has the output: "no feasible solution." Therefore, it is this motivation that requires the adaptive controller to adopt the new operating points.

A reduced-order model based on a selective modal analysis (SMA) is used to reduce the problem size [3,4]. The size of A is reduced from 123 to 30 using SMA. When SMA converges, a reduced-order model preserving low-frequency eigenvalues has been obtained.

Controller Design To strengthen the robustness of the controllers to achieve good performances of controllers, the federated Kalman filters are proposed to identify the operating condition and choose the most suitable controller. In a real power system, it is possible for the operators to predesign a damping controller for different contingencies, which are not limited to the cases shown in Table 7.2. The number of Kalman filters in the power system is able to reach hundreds or thousands and join these filters to form the adaptive controller for various operating conditions. First, a damping controller is designed using H_∞/H_2 robust control for each case. Second, the Kalman filter is also formed for each case. Then the Kalman filters are federated using a confidence interval based on chi-square variables about measurement residuals. Finally, the federated Kalman filters will give the operator the most suitable controller to put into service.

Construction of the control scheme is shown in Figure 7.4. The signals are sent to the federated Kalman filters from the phasor measurement unit (PMU) and then the most suitable controller is chosen either due to having the least chi-square variables or in the case where there is no match between the operating condition and the predesigned cases. The controller is then chosen to form a wide-area damping controller (WADC) through dedicated communication links. The WADC calculates modulation signals and sends them to three DCs and three ESDs, as shown in Figure 7.4. The sampling rate is 30 samples per second for Kalman filters.

In Figure 7.4, the power system is operating precisely at the ith case inside the case bank. The variable $\mathbf{L}$ can be obtained using the Kalman filters and the chi-square variable $\mathbf{L}$ of the ith case will lie in the confidence interval. Then $\mathbf{K}_i$ will be chosen as the controller, as shown in the Figure 7.4. If there is no matching controller, the controller should be combined using (7.24) and (7.25). All Kalman

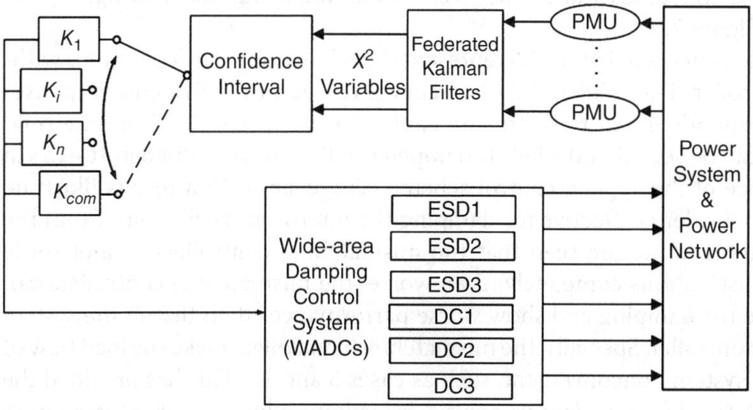

Figure 7.4 Construction of the control scheme.

filters work in parallel. Once the Kalman filters and confidence interval can detect that the operating point is outside the case bank, the controllers will be combined and stopped. Furthermore, the process of combination will be triggered when the Kalman filters detect the changes.

The measurements are the frequency differences of the generator external buses from the values at the reference generator 16. The external bus angle δ_i has a derivative $\dot{\delta}_i(t)$ that is approximate to the generator speed, where i is the external generator bus or the center of the coherency group.The measurement redundancy is critical for bad data detection and identification in a state estimation and control system. Different levels of measurement redundancy can be requested for optimal PMU placements [5]. Therefore, the overall control effectiveness can be guaranteed by the measurement redundancy even if some specific PMU fail to provide data. In this paper, assume that all the PMUs can provide the correct data successfully. There are 16 PMUs needed in the test system. The PMUs are placed near the generator buses.

7.2.1.3 Results and Discussions

Frequency-Domain Results and Analysis The frequency-domain analysis of the study system was carried out for the cases shown in Table 7.3. The eigenvalues of the twelve different cases are shown in Figure 7.5.

The dots and circles represent the eigenvalues of the open-loop and closed-loop systems, respectively. The D-stability region is used as the linear matrix inequality (LMI) region for pole placement. It is obvious from Figure 7.5 that all the poles of all closed loops lie in the D-region to guarantee enough damping ratios.

Time-Domain Results and Analysis Nonlinear simulations have been performed for 25 s to further demonstrate the performance of the controllers for each case. The disturbance for case 2 was a temporary three-phase short-circuit fault applied to the tie line between buses 8 and 9, while the disturbance for the other eleven cases was a temporary three-phase short-circuit fault applied to tie line between buses 1 and 2. The faults lasted for 80 ms (approximately 5 cycles). The relative angular separations between machines G1 and G16, G13 and G14, G13 to G15, and G13 to G16 of all the twelve cases following this contingency are shown in Figure 7.6.

The dotted curves in Figure 7.6 represent the relative angular separations without a controller. The solid and dashed curves represent the dynamic responses with corresponding and mismatching controllers, respectively. From the solid curves, it can be seen that the lightly damped oscillations are settled in 10–15 s in the presence of the applied control scheme. These simulation results illustrate that the controller is effective for damping the inter-area oscillations. From the dashed curves, it can be seen that the mismatching controllers cannot settle down the oscillations completely; even worse, the mismatching controllers can deteriorate the damping and show worse performances than that of the system without a controller. Specially, the mismatching controllers make the load flow of the power system nonconvergent, such as cases 5 and 12. This has provided the motivation to adopt an adaptive approach for improving the robustness under different operating conditions.

Table 7.3 Damping ratios and frequencies of inter-area modes with the controllers.

Case No.		Mode 1	Mode 2	Mode 3	Mode 4
1	Frequency (Hz)	0.2754	0.4176	0.6549	0.7689
	Damp ratio	0.3201	0.2159	0.1362	0.1147
2	Frequency (Hz)	0.2727	0.4031	0.5749	0.7678
	Damp ratio	0.4373	0.2923	0.2092	0.1266
3	Frequency (Hz)	0.2729	0.4139	0.5486	0.7684
	Damp ratio	0.3098	0.2135	0.1605	0.1133
4	Frequency (Hz)	0.2752	0.4159	0.6440	0.7687
	Damp ratio	0.3219	0.2170	0.1392	0.1148
5	Frequency (Hz)	0.2989	0.4062	0.6458	0.7654
	Damp ratio	0.3879	0.2789	0.1922	0.1193
6	Frequency (Hz)	0.2754	0.4190	0.6557	0.7702
	Damp ratio	0.4309	0.2654	0.1799	0.1178
7	Frequency (Hz)	0.2755	0.4177	0.6444	0.7700
	Damp ratio	0.4321	0.2674	0.1853	0.1179
8	Frequency (Hz)	0.2746	0.4131	0.6246	0.7685
	Damp ratio	0.3248	0.2187	0.1444	0.1149
9	Frequency (Hz)	0.2751	0.4159	0.6237	0.7688
	Damp ratio	0.3227	0.2170	0.1446	0.1149
10	Frequency (Hz)	0.2752	0.4161	0.6265	0.7700
	Damp ratio	0.4392	0.2699	0.1951	0.1182
11	Frequency (Hz)	0.2757	0.4172	0.6151	0.7689
	Damp ratio	0.3295	0.2195	0.1487	0.1150
12	Frequency (Hz)	0.3023	0.4065	0.6176	0.7648
	Damp ratio	0.3865	0.2806	0.1990	0.1196

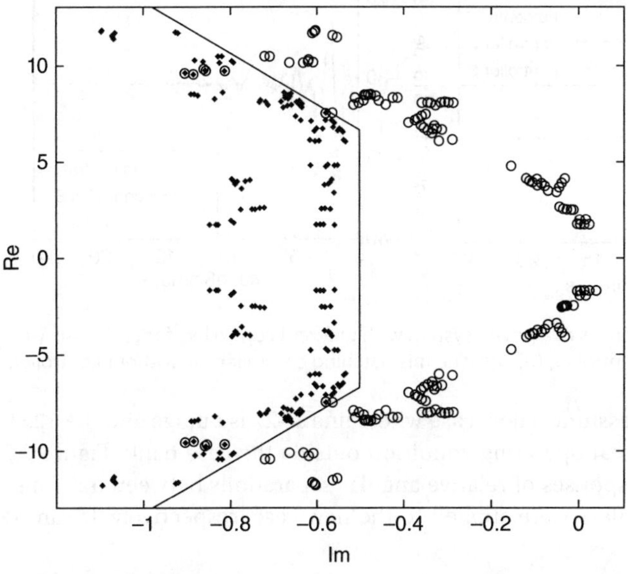

Figure 7.5 Eigenvalues of closed-loop systems for 12 cases shown in Table 7.2.

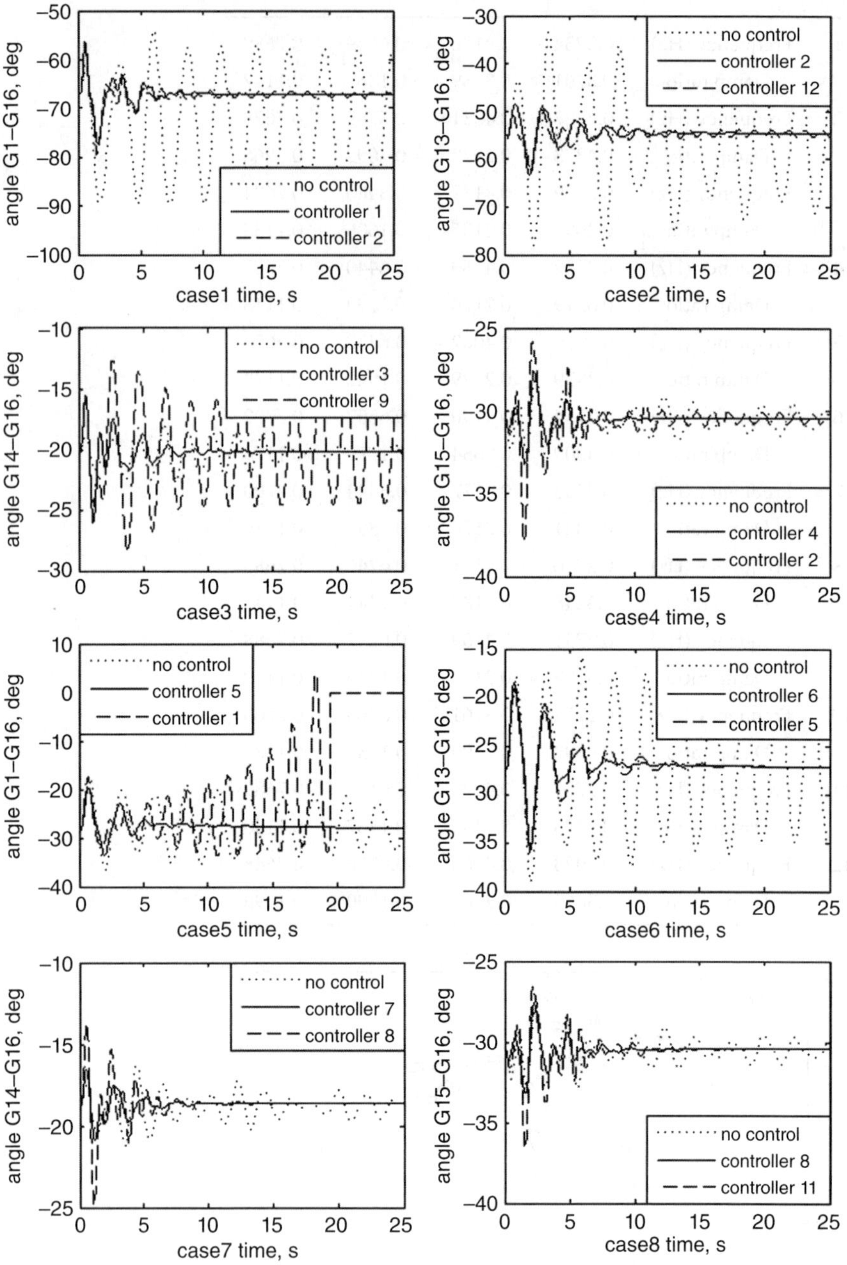

Figure 7.6 Dynamic responses of a power system with different controllers for each case: (-) with the corresponding controller; (--) with the mismatching controller; (···) without controller.

On the other hand, assume a new case where line 1–27 is outage and the 125% load of zone 4 is the real operating condition outside the case bank. Figure 7.7 shows the dynamic responses of relative angular separations between machines when different controllers were applied to the new case, respectively. It can be

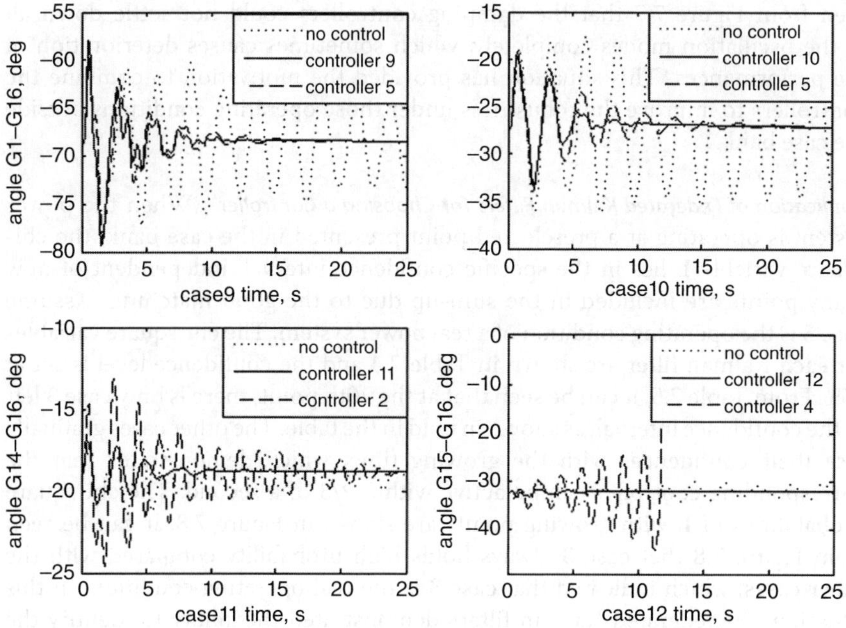

Figure 7.6 (Continued)

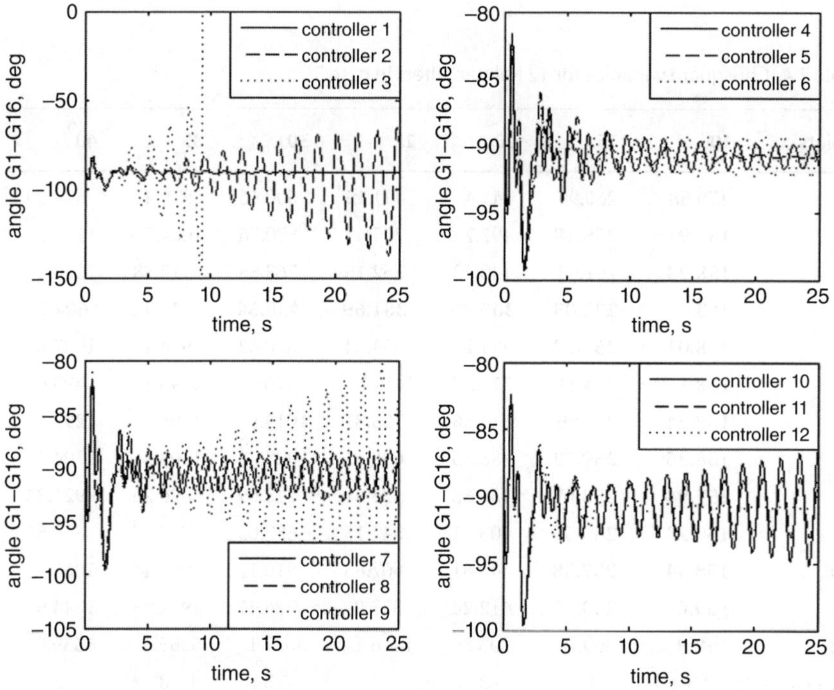

Figure 7.7 Dynamic responses of the closed-loop system using different controllers for the new case.

seen from Figure 7.7 that the damping controllers could not settle down all of the oscillation modes completely, which sometimes causes deterioration of the performances. This situation has provided the motivation to combine the controllers to improve the robustness under those operating conditions outside the case bank.

Application of Federated Kalman Filters for Choosing a Controller When the power system is operating at a preselected point presented in the case bank, the chi-square variable **L** lies in the specific confidence interval, independent of how many points are included in the sum-up due to the good matching. Assume case 3 is the operating condition of a real power system. The chi-square variables for each Kalman filter are shown in Table 7.4 and the confidence level is set at 95%. From Table 7.5, it can be seen that at the 20th point, there is only case 3 left in the confidence interval, as shown in bold in the table. The other cases gradually lose their confidences with the growing time, which demonstrates that the corresponding controller can be active within 2/3 of a second. The chi-square probabilities of **L** with growing points are shown in Figure 7.8. It can be seen from Figure 7.8 that case 3 always holds high probability compared with the other cases, which indicated that case 3 is the real operating condition. In this situation, the federated Kalman filters demonstrated the ability to identify the unknown dynamics and switch to the proper controller within a short time.

The other situation occurs when there is no matched operating condition in the pre-prepared cases. Assume the new case I with line 1–27 is outage and a

Table 7.4 Chi-square variables for 12 Kalman filters in case 3

Points	10	15	20	25	30	35	40
L	179.58	260.99	341.4	421.15	500.46	579.41	658.09
1	**157.91**	**258.18**	497.7	705.41	890.76	1004.7	1118.3
2	**158.23**	261.63	440.42	562.15	763.88	917.88	1024.2
3	**163.47**	**232.03**	**309.35**	**381.69**	**458.54**	**530.12**	**604.7**
4	**158.01**	**256.22**	474.7	654.54	820.62	934.4	1037.8
5	**158.1**	304.21	717.57	1471.2	2990.9	4795.5	5493.6
6	**158.24**	261.69	513.36	755.81	1014.5	1194.1	1297.1
7	**158.35**	**259.72**	488.85	698.13	928.73	1103.3	1204.1
8	**157.94**	**253.67**	436.63	573.61	723.69	826.25	921.38
9	**159.23**	**240.56**	403.21	547.15	683.05	782.77	873.56
10	**158.34**	**257.38**	449.81	607.83	810.12	965.48	1065.4
11	**159.6**	**243.57**	412.24	577.8	774.41	942.98	1044.9
12	**158.9**	309.74	703.39	1416.1	3027.1	5259.8	6359.9
Left cases	1 2 3 4 5 6 7 8 9 10 11 12	1 3 4 7 8 9 10 11	3	3	3	3	3

Table 7.5 Chi-square variables for 12 Kalman filters in the new case I.

Points	10	15	20	25	30	35	40
L	179.58	260.99	341.4	421.15	500.46	579.41	658.09
1	**157.64**	**235.39**	**322.42**	427.3	529.07	644.8	768.72
2	**155.88**	**235.68**	**307.47**	**391.28**	503.83	676.48	861.84
3	**158.83**	**237.97**	**325.1**	427.88	543.45	678.2	811.22
4	**156.88**	**234.14**	**318.29**	**420**	527.03	649.66	774.37
5	**155.94**	**237.39**	**312.75**	**388.21**	**464.78**	**543.87**	**631.08**
6	**159.32**	**236.3**	**320.06**	**417.25**	514.9	621.61	726.49
7	**158.58**	**235.27**	**316.79**	**411.55**	513.91	627.59	733.1
8	**155.83**	**232.88**	**313.67**	411.91	519.78	649.03	778.34
9	**158.05**	**236.22**	**323.24**	426.91	531.93	651.03	775.49
10	**157.49**	**234.21**	**313.23**	**406.58**	510.13	631.39	743.38
11	**159.76**	**237.28**	**321.44**	**418.25**	518.47	628.09	733.32
12	**155.15**	**236.77**	**309.98**	**385.75**	**463.58**	**540.09**	**621.63**
Left cases	1 2 3 4 5 6 7 8 9 10 11 12	1 2 3 4 5 6 7 8 9 10 11 12	1 2 3 4 5 6 7 8 9 10 11 12	2 4 5 6 7 8 10 11 12	5 12	5 12	5 12

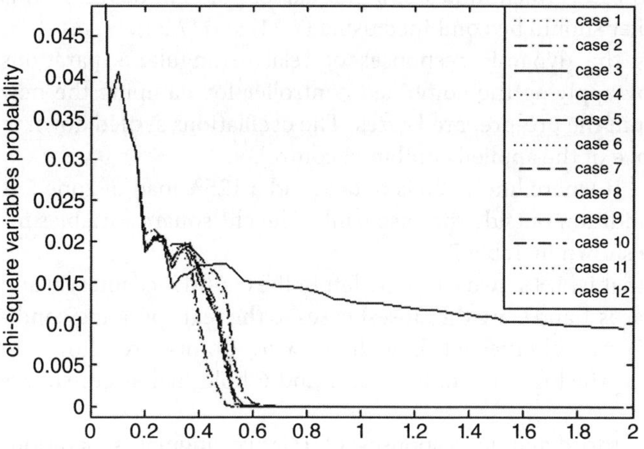

Figure 7.8 Chi-square variables probabilities under case 3.

125% load of zone 4 is the real operating condition outside the case bank mentioned in the previous subsection. The chi-square variables for 12 Kalman filters are shown in Table 7.5. Initially, there were several cases existing in the specific confidence interval. With the increasing number of points, the number of cases lying in the interval decreases. At the 30th point, that is 1 s, two cases are left in the confidence interval and are shown in bold; the results indicate that there was

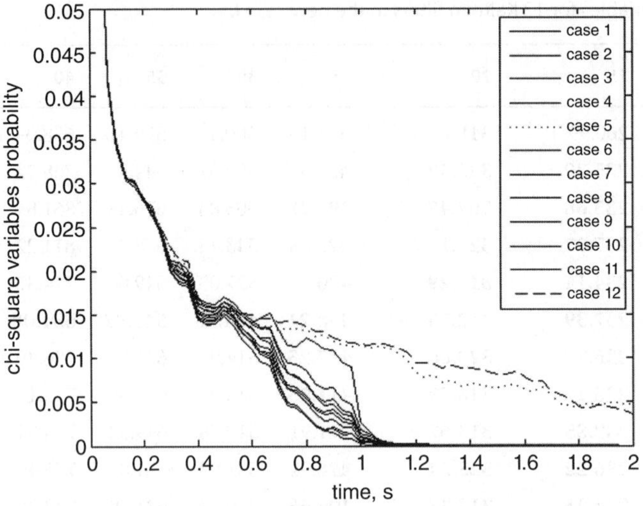

Figure 7.9 Chi-square variables probabilities under the new case I.

no matched case and cases 5 and 12 are the closest cases to the real operating condition. Actually, the chi-square probabilities of **L** with growing points are shown in Figure 7.9. It can be seen that with the growing time, only cases 5 and 12 hold higher chi-square probabilities, and the probabilities of other cases gradually decrease to zero, which indicates a gradual loss of confidence. In this situation, the controller should be combined using (7.24) and (7.25).

Figure 7.10 shows the dynamic responses of relative angular separations between machines by applying the combined controller for damping the new case I, which was not in the pre-prepared cases. The oscillations settled down in 10–15 s in the presence of the applied combined controller.

Assume a new case II where line 1–27 is outage and a 125% load of zone 5 is the real operating condition outside the case bank. The chi-square variables for 12 Kalman filters are shown in Table 7.6.

At the 40th point, that is, 1.3 s, two cases are left in the confidence interval and are shown in bold. Cases 1 and 6 are the closest cases to the real operating condition. The chi-square probabilities of **L** with growing points are shown in Figure 7.11. With the growing time, only cases 1 and 6 hold higher chi-square probabilities.

Figure 7.12 shows the dynamic responses of relative angular separations between machines by applying the combined controller for damping new case II, which was not in the pre-prepared cases.

7.2.2 Classification and Regression Tree Based Adaptive Damping Control of Inter-Area Oscillations

7.2.2.1 Adaptive Control Scheme Based on CART

The controllers can be scheduled adaptively using several independent measurements to deal with the identification of a multiple-input–multiple-output system. The situation is illustrated in Figure 7.13, where p_1 and p_2 denote the changing

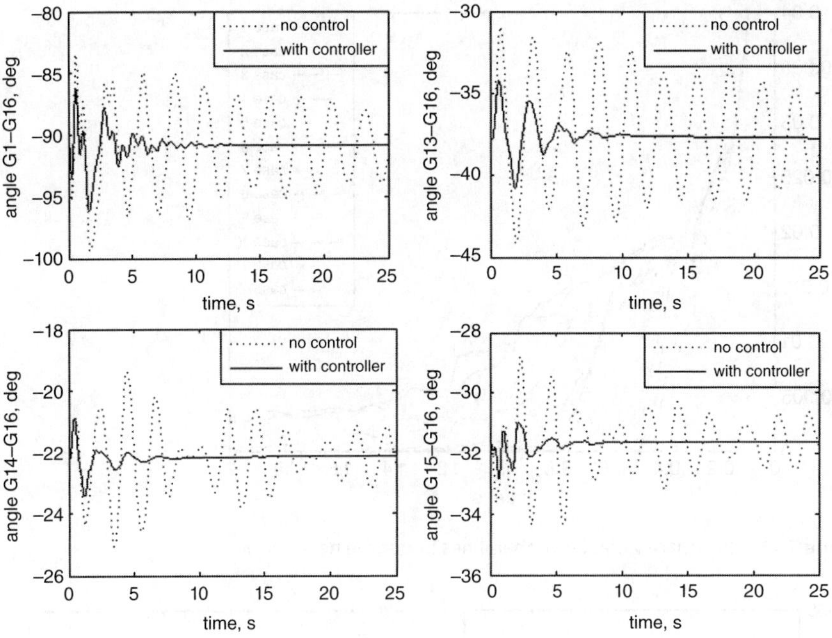

Figure 7.10 Dynamic responses of a closed-loop system with a combined controller for the new case I.

Table 7.6 Chi-square variables for 12 Kalman filters in the new case II.

Points	**10**	**20**	**30**	**40**	**45**	**50**	**55**
L	179.58	341.4	500.46	658.09	736.55	814.82	892.93
1	**159.3**	**302.81**	**471.75**	**639.58**	**726.23**	**807.49**	**885.15**
2	**159.12**	550.76	1419.2	1851.9	2396.6	3592.3	5601.4
3	**170.14**	551.32	1031.5	1292.7	1677.1	2517.2	3820.8
4	**159.34**	**306.01**	**497.33**	667.33	747.08	831.35	923.94
5	**157.24**	560.19	1796.7	3978.3	4352.1	5189.7	7704.1
6	**158.57**	**306.6**	**476.99**	**642.3**	**724.88**	**804.19**	**881.15**
7	**158.65**	**309.03**	**489.94**	662.27	772.9	876.46	966.45
8	**159.2**	**327.43**	594.63	783.17	905.16	1072.7	1300.1
9	**161.56**	**326.56**	529.02	697.43	787	906	1054.3
10	**158.5**	**327.08**	563.6	757.77	959.39	1192.7	1427.7
11	**160.98**	**326.31**	518.89	724.64	875.62	1046.8	1205.9
12	**158.79**	644.04	2023.9	5183.6	5906.6	6606.8	8951.2
Left cases	1 2 3 4 5 6 7 8 9 10 11 12	1 4 6 7 8 9 10 11	1 4 6 7	1 6	1 6	1 6	1 6

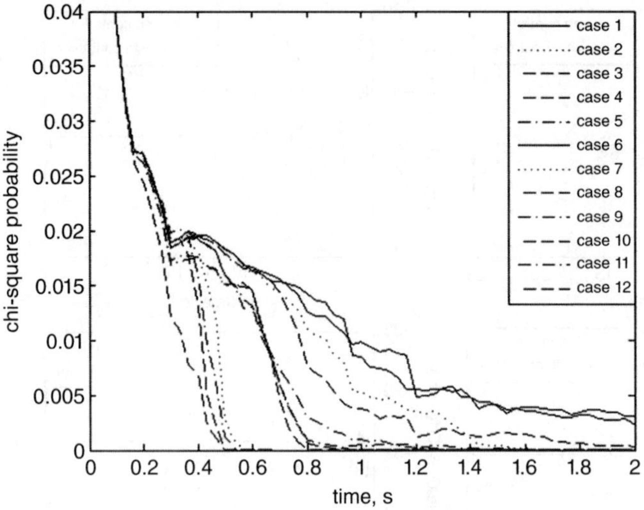

Figure 7.11 Chi-square variables probabilities under the new case II.

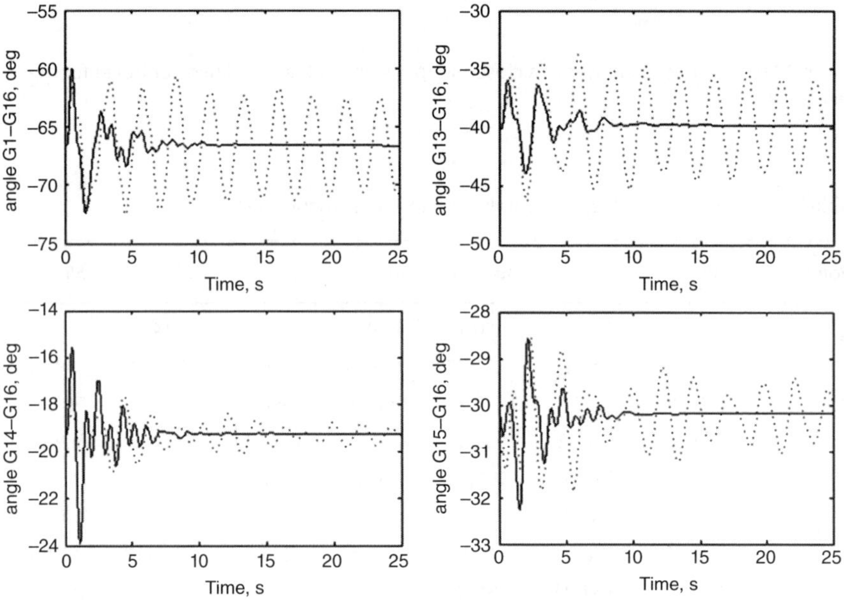

Figure 7.12 Dynamic responses of a closed-loop system with a combined controller for the new case II.

parameters and y_1 and y_2 denote the measurements. The square point represents the initial equilibrium point, the circles represent other equilibrium points, and the star represents the current operating point. Figure 7.13(a) represents the parameter space and Figure 7.13(b) represents the measurement space. The change of the operating point in the measurement space corresponds to the change in the parameter space. Once a disturbance occurs in the power system,

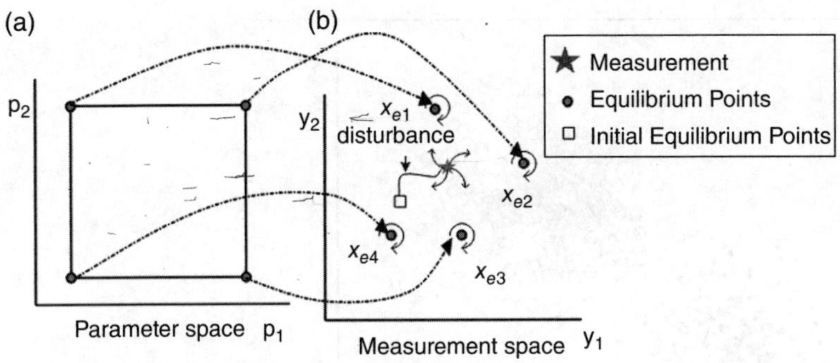

Figure 7.13 Parameter space and measurement space.

the operating point will deviate from the initial point, and move to another equilibrium point or back to the initial point. The track of the operating point indicates that it is part of a domain attracted to some equilibrium point. When the disturbance occurs, the CART is used to identify the attracting equilibrium point.

CART [6–8] is one of the classification data mining algorithms. It is able to partition the prediction space into several subsets recursively according to certain spitting rules. The tree is made up of the root node, middle (trunk) node, and leaf node. Every root node (and middle node) bears two subnodes and every leaf node corresponds to one system equilibrium point. In actual computation, the system starts from the root node and moves to the corresponding leaf node through a continuous yes/no judgment. Generally, the basic design procedure involves 4 steps: (1) attribute selection, (2) data set generation, (3) CART building, and (4) Performance evaluation.

Attribute Selection and Data Set Generation The guiding principle for selection of the attribute set is to select those measurable and controllable variables that are adequate enough in characterizing an equilibrium point. Many researchers have found that the active power generation of a generator carries sufficient information about the class of system oscillation modes [9]. Furthermore, the tie line flow is calculated using the angle shift δ_{ik} and the voltages provided by PMU, as shown by

$$\mathbf{P}_{ik} = \frac{\mathbf{U}_i \mathbf{U}_k}{\mathbf{X}_{ik}} \sin\left(\delta_i - \delta_k\right) = \frac{\mathbf{U}_i \mathbf{U}_k}{\mathbf{X}_{ik}} \sin \delta_{ik} \tag{7.28}$$

where δ_i and $\mathbf{U}_i$ are the voltage angle and amplitude of the ith bus and δ_k and $\mathbf{U}_k$ are the voltage angle and amplitude of the kth bus.

The optimum split of different operating classes using multiple measurements (active power of tie line flows) can be performed by the performance index σ, using Fisher's linear discriminant (FLD), which maximizes the difference between the classes [10]:

$$\sigma = \frac{\left(\mathbf{m}^{\mathrm{T}}\left(\mu_\alpha - \mu_\beta\right)^{\mathrm{T}}\right)^2}{\mathbf{m}^{\mathrm{T}}\left(\Sigma_{z=\alpha} + \Sigma_{z=\beta}\right)\mathbf{m}} \tag{7.29}$$

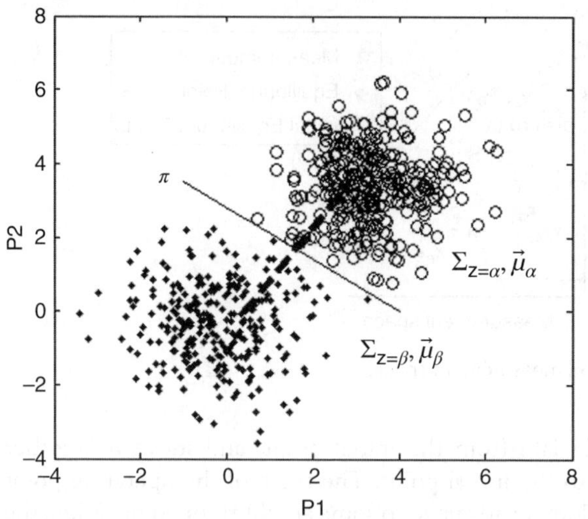

Figure 7.14 The optimum hyperplane for classifying the two different operating classes.

where α and β represent different operating classes, $\Sigma_{z=\alpha}$ and $\Sigma_{z=\beta}$ represent the covariance of measurements of α and β, $\boldsymbol{\mu}_\alpha$ and $\boldsymbol{\mu}_\beta$ represent the means of the measurements of α and β, respectively, as shown in Figure 7.14. Circles and dots denote α and β, respectively, $\mathbf{m}$ is the normal to the hyperplane $\boldsymbol{\pi}$, and σ is the maximum when

$$\mathbf{m} = \left(\Sigma_{z=\alpha} + \Sigma_{z=\beta}\right)^{-1}\left(\boldsymbol{\mu}_\alpha - \boldsymbol{\mu}_\beta\right)^{\mathrm{T}} \tag{7.30}$$

In addition, the middle point $\boldsymbol{\mu}_c = (\boldsymbol{\mu}_\alpha + \boldsymbol{\mu}_\beta)/2$ between $\boldsymbol{\mu}_\alpha$ and $\boldsymbol{\mu}_\beta$ is on the hyperplane. The distance vector $\mathbf{d}$ from any operating point $\mathbf{h}$ at α or β to the hyperplane is calculated by

$$\mathbf{d} = \mathrm{Pr}\, j_m\left(\mathbf{h} - \boldsymbol{\mu}_c\right) = \left(\mathbf{h} - \boldsymbol{\mu}_c\right) \cdot \mathbf{m} \tag{7.31}$$

where $\mathbf{d}$ is employed as the node splitting rule for CART. If $\mathbf{d} > 0$, $\mathbf{h}$ is identified to be at α. Otherwise, $\mathbf{h}$ is identified to be at β; $\cdot$ means the dot product of two vectors. Apply (7.30) to (7.31) and $\mathbf{d}$ is given by

$$\mathbf{d} = \left(\mathbf{h} - \boldsymbol{\mu}_c\right) \cdot \left(\Sigma_{z=\alpha} + \Sigma_{z=\beta}\right)^{-1}\left(\boldsymbol{\mu}_\alpha - \boldsymbol{\mu}_\beta\right)^{\mathrm{T}} \tag{7.32}$$

Then a regression is performed by CART to learn where the current operating point is in the parameter space. A new distance vector $\mathbf{d}'$ from current operating point $\mathbf{h}'$ to the hyperplane $\boldsymbol{\pi}$ is dropped down the tree, and the decision to alter the current operating point is made upon reaching a terminal node. Finally, the controller corresponding to the terminal node is selected and put into service for adaptive damping control.

There are two cases to be discussed in the selection of the most suitable damping controller for the changing power system operating conditions. The first case is when the current operating condition has a well-fitted counterpart in the previously set cases. In this case, the output trajectory of CART will be a specific case number in the overwhelming majority. The corresponding controller can then perform well enough to damp the oscillations.

The second case is when there is no equivalent to the current operating condition in the previously set cases. In this case, the output trajectory of CART will be uncertain numbers of several cases. Each case in the output of CART appears with probability *P*. Assume that in total *s* cases appear in the output trajectory of CART; then a combined control can be formed as

$$\mathbf{K}_{com} = \beta_1\mathbf{K}_1 + \beta_2\mathbf{K}_2 + \cdots + \beta_s\mathbf{K}_s \tag{7.33}$$

The combination factors β_1, β_2, ..., β_s are determined by

$$\beta_i = \sum_{j=1}^{Li} P_j / \sum_{i=1}^{s}\sum_{j=1}^{Li} P_j \tag{7.34}$$

where Li is the number of times case i appears. Combination factors β_1, β_2, ..., β_s represent the weights for different predesigned controllers. From (7.34), it can be seen that the larger Li is, the larger is the combination factor of case i, which indicates a higher matching degree of case i with the current operating conditions.

The data set is mainly decided by system operation experience and can be expressed as a set of measured vectors: $\mathrm{X}=[x_1,x_2,\ldots,x_m]^T$, where x_i^T is an n-dimension row vector (n is the number of measurements) and m is the number of cases in CART. By applying the sampling values of the PMU to (7.28), x_i^T can be calculated. Then by applying x_i^T to (7.32), the attribute value **d** can be gained.

At the same time, in order to improve the robustness of CART, a 1% disturbance is added to the initial value of every center operating state. Thus, x_i^T actually refers to the attribute effective range of case i, denoted as: $x_i^T = [x_{i1}, x_{i2}, \ldots, x_{i100}]^T$, where x_{i1}^T corresponds to the attribute value of the center operating state and x_{ij}^T corresponds to the attribute value of the disturbed center operating state. In the computation, if the actual operating point corresponds to any attribute value in x_i^T, then it could be classified as belonging to case i. Since the CART algorithm is a detective learning method, the classification of all the measured vectors x_i^T should be known.

CART Building Algorithm and Performance Evaluation The CART intent is to partition the space into disjoint subsets in order to increase the "purity" of such subsets. In other words, maximal class separation is equivalent to reduction in the misclassification of cases by node t. Suppose $p(j/t), 0 \le p(j/t) \le 1$, where $j = 1,2,\ldots,J$ and $p(j/t)$ is the probability that a condition is in case j given that it is in node t. J is the number of cases in the system. According to the Gini Index formula, the following misclassification rate could be gained:

$$i(t) = \sum_{j=1}^{J}\sum_{\substack{i=1 \\ i\neq j}}^{J} p(j/t)\,p(i/t) \tag{7.35}$$

On the other hand, the operation performance is closely related to the complexity of the tree, which refers to the number of leaf nodes in the tree. For trees whose number of equilibrium points equal the number of leaf nodes, the resub error (the proportion of cases misclassified) is zero.

For classification performance on data not explicity used to construct the classifier, the resub error is biased low. A tree with one node, however, minimizes complexity and classifies by the plurality rule: the resub error is biased high. The right-sized tree is one with (possibly) greater than one and fewer than N terminal nodes and which balances complexity and unbiased performance estimates.

The resub error of the whole tree is

$$R(T)=\sum_{t\in\tilde{T}}r(t)p(t) \tag{7.36}$$

$$r(t)=1-\max\left(p(j/t)\right),j=1,2,\ldots,J \tag{7.37}$$

$$p(t)=N(t)/N \tag{7.38}$$

where $\tilde{T}$ is the set of leaf nodes in tree T. The quantity $\max(p(j/t))$ gives the proportion of total cases for the majority class in node t, $r(t)$ is the resub error of node t, $N(t)$ represents the number of cases at node t, N is the total number of cases and $p(t)$ measures the contribution of each node to the misclassification rate.

It can be seen from (7.36) to (7.38) that the value of $R(T)$ is dependent on the complexity of the tree. In order to get the best tree size, the cost complexity function $R_\alpha(T)$ is introduced as a criterion to identify the best performing tree under the same tree complexity. $R_\alpha(T)$ is expressed as

$$R_\alpha(T)=R(T)+\alpha|T|,\alpha>0 \tag{7.39}$$

where $|T|$ is the number of leaf nodes in tree T and α is the complexity coefficient. It can be seen from (7.39) that, when the misclassification rate of the tree or the number of leaf nodes increases, the cost complexity will also increase. Therefore, the minimum cost complexity corresponds to the best tree size.

7.2.2.2 An Example Test System

Test System A 16-machine 68-bus system, which is a reduced-order model of the New England–New York interconnected system, is employed to test the performance of the proposed control method. In the single line diagram of the system, zone 4 contains machines that are a simplified representation of the New England system, while zone 5 represents the New York system. The thick lines indicate the tie lines, which are the root cause of the low-frequency inter-area oscillations. The most important connections in this system model are those between groups 4 and 5 (bus 1 to bus 2, bus 1 to bus 27, and bus 8 to bus 9). The dashed lines indicate the coherent group boundaries. In order to examine the performance of the adaptive control in damping inter-area modes, TCSC, SVC, and ESD are added to the system model. IEEE type DC1 excitation systems are installed in all machines. The load applies the mixed-load model. The active power is 15% constant power, 25% constant current, and 60% constant impedance.

Table 7.7 Inter-area modes of the test system.

Mode	Frequency (Hz)	Damping ratio	Participating generators
1	0.40544	−0.0012765	G1–G13 versus G14–G16
2	0.50915	0.0080492	G14 versus G16
3	0.64127	0.029791	G12–G13 versus G1–G9
4	0.79172	0.032764	G13 versus G14

The reactive power is 15% constant power, 25% constant current, and 60% constant admittance.

The geometric measures of controllability (GMC) and observability (GMO) are carried out to identify the locations and input stabilizing signals for the actuators. The most suitable location for TCSC is line 41–42, and bus 22 for ESD.The effective input stabilizing signals with the highest GMO are the bus frequencies $\omega_{5\text{-}16}$, $\omega_{13\text{-}16}$, $\omega_{14\text{-}16}$, and $\omega_{15\text{-}16}$, which denote the differences of the generator external bus frequencies from the reference generator 16 frequency.A transfer function of the frequency measurement (IEEE C37.118 class M) is also used in the design to account for the PMUs frequency measurement dynamics.

Apart from enhancing the transfer capacity of the transmission system, the ESD can also improve the system damping for undesirable inter-area oscillations. There are two control loops in this model, The inner loop is the Q control loop, which can guarantee the voltage stability of the bus where the ESD is installed using local signals. The external control loop is the P control loop, which can guarantee the active power stability using wide-area signals. The wide-area signals based on a combination of bus frequencies serve as input to the P loop as supplementary modulation. The inter-area modes obtained from the eigenvalue analysis are displayed in Table 7.7. There are four critical inter-area modes. The lowest oscillatory mode has the generators in groups 3, 4, and 5 oscillating against the generators in groups 1 and 2. In the second mode, generator 14 is oscillating against generator 16. The third mode has the generators in group 5 oscillating against the generators in group 4. The fourth mode has generator 13 oscillating against generator 14.

Design of Adaptive Control Scheme The structure of the control scheme is shown in Figure 7.15. The control system is composed of a family of multi-input–multi-output (MIMO) controllers $\mathbf{K}_1,\cdots\mathbf{K}_i,\cdots\mathbf{K}_n$, and the CART is used as the decision maker to choose the most suitable controller.

In Figure 7.15, if the system is operating precisely at case i in the case bank, the output trajectory of the CART will be the number of case i. Then $\mathbf{K}_i$ will be chosen as the matching controller. If there is no matching controller, the controllers would be combined according to (7.33) and (7.34). Once the CART detects that the operating point is outside the case bank, the controllers will be combined.

The MIMO controllers $\mathbf{K}_1,\ldots,\mathbf{K}_i,\ldots,\mathbf{K}_n$ are designed off-line. The test system has 141 operating states in open-loop. This number will increase as the loop is closed, that is, in the output feedback design. Therefore, the order of the test

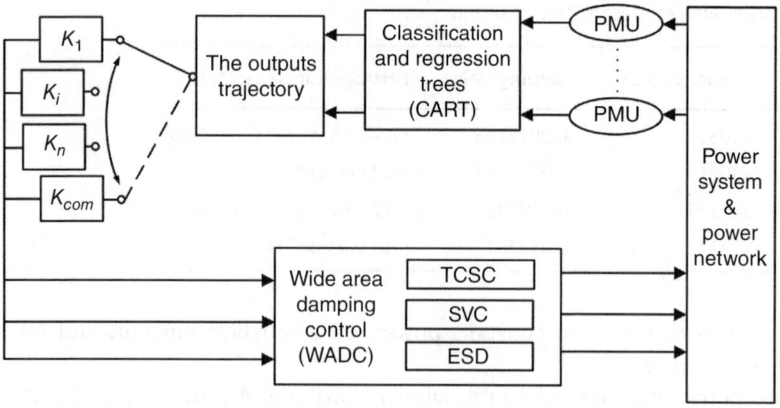

Figure 7.15 The structure of the control scheme.

system has to be reduced first before the controller design is carried out. The Schur model reduction is used to reduce the model order, which is able to reserve the outputs and inputs of the original system. It is demonstrated that the 10th reduced-order model has a similar input–output characteristic in the desired frequency range to the full-order system. Based on the reduced model, the multiple-input–multiple-output feedback controllers are designed using H_2/H_∞ robust control for each case in the case bank.

Test Cases In this section, 10 typical cases, which are greatly different from each other (see Table 7.8), are used to test the effectiveness of the proposed adaptive damping method. The load changes from 40% to 130% under both situations of all lines in service and line 1–2 outage. When the load changes from 40% to 130% or line 1–2 is cut off, it is difficult to find a single optimization feedback gain matrix for such a wide range of changes. The singular value plots of four relatively

Table 7.8 Case bank.

No.	Line outages	Load of zone 4	Load of zone 5
1	No outage	100%	100%
2	No outage	40%	40%
3	No outage	70%	70%
4	No outage	130%	130%
5	Line 1–2 outage	100%	100%
6	Line 1–2 outage	40%	40%
7	Line 1–2 outage	70%	70%
8	Line 1–2 outage	130%	130%
9	Exciter of G13 outage	100%	100%
10	Exciter of G15 outage	100%	100%

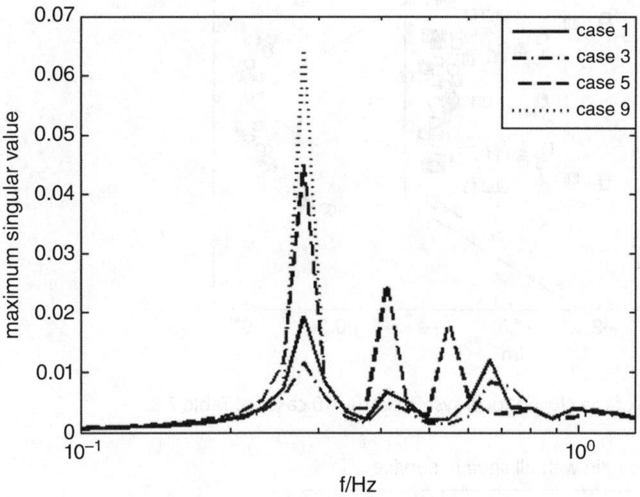

Figure 7.16 Maximum singular value plots for different cases.

close cases are displayed in Figure 7.16. It can be seen that the operating conditions are far apart from each other.

More actuators mean more feasibility for more critical cases. In Table 7.8, cases 5–8 are double-circuit line outage, which are more critical than only one of the double circuit lines outage. There are $N-2$ contingencies that make the simulation even more critical. The simulation environment with a mixed load model is also more demanding than that with a constant impedance load model. In view of these serious contingencies, it is impossible to find a feasible feedback controller using only one or two actuators.

Frequency domain analysis of the test system is carried out concerning the ten cases shown in Table 7.8. The eigenvalues of the closed-loop matrices in the ten cases are shown in Figure 7.17. The circles and squares represent eigenvalues of the open-loop and closed-loop system, respectively. The D-stability region is used as the LMI region for pole placement. It is obvious that all closed-loop poles lie in the D-region using TCSC, SVC, and ESD.

Figure 7.18 shows the stability regions of controllers 1–8, with the damping ratio being 12.5% and the loads in zone 4 and zone 5 changing from 0% to 200%. It can be seen that the stability regions have overlapped each other, which demonstrates the smooth switch between different controllers for different operating conditions. With more predesigned controllers, the whole area could be covered more thoroughly for a wider range of changes in the operating points.

Different time delays in wide-area signals are considered and tested. Nonlinear simulation tests with 150 ms and 300 ms time delays for case 1 are carried out to examine the control performance, shown in Figure 7.19. The simulation results show that the relative angular separations between G13 and G16 and G13 and G14 can be settled within the time limit under different time delays, though the settling time is a little longer due to time delays.

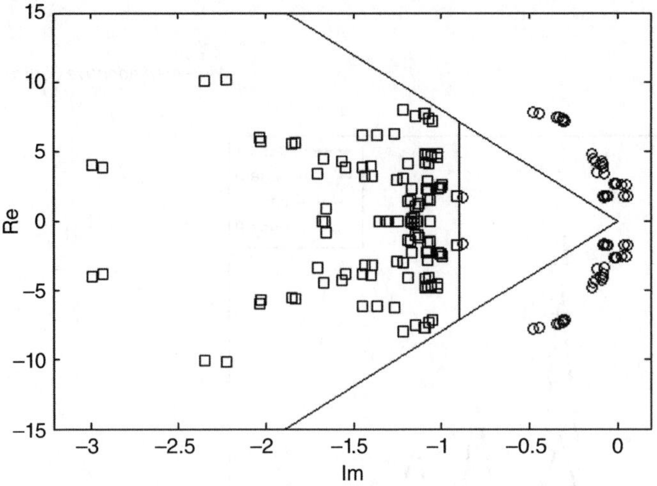

Figure 7.17 Eigenvalues of the closed-loop system in the 10 cases of Table 7.8.

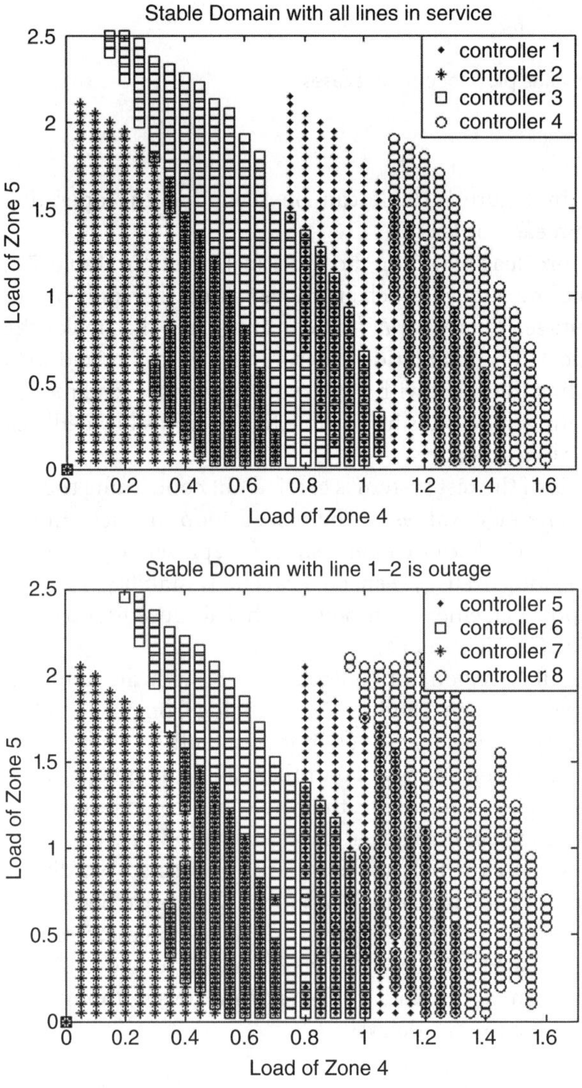

Figure 7.18 Stabilily regions of different controllers under different operating conditions.

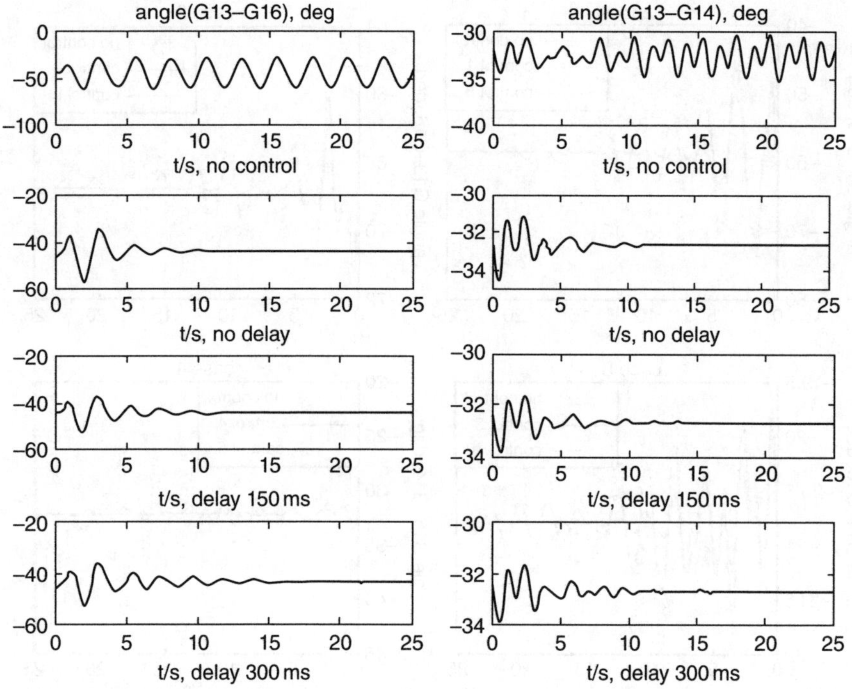

Figure 7.19 Dynamic responses of power systems with different time delays.

A nonlinear simulation test with a time duration of 25 s is carried out to examine the performance of the controllers. An 80 ms (approximately 5 cycles) temporary three-phase short-circuit fault occurs on the tie line between buses 1 and 2, while the disturbance for cases 5–8 is a temporary three-phase short-circuit fault occurring on the tie line between buses 1 and 27. Ten typical cases in the case bank are used to examine the performance of the controllers. The dynamic responses of the relative angular separations between G1 and G16, G13 and G16, G14 and G16, and G13 and G14 are shown in Figure 7.20. The dotted curves represent the dynamic responses without controllers. The solid and dashed curves represent the relative angular separations with corresponding and mismatching controllers put into service, respectively. The solid curves in 12–15s demonstrate the good performance of the corresponding controllers in damping inter-area oscillations. The dashed curves show that, when the controller does not match with the operating point, the oscillations cannot be damped effectively.

7.2.2.3 Results and Discussions

Formation of Classification and Regression Trees (CART) for Adaptive Control Scheme The changing active power of line 1–2, line 8–9, line 41–42, and line 50–51 are four measurements to select the suitable controller. The sampling rate of each of these measurements is 30 samples per second. To meet the statistical test requirement, the learning sample for the classification tree is generated by

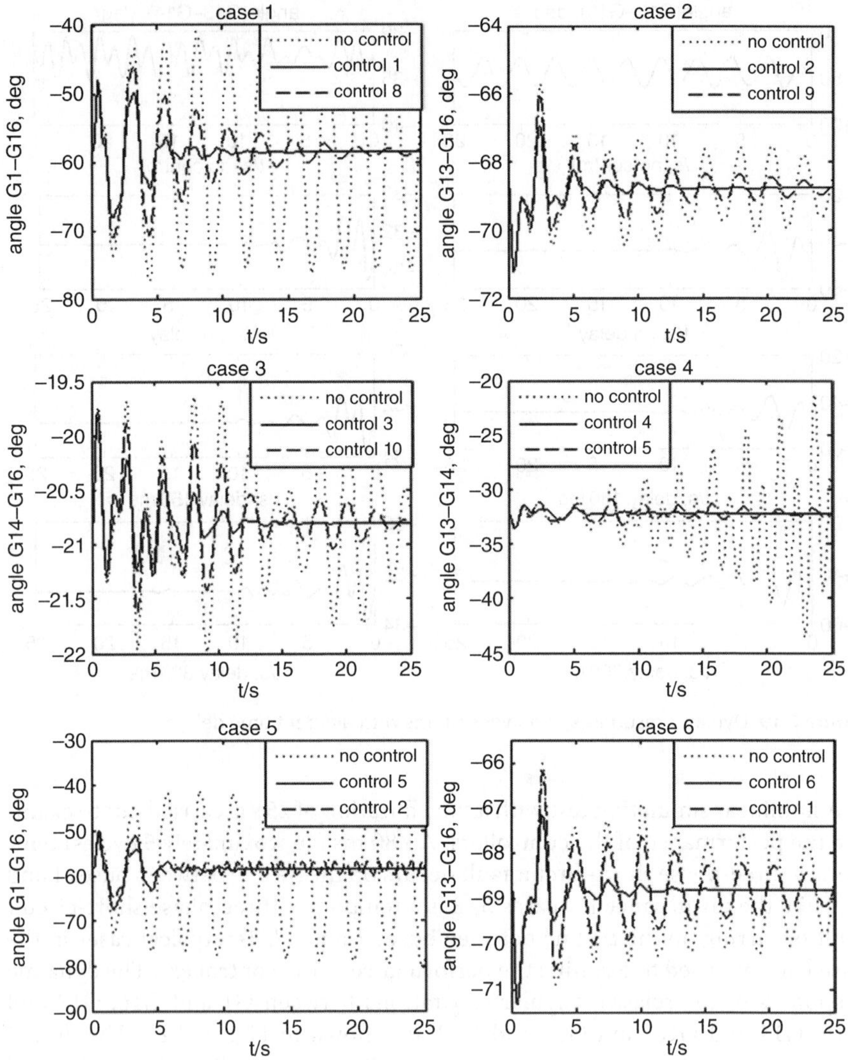

Figure 7.20 Dynamic responses of the power system with different controllers for each case: (-) with the corresponding controller; (--) with the mismatching controller; and (···) without the controller.

simulating the 16-bus system 1000 times with frequently occurring disturbances around the equilibrium point for each case.Therefore, in total 10 000 stimulation results for 10 cases are used as the learning sample for CART. If there is only one measurement, the total learning sample for CART is 10 000 rows of 30 samples. For four measurements, the total learning sample for CART is 10 000 rows of 120 samples. Therefore, each hyperplane π is formed in 120 dimensions according to (7.31). Due to a combinatorial number of 2 out of 10 cases, there are 45 hyperplanes between each of two cases for classification using the CART algorithm to grow DTs. Then regression is performed by CART to learn where

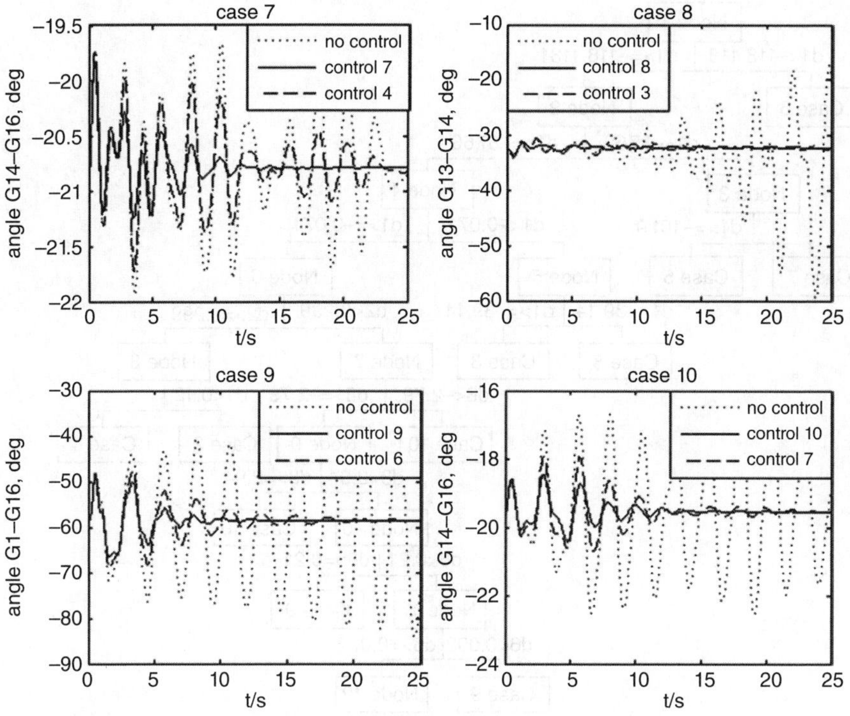

Figure 7.20 (Continued)

the current operating point is in the parameter space. A new distance vector $\mathbf{d}'$ from the current operating point $\mathbf{h}'$ to the hyperplane $\boldsymbol{\pi}$ is dropped down the tree using (7.32), and the decision to alter the current operating point is made upon reaching a terminal node.

Based on the learning sample, a sequence of minimal cost-complexity subtrees is grown. The selection of a correctly sized tree is made based on classification accuracy and tree complexity. Figure 7.20 shows the estimated misclassification rate of subtrees. It can be seen that the proper size for the subtree is 13. The misclassification rate of the su-tree is 0.026, which means a 97.4% probability of choosing the matched case.

The chosen subtree is shown in Figure 7.21. It has 13 terminal nodes that include 10 different operating points. The distances from some operating points to the hyperplanes are expressed as d1, d2,..., d45. The splitting rule for node 1 is its distance to the first hyperplane. If its distance to the first hyperplane is less than −118.118, then the current operating condition is identified to be case 6. Otherwise, the current operating point moves down to node 2 for further determination. The rest of the nodes can be split in the same manner.

Test Cases under Situation I

1) Situation Ia. Situation I represents the situation when the power system is operating precisely at a previously set operating point. For example, at t=1s, a

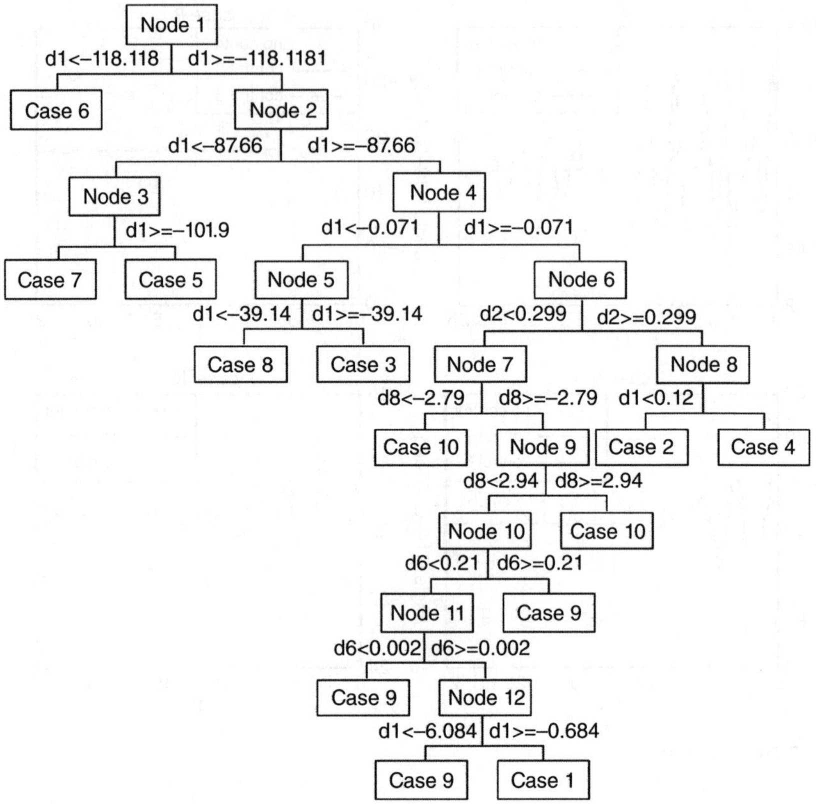

Figure 7.21 Classification and regression trees to distinguish different operating conditions.

three-phase short-circuit fault occurs on the tie line between buses 1 and 2 and lasts for 50 ms, and then this tie line is cut off. The postdisturbance dynamic change is captured by case 5. These two cases are both in the case bank. This situation is denoted as Situation Ia. Measurements obtained from PMUs are sent to CART.

The output of CART is shown in Figure 7.22. It can be seen that, at the first stage after the disturbance occurs, the CART is not able to identify the exact operating condition. The estimation at one point cannot be trusted. With more sampling points obtained, the identification result of CART is closer to the real operating condition. In Figure 7.22, case 5 is in the overwhelming majority and the time taken by CART to identify the current operating condition is less than 2 s. Finally, the postdisturbance operating condition is identified to be case 5 and the corresponding controller 5 is active to damp the oscillation modes at t=3 s.

The dynamic responses of the relative angular separations between G1 and G16, G13 and G16, G14 and G16, and G13 and G14 are shown in Figure 7.23. The dotted curves represent the dynamic responses without controllers. The solid curves represent the relative angular separations with corresponding controllers. It can be seen that the adaptive control scheme is able to

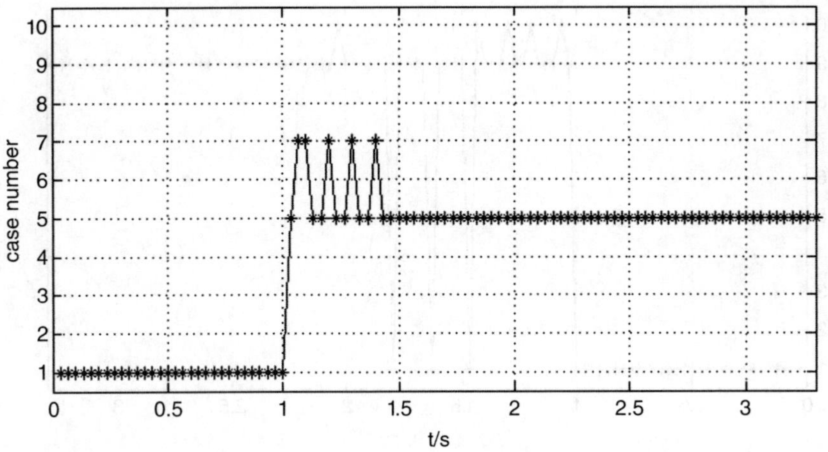

Figure 7.22 The output trajectory of CART in Situation Ia.

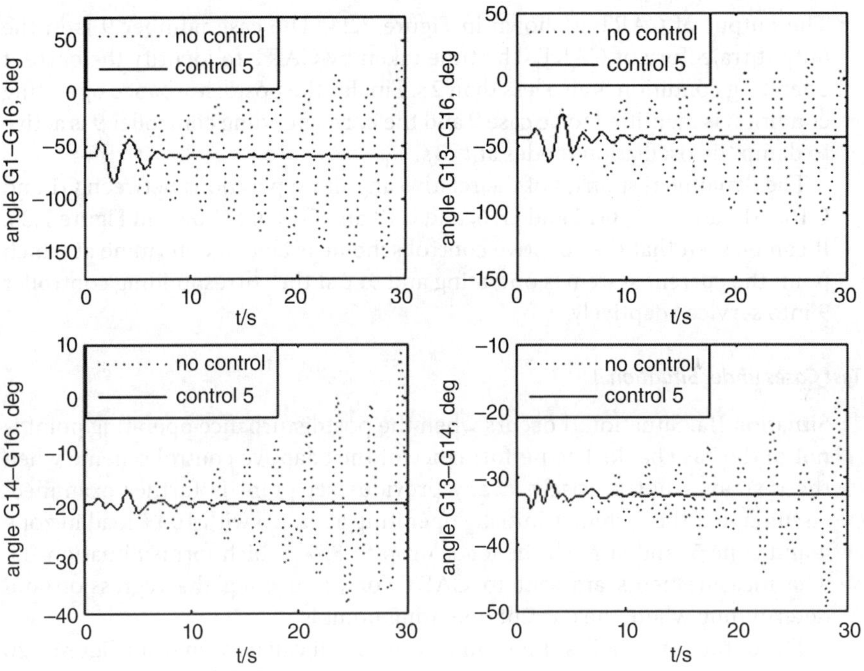

Figure 7.23 Relative angular separations of the closed-loop system with control 5.

determine at which point the current system is operating and to put the corresponding controller 5 into service adaptively.

2) Situation Ib. Situation Ib is operating at the following condition. Initially, the power system is operating at case 1, and at t=1s, the exciter of generator 13 goes out of service suddenly. The postdisturbance dynamic change is captured by case 9. Three paths from the top to the bottom lead to case 9.

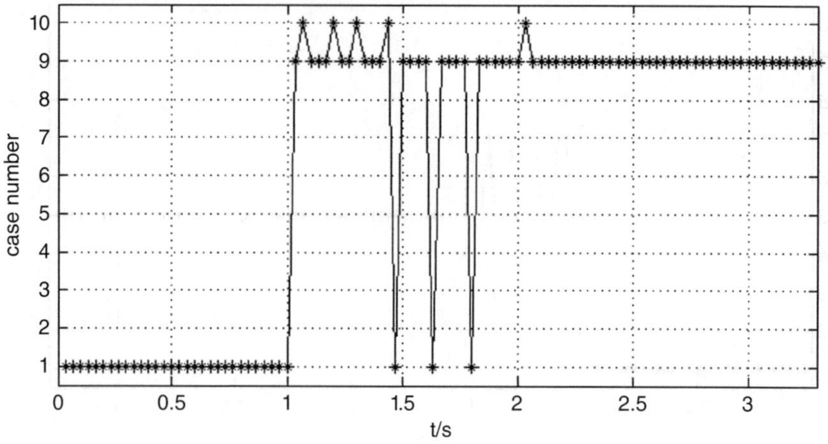

Figure 7.24 The output trajectory of CART in Situation Ib.

The output of CART is shown in Figure 7.24. The case number 9 is in the output trajectory of CART. The time taken by CART to identify the current operating condition is also less than 2 s. Finally, the postdisturbance operating condition is identified to be case 9 and the corresponding controller 9 is active to damp the oscillation modes at t=3 s.

The dynamic responses of the relative angular separations between G1 and G16, G13 and G16, G14 and G16, and G13 and G14 are shown in Figure 7.25. It can be seen that the adaptive control scheme is able to determine at which point the current system is operating and to put the corresponding controller 9 into service adaptively.

Test Cases under Situation II

1) Situation IIa. Situation II occurs when the postdisturbance operating point is not in the case bank. The performance of the adaptive control scheme when the system is not precisely at any previous set point is further examined. Assume that the system is initially operating at case 5 with 100% load in zone 4 and zone 5, and at t=1s, the loads drop to 85%, which forms Situation IIa. The measurements are sent to CART for performing the regression and determining where the current operating point is.

   The output of CART is shown in Figure 7.26. It can be seen from Figure 7.26 that, in the current operating state, the output of CART does not converge to any given case, but oscillates between case 5 and case 7. Therefore, any given controller in the controller bank is not suitable for the control of the current operating state. The combination of the controllers corresponding to case 5 and case 7 is a good way to provide effective control for the current operating state. The combination manner could be gained by solving (7.33) and (7.34). The combination factors in the dynamic process are shown in Figure 7.27. It can be seen that the combination factors converge in less than 3 seconds. Therefore, controllers 5 and 7 are combined according to (7.33) and (7.34)

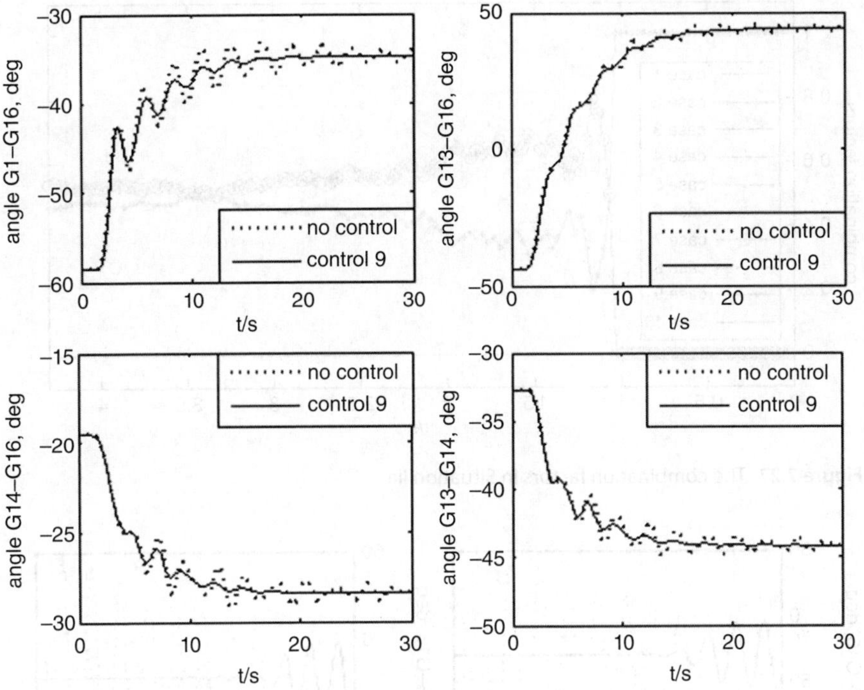

Figure 7.25 Relative angular separations of a closed-loop system with control 9.

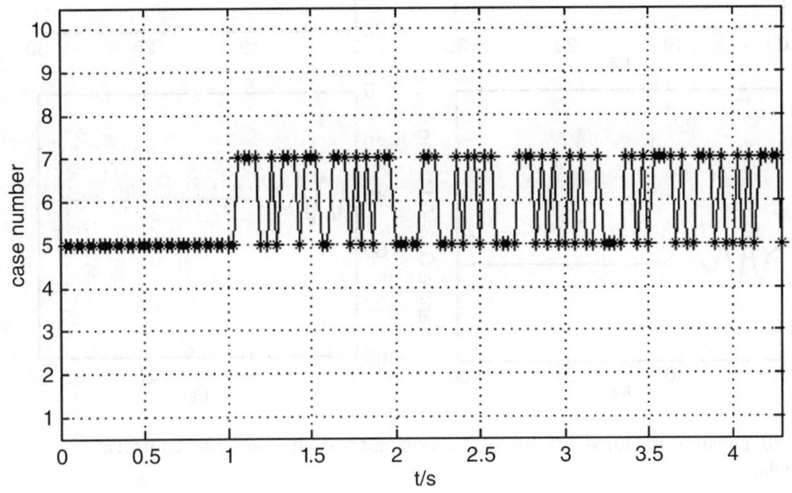

Figure 7.26 The output trajectory of CART in Situation IIa.

and put into service at t=4 s. Figure 7.28 shows the dynamic responses of the relative angular separations between generators in Situation IIa. From Figure 7.28, it can be seen that Situation IIa is in the overlapped stability region of controllers 5 and 7.

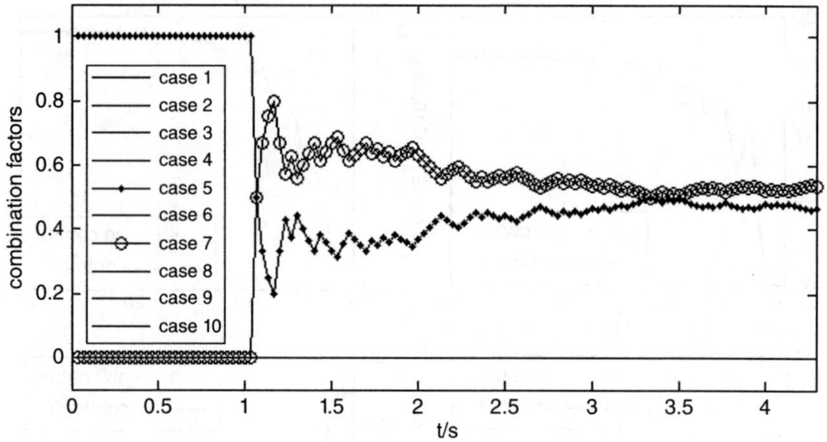

Figure 7.27 The combination factors in Situation IIa.

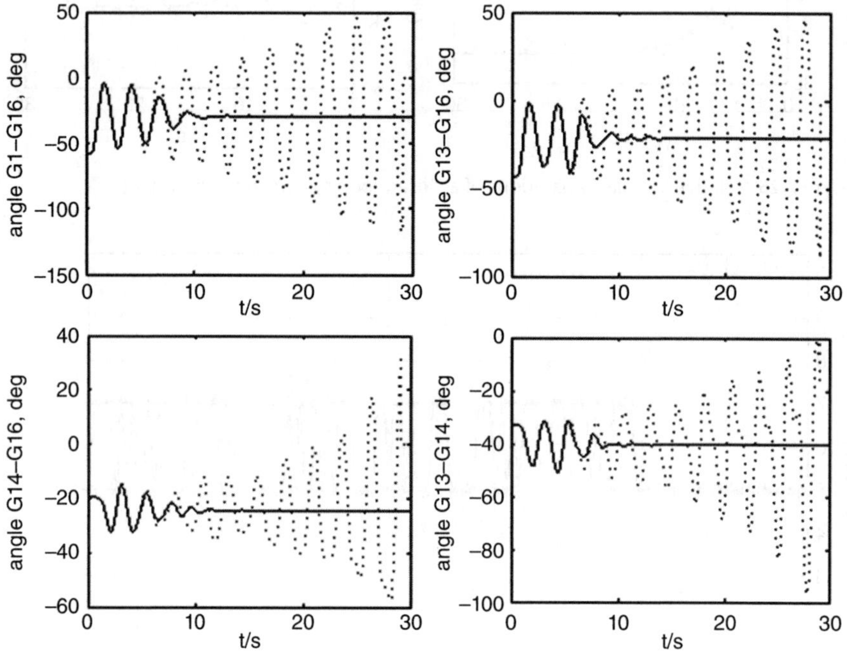

Figure 7.28 Dynamic responses of the closed-loop system with combined control in Situation IIa.

2) Situation IIb. Assume that the system is initially operating at case 3 with 70% load in zone 4 and zone 5. At t=1s, the exciter of generator 15 is suddenly removed, which forms Situation IIb.

The output of CART is shown in Figure 7.29. It can be seen from Figure 7.29 that, in the current operating state, the output of CART does not converge to any given case, but oscillates between case 3 and case 10. Therefore, any given

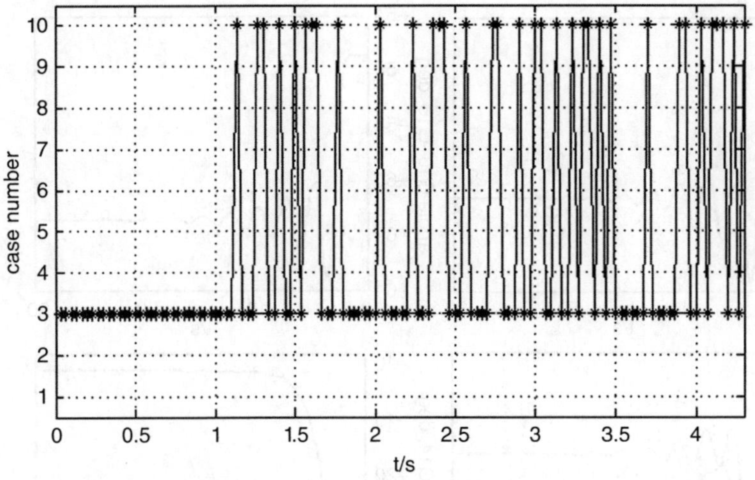

Figure 7.29 The output trajectory of CART in Situation IIb.

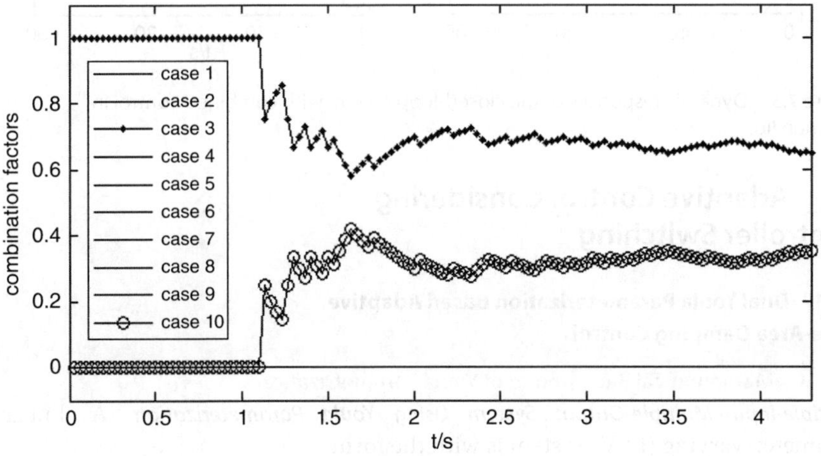

Figure 7.30 The combination factors in Situation IIb.

controller in the controller bank is not suitable for the control of the current operating state. The combination of the controllers corresponding to case 3 and case 10 is a good way to provide effective control for the current operating state. The combination manner could be gained by solving (7.33) and (7.34). The combination factors in the dynamic process are shown in Figure 7.30 and converge in less than 3 seconds. At t=4 s controllers 3 and 10 are combined and put into service according to (7.33) and (7.34). Figure 7.31 shows the dynamic responses of the relative angular separations in Situation IIb. The oscillations are also damped within 20 s.

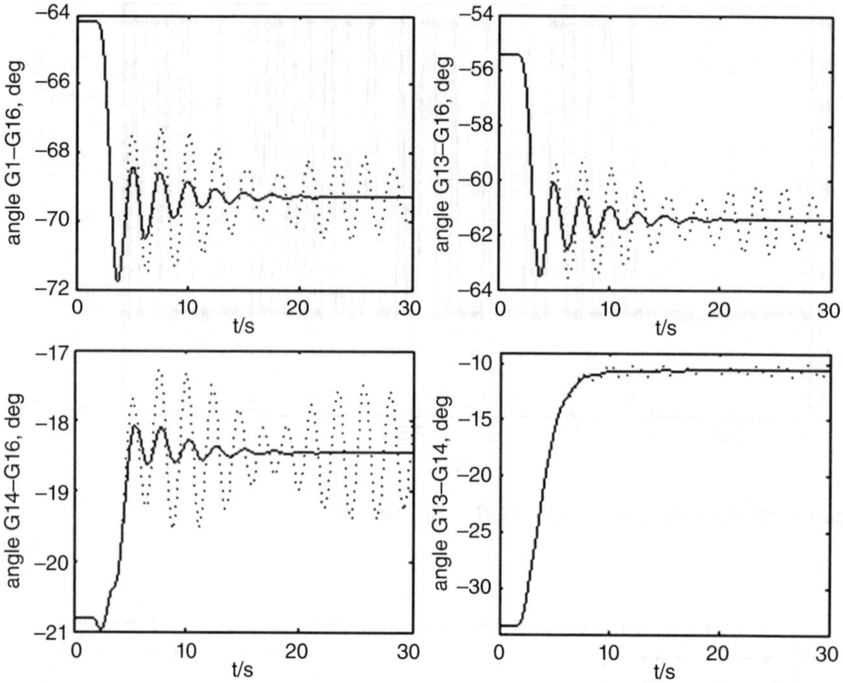

Figure 7.31 Dynamic responses of the closed-loop system with combined control in Situation IIb.

7.3 Adaptive Control Considering Controller Switching

7.3.1 Dual Youla Parameterization Based Adaptive Wide-Area Damping Control

7.3.1.1 Mathematical Background of Youla Parametrization

Multiple-Input–Multiple-Output System Using Youla Parameterization A linear parameter-varying (LPV) system is with the form

$$\begin{cases} \dot{\mathbf{x}}(t) = \mathbf{A}(\rho(t))\mathbf{x}(t) + \mathbf{B}(\rho(t))\mathbf{u}(t) \\ \mathbf{y}(t) = \mathbf{C}(\rho(t))\mathbf{x}(t) + \mathbf{D}(\rho(t))\mathbf{u}(t) \end{cases} \tag{7.40}$$

where $\boldsymbol{\rho}(t)$ is a vector of time-varying system parameters and **A**, **B**, **C**, and **D** are fixed continuous functions of $\boldsymbol{\rho}(t)$. When $\boldsymbol{\rho}(t)$ is assumed to be fixed value, the transfer function from inputs to outputs is expressed as

$$\mathbf{G}(s) = \mathbf{C}(s\mathbf{I} - \mathbf{A})^{-1}\mathbf{B} + \mathbf{D} \tag{7.41}$$

The dynamic system in (7.40) could be controlled by the following stabilizing feedback controller:

$$\mathbf{u} = \mathbf{K}(s)\mathbf{y} \tag{7.42}$$

Matrices with a tilde denote the left coprime factors, for example, $\tilde{\mathbf{M}}$, $\tilde{\mathbf{N}}$, $\tilde{\mathbf{V}}$, $\tilde{\mathbf{U}}$, and matrices without a tilde denote the right coprime factors, for example, **M**, **N**, **V**, **U**. A coprime factorization of the system **G**(s) from (7.41) and a stabilizing controller **K**(s) from (7.42) is given by

$$\begin{cases} \mathbf{G}=\mathbf{NM}^{-1}=\tilde{\mathbf{M}}^{-1}\tilde{\mathbf{N}}, & \mathbf{N},\mathbf{M},\tilde{\mathbf{M}},\tilde{\mathbf{N}}\in RH_\infty \\ \mathbf{K}=\mathbf{UV}^{-1}=\tilde{\mathbf{V}}^{-1}\tilde{\mathbf{U}}, & \mathbf{U},\mathbf{V},\tilde{\mathbf{V}},\tilde{\mathbf{U}}\in RH_\infty \end{cases} \tag{7.43}$$

where the matrices $\mathbf{N},\mathbf{M},\tilde{\mathbf{M}},\tilde{\mathbf{N}}$ and $\mathbf{U}$, $\mathbf{V}$, $\tilde{\mathbf{V}}$, $\tilde{\mathbf{U}}$ in (7.43) must satisfy the double Bezout equation given by

$$\begin{aligned} \begin{pmatrix} \mathbf{I} & \mathbf{0} \\ \mathbf{0} & \mathbf{I} \end{pmatrix} &= \begin{pmatrix} \tilde{\mathbf{V}} & -\tilde{\mathbf{U}} \\ -\tilde{\mathbf{N}} & \tilde{\mathbf{M}} \end{pmatrix} \begin{pmatrix} \mathbf{M} & \mathbf{U} \\ \mathbf{N} & \mathbf{V} \end{pmatrix} \\ \begin{pmatrix} \mathbf{I} & \mathbf{0} \\ \mathbf{0} & \mathbf{I} \end{pmatrix} &= \begin{pmatrix} \mathbf{M} & \mathbf{U} \\ \mathbf{N} & \mathbf{V} \end{pmatrix} \begin{pmatrix} \tilde{\mathbf{V}} & -\tilde{\mathbf{U}} \\ -\tilde{\mathbf{N}} & \tilde{\mathbf{M}} \end{pmatrix} \end{aligned} \tag{7.44}$$

Also taking the disturbances into account and considering a generalized nominal 2×2 system, and assuming that **G**(s) has a stable and detectable realization, gives

$$\mathbf{G}(s)=\begin{bmatrix} \mathbf{G}_{vw}(s) & \mathbf{G}_{vu}(s) \\ \mathbf{G}_{yw}(s) & \mathbf{G}_{yu}(s) \end{bmatrix} = \left[\begin{array}{c|cc} \mathbf{A} & \mathbf{B}_w & \mathbf{B}_u \\ \hline \mathbf{C}_v & \mathbf{D}_{vw} & \mathbf{D}_{vu} \\ \mathbf{C}_y & \mathbf{D}_{yw} & \mathbf{D}_{yu} \end{array}\right] \tag{7.45}$$

where $w\in\Re^r$ is the disturbance input vector, $u\in\Re^m$ is the control input signal vector, $v\in\Re^q$ is the external output signal vector to be controlled, and $y\in\Re^p$ is the measurement vector.

Let the doubly coprime factorizations of $\mathbf{G}_{yu}$ be chosen as

$$\begin{bmatrix} \mathbf{M} & \mathbf{U} \\ \mathbf{N} & \mathbf{V} \end{bmatrix} = \left[\begin{array}{c|cc} \mathbf{A}+\mathbf{B}_u\mathbf{F} & \mathbf{B}_u & -\mathbf{L} \\ \hline \mathbf{F} & \mathbf{I} & \mathbf{0} \\ \mathbf{C}_y+\mathbf{D}_{yu}\mathbf{F} & \mathbf{D}_{yu} & \mathbf{I} \end{array}\right] \tag{7.46}$$

$$\begin{bmatrix} \tilde{\mathbf{V}} & -\tilde{\mathbf{U}} \\ -\tilde{\mathbf{N}} & \tilde{\mathbf{M}} \end{bmatrix} = \left[\begin{array}{c|cc} \mathbf{A}+\mathbf{LC}_y & \mu-(\mathbf{B}_u+\mathbf{LD}_{yu}) & \mathbf{L} \\ \hline \mathbf{F} & \mathbf{I} & \mathbf{0} \\ \mathbf{C}_y & -\mathbf{D}_{yu} & \mathbf{I} \end{array}\right] \tag{7.47}$$

where **F** and **L** are chosen such that $\mathbf{A}+\mathbf{B}_u\mathbf{F}$ and $\mathbf{A}+\mathbf{LC}_y$ are both stable. The controller **F** and observer **L** are designed using the H_2/H_∞ robust control method with a pole-placement region to guarantee the good performance for nominal operating conditions [11,12].

In this manner, the nominal controller $\mathbf{K}_0 := \mathbf{U}_0\mathbf{V}_0^{-1}$ is an observer-based H_2/H_∞ robust control scheme with

$$\mathbf{K}_0=\left[\begin{array}{c|c} \mathbf{A}+\mathbf{B}_u\mathbf{F}+\mathbf{LC}_y+\mathbf{LD}_{yu}\mathbf{F} & -\mathbf{L} \\ \hline \mathbf{F} & \mathbf{0} \end{array}\right] \tag{7.48}$$

Taking the varying parameter $\rho(t)$ into account, based on the coprime factorization of the system $\mathbf{G}_{yu}(s)$ and the controller $\mathbf{K}(s)$, then the parameterization of all controllers that stabilizes the system in terms of a stable parameter $\mathbf{Q}(s)$ can be obtained; i.e. all stabilizing controllers are given by

$$\mathbf{K}(\mathbf{Q}) = \mathbf{U}(\mathbf{Q})\mathbf{V}(\mathbf{Q})^{-1} = \tilde{\mathbf{V}}(\mathbf{Q})^{-1}\tilde{\mathbf{U}}(\mathbf{Q}) \tag{7.49}$$

where $\mathbf{U}(\mathbf{Q}) = \mathbf{U} + \mathbf{MQ}$, $\mathbf{V}(\mathbf{Q}) = \mathbf{V} + \mathbf{NQ}$, $\tilde{\mathbf{U}}(\mathbf{Q}) = \tilde{\mathbf{U}} + \mathbf{Q}\tilde{\mathbf{M}}$, $\tilde{\mathbf{V}}(\mathbf{Q}) = \tilde{\mathbf{V}} + \mathbf{Q}\tilde{\mathbf{N}}$, where $\mathbf{Q} \in RH_\infty$. Using the Bezout equation, the controller given by (7.49) is able to be expressed as a linear fractional transformation (LFT) in the parameter $\mathbf{Q}$:

$$\mathbf{K}(\mathbf{Q}) = \mathbf{F}_l(\mathbf{J}_\mathbf{K}, \mathbf{Q}) \tag{7.50}$$

$$\mathbf{J}_\mathbf{K} = \begin{pmatrix} \mathbf{UV}^{-1} & \tilde{\mathbf{V}}^{-1} \\ \mathbf{V}^{-1} & -\mathbf{V}^{-1}\mathbf{N} \end{pmatrix} = \begin{pmatrix} \tilde{\mathbf{V}}^{-1}\tilde{\mathbf{U}} & \tilde{\mathbf{V}}^{-1} \\ \mathbf{V}^{-1} & -\mathbf{V}^{-1}\mathbf{N} \end{pmatrix} \tag{7.51}$$

where $\mathbf{F}_l$ represents the lower LFT. It is clear from (7.51) that $\mathbf{K}(\mathbf{Q}) = \mathbf{K}_0$ when $\mathbf{Q} = 0$.

In the same way, taking the variations of the power system into account, the parameterization in terms of a stable parameter $\mathbf{S}$ of all system that are stabilized by one controller gives

$$\mathbf{G}(\mathbf{S}) = \mathbf{N}(\mathbf{S})\mathbf{M}(\mathbf{S})^{-1} = \tilde{\mathbf{M}}(\mathbf{S})^{-1}\tilde{\mathbf{N}}(\mathbf{S}) \tag{7.52}$$

where $\mathbf{N}(\mathbf{S}) = \mathbf{N} + \mathbf{VS}$, $\mathbf{M}(\mathbf{S}) = \mathbf{M} + \mathbf{US}$, $\tilde{\mathbf{N}}(\mathbf{S}) = \tilde{\mathbf{N}} + \mathbf{S}\tilde{\mathbf{V}}$, $\tilde{\mathbf{M}}(\mathbf{S}) = \tilde{\mathbf{M}} + \mathbf{S}\tilde{\mathbf{U}}\ \mathbf{Q} \in RH_\infty$. Using the Bezout equation, the system given by (7.52) is able to be realized as a linear fractional transformation (LFT) in the parameter $\mathbf{S}$:

$$\mathbf{G}(\mathbf{S}) = \mathbf{F}_u(\mathbf{J}_\mathbf{G}, \mathbf{S}) \tag{7.53}$$

$$\mathbf{J}_\mathbf{G} = \begin{pmatrix} \mathbf{NM}^{-1} & \tilde{\mathbf{M}}^{-1} \\ \mathbf{M}^{-1} & -\mathbf{M}^{-1}\mathbf{U} \end{pmatrix} \tag{7.54}$$

where $\mathbf{F}_u$ represents the upper LFT. It is clear from (7.51) that $\mathbf{G}(\mathbf{S}) = \mathbf{G}_0$ when $\mathbf{S} = 0$.

Due to the varying parameter $\rho(t)$, the connection between the real system described as $\mathbf{G}_{yu}(\rho(t))$ and the $\mathbf{S}$ parameterization is deduced as follows:

$$\mathbf{S} = \left(\mathbf{V} - \mathbf{G}_{yu}(\rho(t))\mathbf{U}\right)^{-1}\left(\mathbf{G}_{yu}(\rho(t))\mathbf{M} - \mathbf{N}\right) \tag{7.55}$$

From equation (7.55), it can be seen that the varying parameter $\rho(t)$ is coupled with the Youla parameter $\mathbf{S}$. It is therefore necessary to give a more explicit connection between $\rho(t)$ and $\mathbf{S}$, and the upper LFT can realize the connection between $\rho(t)$ and $\mathbf{G}_{yu}(\rho(t))$ using (7.45) as follows:

$$\mathbf{G}_{yu}(\rho(t)) = \mathbf{F}_u(\mathbf{G}, \rho(t)) \tag{7.56}$$

Figure 7.32 Closedloop system with a Youla parameterization-based control scheme.

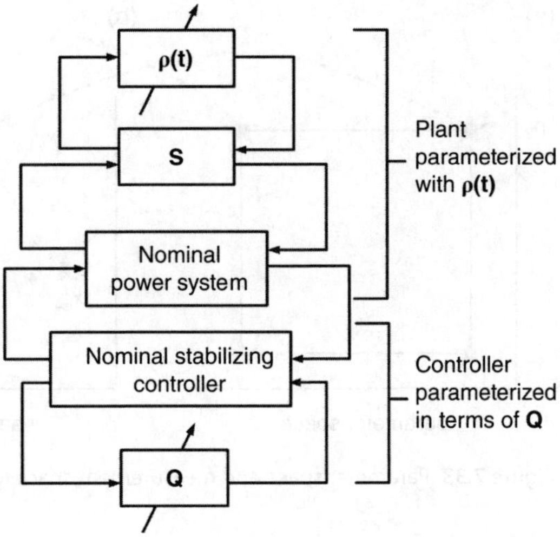

Substituting (7.56) into (7.55) yields

$$\mathbf{S}\big(\rho(t)\big)=\tilde{\mathbf{M}}\mathbf{G}_{yw}\rho(t)\Big(\mathbf{I}-\big(\mathbf{G}_{zw}+\mathbf{G}_{zu}\mathbf{U}\tilde{\mathbf{M}}\mathbf{G}_{yw}\big)\rho(t)\Big)^{-1}\mathbf{G}_{zu}\mathbf{M} \tag{7.57}$$

In this manner, the Youla parameterization-based control scheme design consists of two stages. The first stage is the design of a stabilizing nominal controller **K** for a nominal operating condition **G,** which is regarded as an off-line design. The second stage, regarded as an enhancement stage, utilizes on-line measurement to identify the changing dynamics through the updated Youla parameter **S** by (7.57). The augmented control **Q** is then designed based on the updated **S** using the loop transfer recovery [10]. The closed-loop system is shown in Figure 7.32. The measurements $\boldsymbol{\rho}(t)$ used to sense the changes of the operating condition are the frequency differences of the generator buses with reference to generator 16, since all the information of variation of system load/generation patterns or topology change is involved in the buses frequencies.

The adaptive control scheme is proposed to be scheduled adaptively using several measurements that are independent of each other to deal with the adaptability using Youla parameterization. The situation is illustrated in Figure 7.33, where p_1 and p_2 denote the changing parameters and y_1 and y_2 denote the measurements. The stars represent the equilibrium points in parameter space and the dots represent the equilibrium points in measurement space. Figure 7.33(a) represents the parameter space and Figure 7.33(b) represents the measurement space. The main idea in this paper is to make use of the measurement to schedule the controller adaptively. The changes of the operating point in the measurement space are caused by that in the parameter space. As long as the disturbance occurs in the power system, the operating point will deviate from the initial point and move toward another equilibrium point or back to the initial point. The changes in space parameter are suggested by the changes in measurement space. The engineering significance of the Youla parameterization is to find **Q** such that the Youla-based adaptive control **K(Q)** is obtained as shown in Figure 7.32.

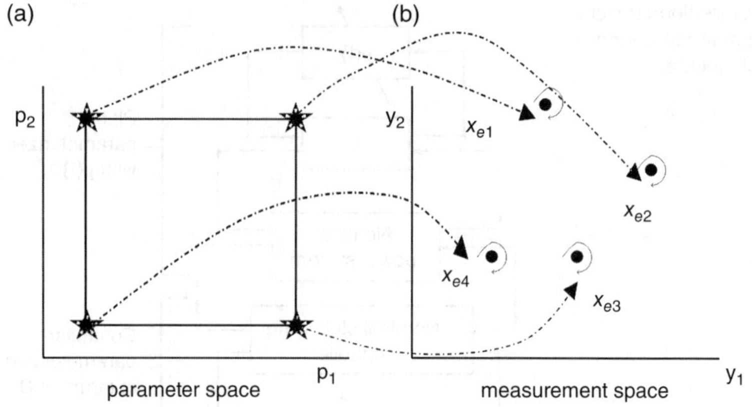

Figure 7.33 Parameter space and measurement space for the dynamic power system.

7.3.1.2 An Example Test System

Test System A 127-bus equivalent of the Western Electricity Coordinating Council (WECC) system with three additional controllers of an energy storage system is employed for testing the performance of the control method in this study. A single line diagram of the system is shown in Figure 7.34.

In Figure 7.34 the dashed lines indicate the coherent group boundaries. Groups 1 and 2 are connected by thick lines, which are the root cause of the low-frequency inter-area oscillations. The most important connections between these two areas are lines from bus 12 to bus 15. In order to damp inter-area modes and investigate the performances of an adaptive control scheme, this system is modified by adding ESDs (energy storage devices). The capacity of ESD is 10 MW, which is a feasible capacity. Three ESDs are installed at buses 37, 44, and 52, respectively, due to the highest geometric measures of controllability (GMC). The effective input stabilizing signals with the highest GMO are the bus frequencies, ω_1, ω_2, ω_6 and ω_7. Here ω_1, ω_2, ω_6 and ω_7 denote the generator external buses frequency differences from that of the reference generator 29.IEEE type DC1 excitation systems are installed in all the generators. The mixed load model is assumed at all load buses. The real power is 15% constant current and 15% constant power, and all loads support the remaining 70% as constant impedance.

The inter-area modes are obtained from the eigenvalue analysis and are displayed in Table 7.9. There are four critical inter-area modes. In the lowest oscillatory mode shape, generator 1 and 2 in group 1 is oscillating against generator 7 in group 2. In the second mode shape, generator 18 in group 3 is oscillating against the generator 20 in group 5. The third mode shape has the generators in group 2 oscillating against the generators in group 3. The final mode shape has generator 8 and 9 oscillating against generators 10–14 in group 4.

Youla-Based Adaptive Control Design The test system has 232 states in the open loop. In output feedback design, this number will increase as the loop is closed, depending on the number of states associated with the controllers of these devices. Therefore, the full-order model of the power system has to be reduced before the controller design is carried out. First, the original system is reduced to

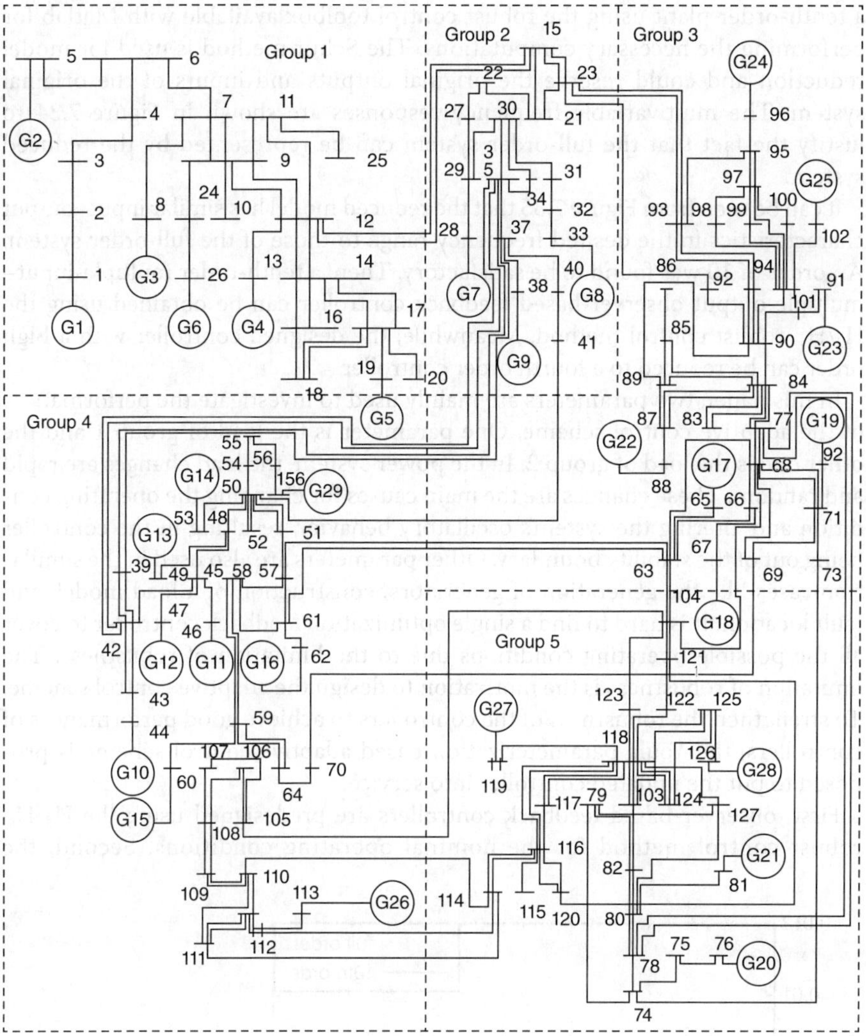

Figure 7.34 WECC system.

Table 7.9 Interarea modes of the study system.

Mode	Frequency(Hz)	Damping ratio	Dominant participated generators
1	0.43234	−0.008543	G1–G2 versus G7
2	0.5537	0.01575	G18 versus G20
3	0.6551	0.03842	G7–G9 versus G22–G23
4	0.8214	0.04531	G8–G9 versus G10–G14

a tenth-order plant using the robust control toolbox available with Matlab for performing the necessary computations. The Schur method is used for model reduction and could reserve the original outputs and inputs of the original system. The multivariable frequency responses are shown in Figure 7.34 to justify the fact that the full-order system can be represented by the reduced system.

It can be seen from Figure 7.35 that the reduced model has similar input–output characteristics in the desired frequency range to those of the full-order system. An order of 10 was found to be satisfactory. Then, a tenth-order multiple-input–multiple-output observer-based feedback controller can be obtained using the H_2/H_∞ robust control method. Meanwhile, the designed controller with a high order can be reduced to a fourth order controller.

In this paper, two parameters are mainly used to investigate the performances of the adaptive control scheme. One parameter is the load of group 1 and the other one is the load of group 2. In the power system, the load changes are rapid and random. These changes are the main causes for changing the operating condition and altering the system's oscillatory behavior, resulting in the controller being out of the stability boundary. Other parameters are also used in the simulation cases like the generation of generators, construction of a load model, and fault location. It is hard to find a single optimization feedback controller to cover all the possible operating conditions due to the limitation of robustness. The limitation of robustness is the motivation to design the adaptive control scheme. To strengthen the robustness of the controllers to achieve good performances of controllers, the Youla parameterization based adaptive control scheme is proposed to put the updated controller into service.

First, observer-based feedback controllers are predesigned using the H_2/H_∞ robust control method for the nominal operating conditions. Second, the

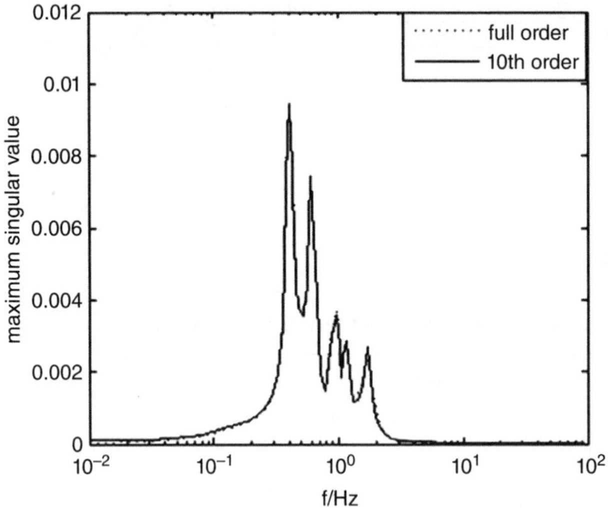

Figure 7.35 Multivariable frequency response plots of the input–output characteristics of the power system. (-) reduced; (···) full order.

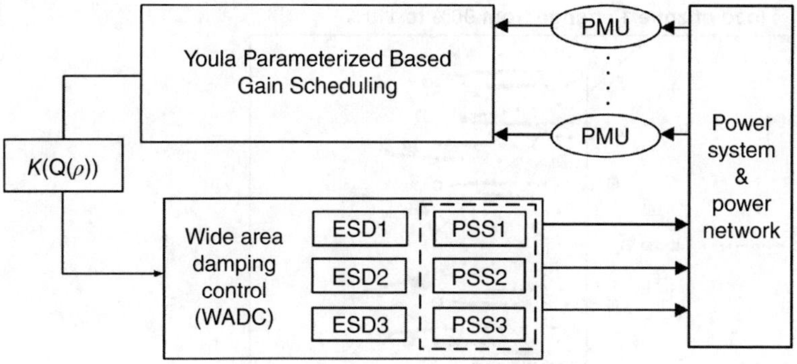

Figure 7.36 The construction of an adaptive control scheme.

controllers are parameterized using the Youla method. Finally, the Youla-based gain scheduling variables will adjust the parameterized controllers to adapt to changing operating conditions through the updated Youla parameters and put the updated controller into service. The construction of the control scheme is shown in Figure 7.36. The signals are sent to the Youla parameterized control scheme from the phasor measurement unit (PMU). The process of Youla parameterized control and **K(Q)** are playing the centralized role of an adaptive control scheme. Three ESDs or PSSs are playing the decentralized role of actuators. Then the wide-area damping control (WADC) is formed through dedicated communication links. The WADC calculates modulation signals and sends them to three ESDs or PSSs, as shown in Figure 7.36.

7.3.1.3 Results and Discussions

Frequency-Domain Results and Analysis The frequency-domain analysis of the study system was carried out for different operating conditions in Figure 7.37. The trajectory is given of eigenvalues of open loops and closed loops with the changing loads of groups 1 and 2, respectively. The circles and dots represent the eigenvalues of open-loop and closed-loop systems, respectively. The D-stability region is used as the LMI region for pole placement. Figure 7.37(a) and (c) represents the eigenvalue trajectory with the application of the fixed observer-based H_2/H_∞ robust control scheme (abbreviated to the fixed control scheme in this paper) designed for the specific equilibrium operating point. It can be seen that the performances of the fixed control scheme demonstrate the limitation of robustness. The fixed control scheme shows poor performance as long as the operating point is beyond the boundary. Even worse, when the loads changed to 110% or 90%, the fixed control scheme deteriorated the performances. On the contrary, the adaptive control scheme based on Youla parameterization in Figure 7.37 (b) and (d) demonstrated excellent performances, independent of the parameters changing. It is obvious from the figures that all the closed-loop poles lay in the D-region with the applications of the Youla-based adaptive control scheme. The minimum and maximum loads of group 1 are 56.889% and 132%, respectively, which can guarantee that the load flows of the power system

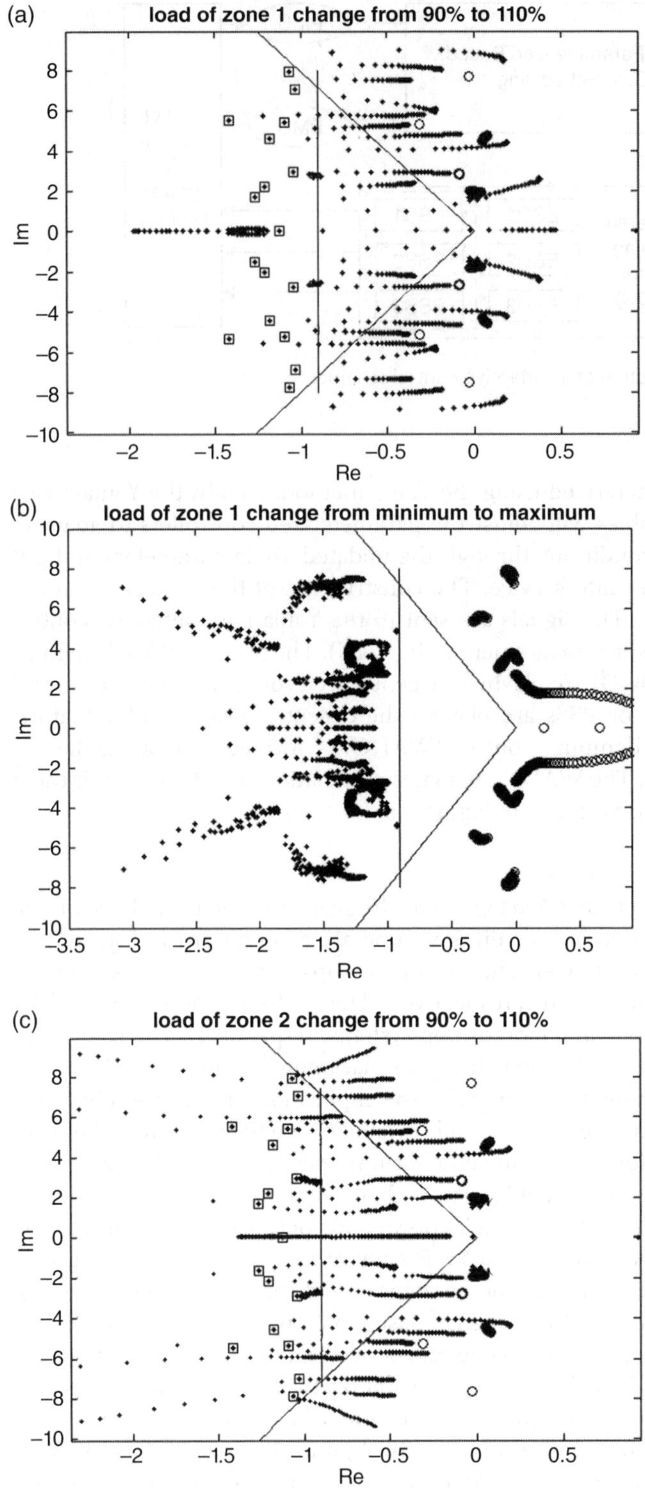

Figure 7.37 The trajectory of eigenvalues of open loops and closed loops with the changing loads of groups 1 and 2, respectively.

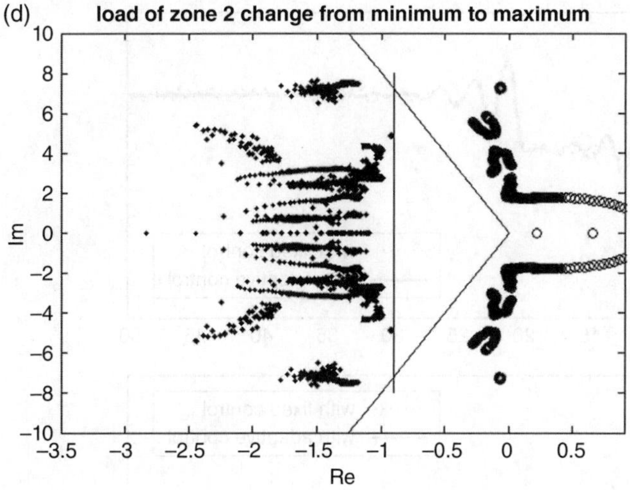

Figure 7.37 (Continued)

converge. The minimum and maximum loads of group 2 are 48.322 and 145%, respectively, which also guarantees that the load flows of the power system converge.

Time Domain Results and Analysis In order to verify the effectiveness of the adaptive scheme, the following seven cases involving different parameters are used to investigate the performance of the system under different disturbances, as shown in Table 7.10.

1) Test cases I. The disturbances for simulations in Figure 7.38 are due to continuously changing loads of zone 1. The load of zone 1 dropped to 90%, 80%, and 70% at t=1 s, t=10 s, and t=30s, respectively. Figure 7.37 shows the relative angular separations between machines G1 to G7 and G7 to G22. The dotted curves represent the relative angular separations with a fixed control scheme. The solid curves represent the dynamic responses with the Youla-based adaptive

Table 7.10 Cases used in the time-domain simulation.

Case	Main criteria used in each test
1	Load of zone 1 changing from 100% to 70%
2	Load of zone 2 changing from 100% to 70%
3	Load of zone 1 and zone 2 changing from 100% to 70% simultaneously
4	Load of zone 1 changing from 100% to 70% while load of zone 2 changing from 100% to 130%
5	Change of load construction
6	Different fault type
7	Generation of generator 7 changing from 100% to 70%

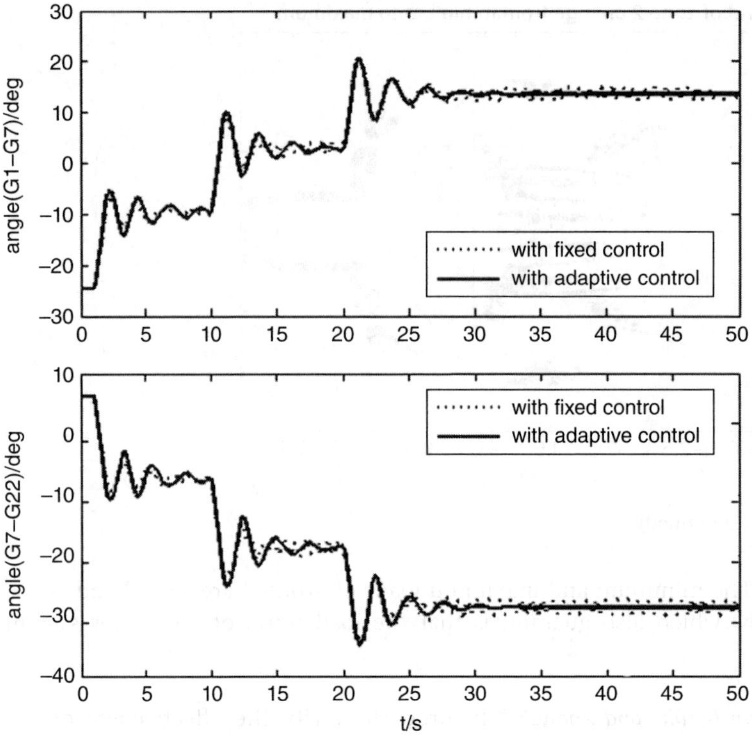

Figure 7.38 Dynamic responses of a power system with fixed and adaptive control schemes in the case of multiple disturbances of the load of zone 1.

control scheme. It can be seen from Figure 7.38 that the dashed curves demonstrate that fixed control scheme and do show some effectiveness for damping the oscillation compared with fixed control. However, with the continued dropped load, the fixed control scheme gradually loses its effectiveness. This has provided the motivation to adopt the adaptive approach. It can be seen that the Youla-based adaptive control scheme improves the robustness and effectiveness greatly under continuous changing of the load of zone 1.

2) Test cases II. The disturbances for simulations in Figure 7.39 are due to continuously changing loads of zone 2. The load of zone 2 dropped to 90%, 80%, and 70% at t=1s, t=10 s, and t=30 s, respectively.

   In Figure 7.39, the relative angular separations between machines G1 to G7 and G7 to G22 are displayed. The dotted curves represent the relative angular separations with a fixed control scheme. The solid curves represent the dynamic responses with a Youla-based adaptive control scheme. From the dotted curves, it can be seen that the fixed control scheme cannot settle down the oscillations completely. However, from the solid curves, it can be seen that the adaptive control scheme shows a better performance by adjusting itself to an increasingly changing dropped load of zone 2.

3) Test cases III. The disturbances for simulations in Figure 7.40 are due to continuously changing loads of zones 1 and 2. At t=1s, the loads of zones

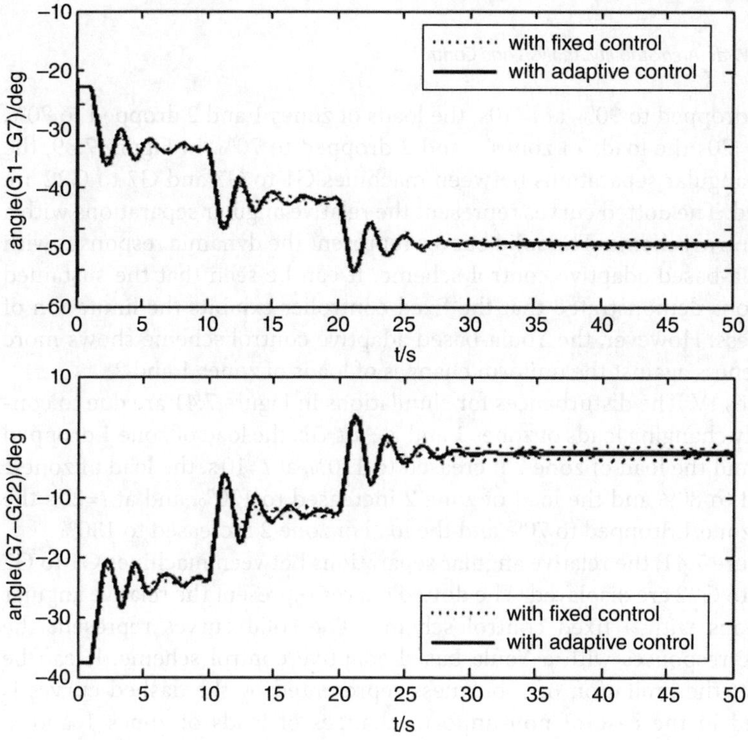

Figure 7.39 Dynamic responses of a power system with fixed and adaptive control schemes in the case of multiple disturbances of the load of zone 2.

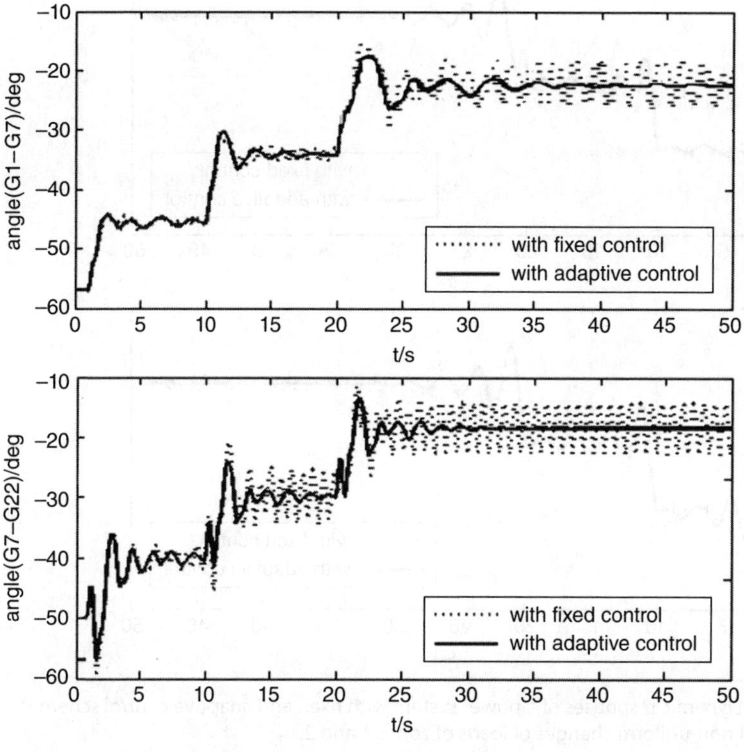

Figure 7.40 Dynamic responses of a power system with fixed and adaptive control schemes in the case of uniform changes of loads of zones 1 and 2.

1 and 2 dropped to 90%, at t=10s, the loads of zones 1 and 2 dropped to 80%, and at t=20s the loads of zones 1 and 2 dropped to 70%. In Figure 7.39, the relative angular separations between machines G1 to G7 and G7 to G22 are displayed. The dotted curves represent the relative angular separations with a fixed control scheme. The solid curves represent the dynamic responses with the Youla-based adaptive control scheme. It can be seen that the sustained oscillations demonstrated that the fixed controller exhibits the limitation of robustness. However, the Youla-based adaptive control scheme shows more effectiveness against the uniform changes of loads of zones 1 and 2.

4) Test cases IV. The disturbances for simulations in Figure 7.41 are due to continuously changing loads of zones 1 and 2. At t=1s, the load of zone 1 dropped to 90% and the load of zone 2 increased to 110%, at t=10 s, the load of zone 1 dropped to 80% and the load of zone 2 increased to 120%, and at t=20 s the load of zone 1 dropped to 70% and the load of zone 2 increased to 130%.

In Figure 7.41, the relative angular separations between machines G1 to G7 and G7 to G22 are displayed. The dotted curves represent the relative angular separations with a fixed control scheme. The solid curves represent the dynamic responses with a Youla-based adaptive control scheme. It can be seen that the limitation of robustness represented by the dashed curves is exhibited in the case of non-uniform changes of loads of zones 1 and 2.

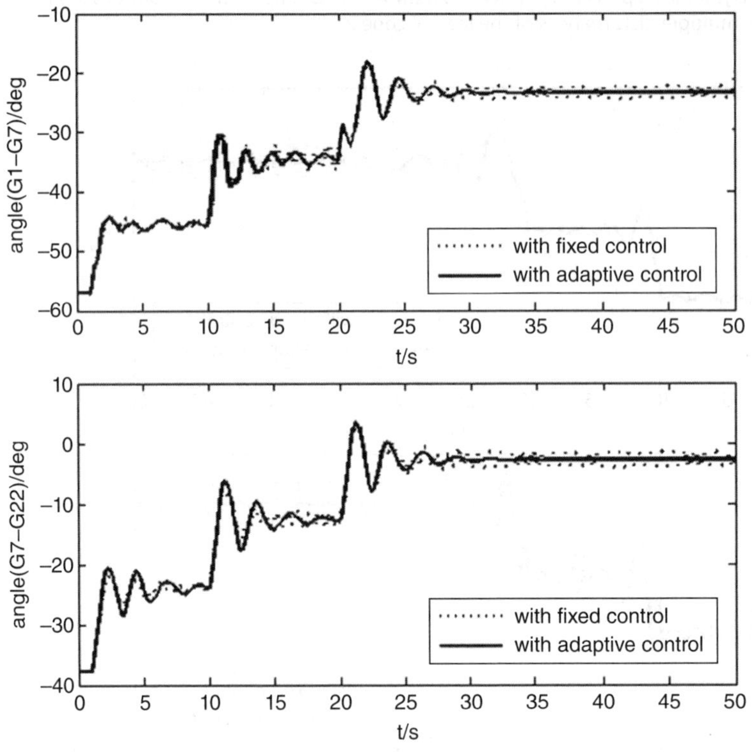

Figure 7.41 Dynamic responses of a power system with fixed and adaptive control schemes in the case of non-uniform changes of loads of zones 1 and 2.

However, the lightly damped oscillations are settled in the presence of the adaptive control scheme.

5) Test cases V. The disturbances for simulations in Figure 7.41 are that, at t=1s, the load model changed to 20% constant current and 20% constant power, at t=10 s, the load model changed to 25% constant current and 25% constant power, and at t=20 s the load model changed to 30% constant current and 30% constant power. In Figure 7.42, the relative angular separations between machines G1 and G7 and G7 and G22 are displayed. The dotted curves represent the relative angular separations with a fixed control scheme. The solid curves represent the dynamic responses with the Youla-based adaptive control scheme. The sustained oscillation represented by the dashed curve indicates the limitation of robustness. However, the adaptive control scheme shows a better performance by adjusting itself to uniform changes in the load model.
6) Test case VI. Wide-area PSSs instead of the ESD are employed in the following simulations. In the same manner, the optimum locations and feedback signals are chosen by the highest GMC and GMO. The disturbances for simulations in Figure 7.43 are continuous changing of topology. Assume lines 34–35 and 15–21 are both double circuit lines. At t=1s, line 11–12 is outage, at t=10 s, one of the double circuit lines, 34–35, is outage and at t=20 s the

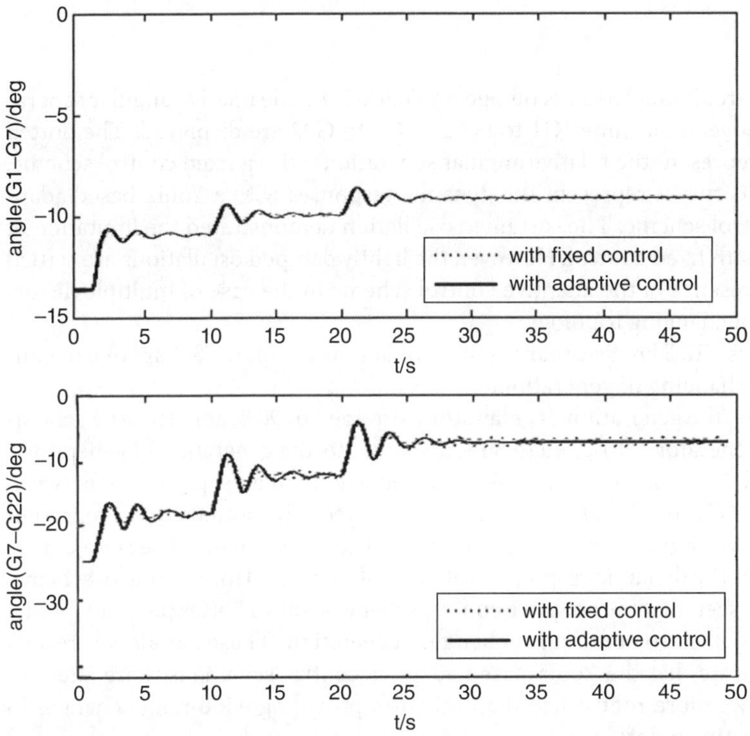

Figure 7.42 Dynamic responses of a power system with fixed and adaptive control schemes in the case of uniform changes of the load model.

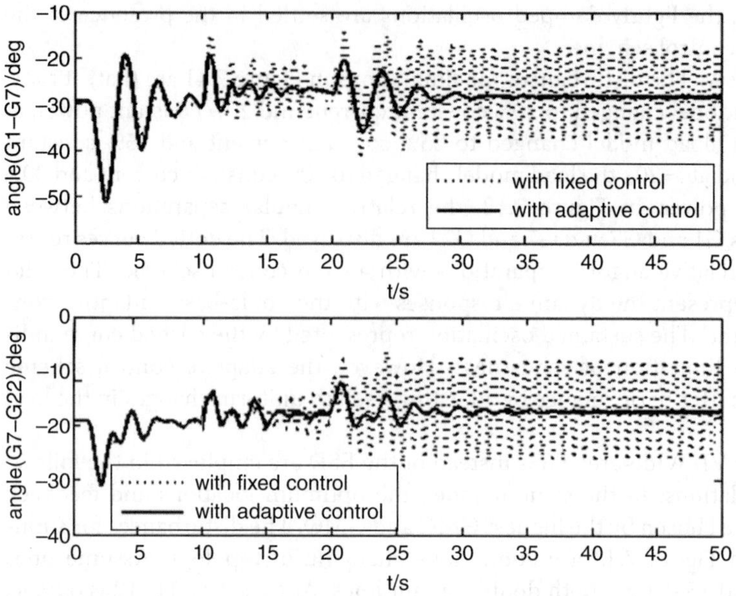

Figure 7.43 Dynamic responses of a power system without control and with a fixed control scheme in the case of multiple disturbances of changing topology.

double circuit line 15–21 is outage. In Figure 7.42, the relative angular separations between machines G1 to G7 and G7 to G22 are displayed. The dotted curves represent the relative angular separations with a fixed control scheme. The solid curves represent the dynamic responses with a Youla-based adaptive control scheme. The sustained oscillation demonstrated the limitation of robust with fixed control. However, the lightly damped oscillations are settled in the presence of the adaptive control scheme in the case of multiple disturbances of changing topology.

7) Test cases VII. The disturbances for simulations in Figure 7.44 are due to continuous changing of generation.

   At t=1s, the generation of generator 7 dropped to 90%, at t=10 s, the generation of generator 7 dropped to 80%, and at t=20 s the generation of generator 7 dropped to 70%. In Figure 7.44, the relative angular separations between machines G1 to G7 and G7 to G22 are displayed. The dotted curves represent the relative angular separations with a fixed control scheme. The solid curves represent the dynamic responses with a Youla-based adaptive control scheme. It can be seen that the adaptive control scheme shows a better performance by adjusting itself to continuous changing generation. These simulation results demonstrate that the Youla-based adaptive control scheme is more effective and shows more robustness than schemes providing wide-range changes in the operating points.

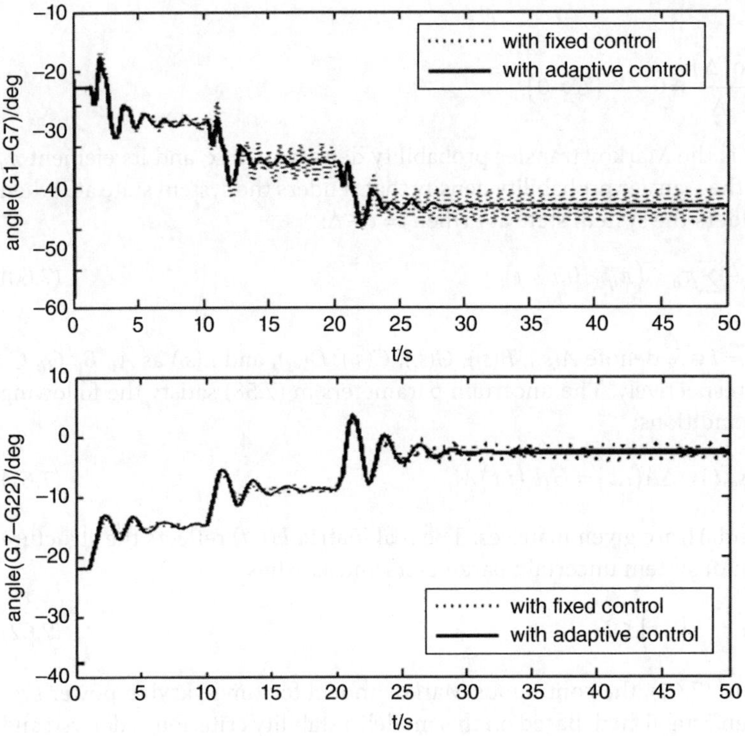

Figure 7.44 Dynamic responses of power system without control and with a fixed control scheme in the case of multiple disturbances of changing generation.

7.3.2 Continuous Markov Model Based Adaptive Control Strategy for Time-Varying Power System

7.3.2.1 Continuous Markov Model for Time-Varying Power System

The model of a time-varying power system can be expressed using the continuous Markov system as

$$\begin{cases} x_t = \left(A(s_t) + \Delta A(s_t, t)\right)x_t + B(s_t)u_t + G(s_t)\omega_t \\ z_t = C(s_t)x_t + D(s_t)u_t + L(s_t)\omega_t \end{cases} \tag{7.58}$$

where $x_t \in R^n$ is the state vector and $u_t \in R^p$ is the control input vector, ω_t is a white noise sequence that satisfies $E\{d\omega_t\} = 0, E\{d\omega_t^2\} = Wdt(W > 0)$, and $\{s(t), t \geq 0\}$ is a homogeneous Markov process defined in the finite space $S = \{1, 2, \cdots, l\}$, which is discrete in state and continuous in time and describes the variation of system operating conditions with time t. The state transfer probability of the Markov process is

$$p_{ij} = P\left(\eta_{t+\Delta} = j \mid \eta_t = i\right) = \begin{cases} \pi_{ij}\Delta + o(\Delta), & i \neq j \\ 1 + \pi_{ii}\Delta + o(\Delta), & i = j \end{cases} \tag{7.59}$$

where

$$\lim_{\Delta\to\infty}\frac{o(\Delta)}{\Delta}=0 \qquad (\Delta>0)$$

In (7.59), π is the Markov transfer probability density matrix, and its element π_{ij} represents the transfer probability density that renders the system state at i when t=t and renders the system state at j when $t=t+\Delta$:

$$\pi_{ii}=-\sum_{i\neq j}\pi_{ij} \quad (\pi_{ij}\geq 0, i\neq j) \tag{7.60}$$

For each $s_t=i\in S$, denote $A(s_t)$, $B(s_t)$, $G(s_t)$, $C(s_t)$, $D(s_t)$, and $L(s_t)$ as A_i, B_i, G_i, C_i, D_i, and L_i, respectively. The uncertain parameters in (7.58) satisfy the following matching conditions:

$$\Delta A(s_t,t)=\Delta A(i,t)=H_iF(i,t)M_i \tag{7.61}$$

where H_i and M_i are given matrices. The real matrix $F(i, t)$ reflects the structure information of system uncertain parameters and satisfies

$$F^T(i,t)F(i,t)\leq I \tag{7.62}$$

In (7.58) to (7.62), the continuous Markov model for time-varying power system has been established. Based on this model, a stability criterion is derived and an adaptive control strategy is formulated in this paper.

7.3.2.2 Robust Stochastic Stability Criterion Based on the Continuous Markov Power System Model

Definition 7.1 For the system in (7.58), when $u(t)\equiv 0$ and all the initial conditions $x_0\in R^n$ and $s_0\in S$ are satisfied, if

$$\int_0^\infty E\left\{\left\|x(t,x_0,s_0)\right\|^2\right\}dt<\infty \tag{7.63}$$

then the continuous Markov system in (7.58) is stochastic stable.

Lemma 7.1 Suppose D, E, and F are appropriately dimensioned real matrices, and matrix $P>I$, $\|F\|\leq 1$, then

1) For any vector $\varepsilon>0$ and $x,y\in R^n$, $2x^TDFEy\leq\varepsilon^{-1}x^TDD^Tx+\varepsilon y^TEE^Ty$.
2) $DFE+E^TF^TD^T\leq\varepsilon DD^T+\varepsilon^{-1}E^TE$, where I is the unit matrix.

Theorem 7.1 For given positive constant $\gamma>0$, if there exists a set of symmetric positive definite matrices $P_i>0$ $(i\in S)$, which satisfy the following matrix inequalities:

$$\begin{bmatrix} J_i & C_i^TL_i+P_iG_i \\ * & L_i^TL_i-\gamma^2I \end{bmatrix}<0 \tag{7.64}$$

$$J_i = A_i^T P_i + P_i A_i + \sum_{j=1}^{l} \pi_{ij} P_j + C_i^T C_i \tag{7.65}$$

then when $u(t) \equiv 0$, the system in (7.58) is robust stochastic stable and satisfies the disturbance attenuation degree γ, that is,

$$\|z\|_2 \le \gamma \left[\|\omega\|_2^2 + x_0^T P(s_0) x_0 \right]^{\frac{1}{2}} \tag{7.66}$$

Proof:

Construct the following Lyapunov functional:

$$V(x_t, s_t = i) = V(x_t, i) = x_t^T P_i x_i \quad (P_i > 0, i \in S) \tag{7.67}$$

when $\omega(t) = 0$, according to the weak infinitesimal operator of $V(x_t, i)$:

$$\begin{aligned} \ell V(x_t, i) &= \sum_{j=1}^{l} \pi_{ij} V(x_t, i) + x_t^T A_i^T \frac{\partial}{\partial x} V(x_t, i) \\ &= x_t^T \left[\sum_{j=1}^{l} \pi_{ij} P_j + A_i^T P_j + P_j A_i \right] x_t < 0 \end{aligned} \tag{7.68}$$

that is, there exists a set of symmetric positive definite matrices $Q_i > 0 (i \in S)$ that satisfy

$$\ell V(x_t, i) = -x_t^T Q_i x_t \tag{7.69}$$

Thus, according to (7.68) and (7.69),

$$\ell V(x_t, i) \le -\min_{i \in S} \lambda_{\min}[Q_i] x_t^T x_t < 0 \tag{7.70}$$

and then, according to the Dynkin lemma,

$$\begin{aligned} E[V(x_t, i)] - E[V(x_0, s_0)] &= E\left[\int_0^t \ell V(x_r, s_r) dr \right] \\ &\le -\min_{i \in S} \lambda_{\min}[Q_i] E\left[\int_0^t x_r^T x_r dr \middle| (x_0, s_0) \right] \end{aligned} \tag{7.71}$$

It can be seen from (7.71) that

$$\begin{aligned} &\min_{i \in S} \lambda_{\min}[Q_i] E\left[\int_0^t x_r^T x_r dr \middle| (x_0, s_0) \right] \\ &\le E[V(x_0, s_0)] - E[V(x_t, i)] \le E[V(x_0, s_0)] \end{aligned} \tag{7.72}$$

Thus

$$E\left[\int_0^t x_r^T x_r dr \middle| (x_0, s_0) \right] \le \frac{E[V(x_0, s_0)]}{\min_{i \in S} \lambda_{\min}[Q_i]} \tag{7.73}$$

When $t \to \infty$, according to (7.72),

$$E\left[\int_0^{\infty} x_r^T x_r dr \middle| (x_0, s_0)\right] \le \frac{E\left[V(x_0, s_0)\right]}{\min_{i \in S} \lambda_{\min}\left[Q_i\right]} < \infty \tag{7.74}$$

Thus, according to (7.67) to (7.74),

$$A_i^T P_i + P_i A_i + \sum_{j=1}^{l} \pi_{ij} P_j < 0 \tag{7.75}$$

For $i \in S$, $C_i^T C_i \ge 0$. Thus

$$A_i^T P_i + P_i A_i + \sum_{j=1}^{l} \pi_{ij} P_j + C_i^T C_i < 0 \tag{7.76}$$

Therefore, according to the Schur complement theorem, the system in (7.58) is robust stochastic stable.

The disturbance attenuation degree γ is proved as follows. When $\omega(t) \ne 0$, the weak infinitesimal operator of functional $V(x_t, i)$ can be expressed as

$$\ell V(x_t, i) = x_t^T \left[\sum_{j=1}^{l} \pi_{ij} P_j + A_i^T P_j + P_j A_i\right] x_t + 2 x_t^T P_i G_i \omega_t \tag{7.77}$$

Define the index function:

$$J_T = E\left[\int_0^T \left(z_t^T z_t - \gamma^2 \omega_t^T \omega_t\right) dt\right] \tag{7.78}$$

Since

$$\begin{aligned} & z_t^T z_t - \gamma^2 \omega_t^T \omega_t + \ell V(x_t, i) \\ & = x_t^T C_i^T C_i x_t - \gamma^2 \omega_t^T \omega_t + x_t^T C_i^T L_i \omega_t + \omega_t^T L_i^T C_i x_t \\ & + \omega_t^T L_i^T L_i \omega_t + x_t^T \left[\sum_{j=1}^{l} \pi_{ij} P_j + A_i^T P_j + P_j A_i\right] x_t \\ & + 2 \omega_t^T G_i^T P_i x_t + x_t^T P_i G_i \omega_t \\ & = \begin{bmatrix} x_t^T & \omega_t^T \end{bmatrix} \begin{bmatrix} J_i & C_i^T L_i + P_i G_i \\ * & L_i^T L_i - \gamma^2 I \end{bmatrix} \begin{bmatrix} x_t \\ \omega_t \end{bmatrix} \end{aligned} \tag{7.79}$$

According to (7.77) to (7.79),

$$\begin{aligned} J_T & = E\left[\int_0^T \left(z_t^T z_t - \gamma^2 \omega_t^T \omega_t + \ell V(x_t, i)\right) dt\right] \\ & - E\left[\int_0^T \left(\ell V(x_t, i)\right) dt\right] \end{aligned} \tag{7.80}$$

According to the Dynkin lemma,

$$E\left[\int_0^T (\ell V(x_t, i))dt\right] = E\left[V(x_T, s_T)\right] - E\left[V(x_0, s_0)\right] \tag{7.81}$$

Thus, according to (7.80) and (7.81),

$$\begin{aligned} J_T = E\left\{\int_0^T \begin{bmatrix} x_t^T & \omega_t^T \end{bmatrix} \begin{bmatrix} J_i & C_i^T L_i + P_i G_i \\ * & L_i^T L_i - \gamma^2 I \end{bmatrix} \begin{bmatrix} x_t \\ \omega_t \end{bmatrix} dt\right\} \\ - E\left[V(x_T, s_T)\right] + E\left[V(x_0, s_0)\right] \end{aligned} \tag{7.82}$$

Consider (7.75) and $E\left[V(x_T, s_T)\right] \geq 0$,

$$J_T \leq E\left[V(x_0, s_0)\right] \tag{7.83}$$

When $T \to \infty$,

$$\|z_2\| \leq \gamma\left[\|\omega\|_2^2 + x_0^T P(s_0) x_0\right]^{\frac{1}{2}} \tag{7.84}$$

According to (7.77) to (7.84), it is proved that (7.64) could render the system in (7.58) with a disturbance attenuation degree γ.

7.3.2.3 Adaptive Control Strategy Based on the Continuous Markov Power System Model

Design of the Adaptive Controller For all the permitted uncertainties, suppose the continuous Markov model for time-varying power system in (7.58) meets the following conditions:

1) There exists a feedback control matrix K_i that renders the system in (7.58) robust stochastic stable.
2) For the given scalar quantity $\gamma > 0$ and initial states $x_0 \in R^n$ and $s_0 \in S$, the control output z_t satisfies

$$\left[E\int_0^T z_t^T z_t dt \middle| (x_0, s_0)\right]^{\frac{1}{2}} \leq \gamma\left[\|\omega\|_2^2 + M(x_0, s_0)\right]^{\frac{1}{2}} \tag{7.85}$$

3) The steady-state error of each operating condition meets the following constraints:

$$\mathrm{Var}\left[x_t(s_t)\right] := \left[\hat{Q}(s_t)\right]_{ii} < \sigma_i^2(s_t) \tag{7.86}$$

$$\hat{Q}(s_t) = \lim_{t\to\infty} \hat{Q}_t(s_t) = \lim_{t\to\infty} E\left[x_t(s_t) x_t^T(s_t)\right] = \sum_{i=1}^{l} \hat{Q}(i) P_i \tag{7.87}$$

4) where $\sigma_i^2(s_t) > 0 (i = 1, 2, \ldots, l)$ is the upper bound of the acceptable error that meets actual need.

According to (7.61) and (7.65),

$$\begin{bmatrix} J_i & C_i^T L_i + P_i G_i \\ * & L_i^T L_i - \gamma^2 I \end{bmatrix} = \begin{bmatrix} \bar{A}_i^T P_i + P_i \bar{A}_i + \sum_{j=1}^{l} \pi_{ij} P_j & \bar{C}_i^T L_i \\ * & L_i^T L_i - \gamma^2 I \end{bmatrix} + \begin{bmatrix} \bar{C}_i^T \\ L_i^T \end{bmatrix} \begin{bmatrix} \bar{C}_i & L_i \end{bmatrix} < 0 \tag{7.88}$$

According to the Schur complement theorem, (7.88) could be transformed to

$$\begin{bmatrix} \bar{A}_i^T P_i + P_i \bar{A}_i + \sum_{j=1}^{l} \pi_{ij} P_j & P_i G & \bar{C}_i^T \\ * & -\gamma^2 I & L_i^T \\ * & * & -I \end{bmatrix} < 0 \tag{7.89}$$

where $\bar{A}_i = A_i + B_i K_i + H_i F(i,t) M_i$, $\bar{C}_i = C_i + D_i K_i$.

Since (7.89) contains nonlinear items, it needs to be transformed to LMI to facilitate the solution. Set $Q_i = P_i^{-1} (i \in S)$ and multiply both sides of (7.89) by matrix $diag(P_1, I, I)$, so that

$$\begin{bmatrix} Q_i \bar{A}_i^T + \bar{A}_i Q_i + \sum_{j=1}^{l} \pi_{ij} Q_i Q_j^{-1} Q_i & G & Q_i \bar{C}_i^T \\ * & -\gamma^2 I & L_i^T \\ * & * & -I \end{bmatrix} < 0 \tag{7.90}$$

Set $Y_i = K_i Q_i$ so that the following equation is true:

$$\begin{aligned} Q_i \bar{A}_i^T + \bar{A}_i Q_i &= Q_i A_i^T + A_i Q_i + B_i Y_i + Y_i^T B_i^T \\ &\quad + H_i F(i,t) M_i Q_i + Q\left(H_i F(i,t) M_i\right)^T \end{aligned} \tag{7.91}$$

$$Q_i \left[\sum_{j=1}^{l} \pi_{ij} Q_j^{-1} \right] Q_i = \pi_{ii} Q_i + \Im_i(Q) Q_j^{-1} \Im_i^T(Q) \tag{7.92}$$

$$Q_i \left(C_i + D_i K_i\right)^T = Q_i C_i^T + Y_i^T D_i^T \tag{7.93}$$

According to Lemma 7.1, there exists a positive constant $\xi_i > 0$ that satisfies

$$\begin{aligned} & H_i F(i,t) M_i Q_i + Q\left(H_i F(i,t) M_i\right)^T \\ & \quad \le \xi_i H_i H_i^T + \xi_i^{-1} Q_i M_i^T M_i Q_i \end{aligned} \tag{7.94}$$

According to the Schur complement theorem, the LMI of the feedback control matrix when the continuous Markov time-varying system is robust stochastic stable can be gained. In addition, referring to the method used in Theorem 7.1 to verify that the system satisfies attenuation degree γ, a similar method could be used to obtain the following robust stochastic stability control theorem

(Theorem 7.2) for the continuous Markov time-varying power system when uncertainty conditions (7.58) – (7.60) are allowed.

Theorem 7.2 For a given positive constant $\gamma > 0$ and $\sigma_{ij}^2 > 0 (i = 1,\ldots,l;$ $j = 1,\ldots, n)$, if there exist constants $\varepsilon_i > 0$, $\xi_i > 0$, and symmetric positive definite matrices $Q_i > 0$ and matrix $Y_i > 0$, which satisfy the following matrix inequalities:

$$\begin{bmatrix} \Gamma_{1i} & G_i & Q_i C_i^T + Y_i^T D_i^T & Q_i M_i^T & \Im_i(Q) \\ * & -\gamma^2 I & L_i^T & 0 & 0 \\ * & * & -I & 0 & 0 \\ * & * & * & -\xi_i I & 0 \\ * & * & * & * & -\aleph_i(Q) \end{bmatrix} < 0 \tag{7.95}$$

$$\begin{bmatrix} \Gamma_{2i} & Q_i M_i^T \\ * & -\varepsilon_i I \end{bmatrix} < 0 \tag{7.96}$$

$$[Q_i]_{jj} \le \sigma_{ij}^2 \left(i = 1,\ldots,l; j = 1,\ldots,n\right) \tag{7.97}$$

$$\Gamma_{1i} = A_i Q_i + Q_i A_i^T + B_i Y_i + Y_i^T B_i^T + \pi_{ii} Q_i + \xi_i H_i^T H_i \tag{7.98}$$

$$\Gamma_{2i} = A_i Q_i + Q_i A_i^T + B_i Y_i + Y_i^T B_i^T + \pi_{ii} Q_i + \varepsilon_i H_i^T H_i + \sum_{j=1}^{l} \pi_{ij} Q_j + G_i W G_i^T \tag{7.99}$$

$$\Im_i(Q) == \left[\sqrt{\pi_{i1}} I \quad \sqrt{\pi_{i2}} I \quad \cdots \quad \sqrt{\pi_{il}} I\right] \tag{7.100}$$

$$\aleph_i(Q) = diag\left(Q_1, Q_2, \ldots, Q_l\right) \tag{7.101}$$

then there exists the feedback control gain $K_i = Y_i Q_i^{-1}$ that renders the system in (7.58) robust stochastic stable.

Inequalities (7.95) to (7.101) form the LMI of constant variables σ_{ij}, ξ_i and matrix variables Q_i, Y_i. The matrix variables and constant variables that satisfy the LMI in (7.95) to (7.101) constitute a convex set. In order to make the solution to the LMI better to meet the actual needs, the allowed uncertainty conditions in (7.58) to (7.60) could be used to design a controller that meets specific requirements. Thus the following minimum variance robust controller in Theorem 7.3 can be gained.

Theorem 7.3 For a given positive constant $\gamma > 0$ and a set of weighting coefficients $\beta_{ij} > 0 (i = 1,\ldots,l; j = 1,\ldots,n)$ that satisfies $\sum_{i=1}^{l}\sum_{j=1}^{n} \beta_{ij} = 1$, if the following optimization problem has a solution,

$$\min_{Q_i, Y_i, \varepsilon_i, \xi_i} \sum_{i=1}^{l}\sum_{j=1}^{n} \beta_{ij} \sigma_{ij}^2 \quad s.t. \quad (7.95)-(7.101) \tag{7.102}$$

then $u_t = K_i x_t = Y_i P_i^{-1} x_t$ is the robust controller of the system in (7.58) with minimum variance.

The optimization problem in (7.102) could be solved with the method to solve the minimization problem of a linear objective function. If there exists constants $\varepsilon_i > 0$, $\xi_i > 0$ and positive definite symmetric matrices $Q_i > 0$, $Y_i > 0$, then the robust control that satisfies the minimum variance constraint can be expressed as $u_t = K_i x_t = Y_i P_i^{-1} x_t$.

System Identification To ensure that an appropriate controller could be applied to system control each time a system operating condition changes, the current system operating condition needs to be identified on line and then an appropriate combination of the subcontrollers could be formed according to the weighting indexes.

The system multivariable identification model is

$$y(t) = \Phi(t)\vartheta + v(t) \tag{7.103}$$

Define the loss function as

$$J(\vartheta) = \left\| y(t) - \Phi(t)\vartheta \right\|^2 \tag{7.104}$$

Suppose $\mu(t)$ is the step and $\mu(t) = 1/r(t)$, $r(t) = r(t-1) + \left\|\Phi(t)\right\|^2$. Use the negative gradient search method to minimize $J(\vartheta)$; thus the stochastic gradient algorithm of the estimation parameter vector ϑ can be gained:

$$\begin{aligned} \hat{\vartheta}(t) &= \hat{\vartheta}(t-1) - \frac{\mu(t)}{2} grad\left[J\left(\hat{\vartheta}(t-1)\right)\right] \\ &= \hat{\vartheta}(t-1) - \frac{\Phi^T(t)}{r(t)}\left[y(t) - \Phi(t)\hat{\vartheta}(t-1)\right] \end{aligned} \tag{7.105}$$

$$r(t) = r(t-1) + \left\|\Phi(t)\right\|^2, r(0) = 1 \tag{7.106}$$

Define the combined controller as the combination of matching controllers of the subconditions:

$$K = \alpha_1 K_1 + \alpha_2 K_2 + \cdots + \alpha_l K_l \tag{7.107}$$

where the weighting coefficient α_i at $t = k$ can be expressed as

$$J_{ik} = \alpha e_{ik}^2 + \beta \int_0^k e^{-\lambda(k-\tau)} e_{i\tau}^2 d\tau, \quad \alpha, \beta, \lambda \geq 0 \tag{7.108}$$

$$\alpha_i = \frac{\exp\left(-\left|J_i\right|^2\right)}{\sum_{i=1}^{N} \exp\left(-\left|J_i\right|^2\right)}, \quad \sum_{i=1}^{N} \alpha_i = 1 \tag{7.109}$$

$$e_{mj} = y_m - \hat{y}_m = y_m - \Phi(j)\hat{\vartheta} + v(j) \tag{7.110}$$

It can be seen from (7.107) to (7.110) that the output errors of the subconditions and the controlled object are estimated at different instants. The controller combination is then revised and updated adaptively. The more completely the subcondition where the controlled object exists matches the characteristic model of the ith subcondition, the bigger is the corresponding weighting coefficient α_i.

7.3.2.4 Simulation Analysis

Time-Varying Power System Adaptive Control Based on Wind Generator Output Power Fluctuation Take the IEEE 16-machine 68-bus New England–New York interconnected system for example. Substitute G16 in Area 3 with a power source with equal capacity and simulate the variation of the wind generator output power. The time-varying power system is divided into 10 different subconditions according to the wind speed. Thus the state set of the continuous Markov time-varying system model can be formed, denoted as $S = \{1,2,\ldots,10\}$. The cut-in wind speed is 3 m/s, the rated wind speed is 12 m/s, and the cut-out wind speed is 22 m/s.The information of different operating conditions is shown in Table 7.11.

First, use the modal analysis method to obtain the state matrix of the 16-generator system in each subcondition. Second, reduce the order of the state matrix with the SMA method. Then apply the reduced-order state matrix and the following state transfer probability density matrix to (7.90)(the disturbance attenuation degree γ is 0.1), so that the control feedback response matrix K_i corresponding to each subcondition could be gained according to the linear objective function minimization problem-solving method. Two different cases will result according to whether the current system operating condition matches the characteristic model of any subcondition or not.

Table 7.11 Operating condition information.

Subcondition number	Wind speed interval	Characteristic wind speed
Subcondition 1	3 m/s-4 m/s	3.5 m/s
Subcondition 2	4 m/s–5 m/s	4.5 m/s
Subcondition 3	5 m/s–6 m/s	5.5 m/s
Subcondition 4	6 m/s–7 m/s	6.5 m/s
Subcondition 5	7 m/s–8 m/s	7.5 m/s
Subcondition 6	8 m/s–9 m/s	8.5 m/s
Subcondition 7	9 m/s–10 m/s	9.5 m/s
Subcondition 8	10 m/s–11 m/s	10.5 m/s
Subcondition 9	11 m/s–12 m/s	11.5 m/s
Subcondition 10	>12 m/s	12.5 m/s

$$
\pi = \begin{bmatrix}
-0.2414 & 0.2413 & 0.0001 & 0 & 0 & 0 & 0 & 0 & 0 & 0 \\
0.2061 & -0.4255 & 0.2184 & 0.001 & 0 & 0 & 0 & 0 & 0 & 0 \\
0.0469 & 0.2205 & -0.5102 & 0.2388 & 0.004 & 0 & 0 & 0 & 0 & 0 \\
0.0079 & 0.0520 & 0.2102 & -0.5481 & 0.2694 & 0.0086 & 0 & 0 & 0 & 0 \\
0.0012 & 0.0103 & 0.0553 & 0.202 & -0.5736 & 0.2901 & 0.0147 & 0 & 0 & 0 \\
0.0001 & 0.0013 & 0.01 & 0.0523 & 0.1953 & -0.5678 & 0.2975 & 0.0113 & 0 & 0 \\
0 & 0.0001 & 0.0011 & 0.0086 & 0.0483 & 0.1979 & -0.5319 & 0.2728 & 0.0031 & 0.0001 \\
0 & 0 & 0.0001 & 0.0009 & 0.0073 & 0.045 & 0.2057 & -0.4779 & 0.2189 & 0.0003 \\
0 & 0 & 0 & 0 & 0.0005 & 0.0046 & 0.0357 & 0.204 & -0.3848 & 0.14 \\
0 & 0 & 0 & 0 & 0 & 0 & 0 & 0 & 0.0049 & -0.0049
\end{bmatrix}
$$

a) The current operating condition matches with the characteristic model of a certain subcondition.

For the 16-machine system, suppose that the wind speed changes from Subcondition 1 to Subcondition 3 and then to Subcondition 5 (7.5 m/s). Take the change from Subcondition 3 to Subcondition 5, for example, and analyze the control effect of the proposed control strategy. First, the stochastic gradient method is used to identify the system operating condition and, according to (7.107) to (7.110), the weighting coefficient of each subcontroller is calculated, shown in Table 7.12. It can be seen that, except for subcontroller 5, the weighting coefficients of all the other subcontrollers are very small and negligible, and the current system operating condition matches with the characteristic model of subcontroller 5. In this case, examine the dynamic responses of the G1–G8 relative power angle and the G1–G13 relative power angle when: (1) only subcontroller 5 (the matching controller) is used; (2) a combined controller is used (i.e. all the subcontrollers combined according to the weighting coefficients); and (3) a nonmatching subcontroller is used. The simulation results are shown in Figure 7.45, where the black dashed line represents the simulation result when the controller is not applied and the red solid line

Table 7.12 Weighting coefficients of subcontrollers in the 16-machine system after the wind speed varies to 7.5 m/s

Subcontroller number	Weighting coefficient
Subcontroller 1	0
Subcontroller 2	0
Subcontroller 3	0
Subcontroller 4	0.0096
Subcontroller 5	0.9871
Subcontroller 6	0.0033
Subcontroller 7	0
Subcontroller 8	0
Subcontroller 9	0
Subcontroller 10	0

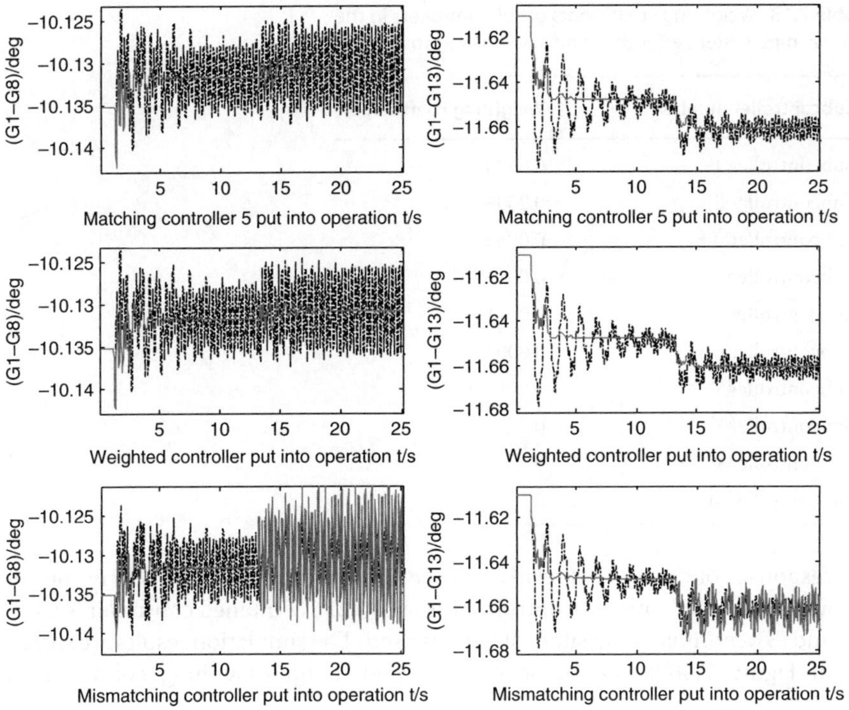

Figure 7.45 Dynamic responses of the G1–G8 relative power angle and the G1–G13 relative power angle after different controllers are put into operation in the new system operating condition.

represents the simulation result after the controller is applied. It can be seen that, when the real system operating condition matches with the characteristic model of a certain subcontroller, the weighting coefficient of this subcontroller is much bigger than that of other subcontrollers. When only the matching controller is used, the control effect is almost the same as that of using the combined controller. However, when a nonmatching subcontroller is used, effective control is not possible and system oscillation even deteriorates.

b) The current operating condition does not match with the characteristic model of any subcondition.

For the 16-machine system, suppose that the wind speed changes from Subcondition 3 to Subcondition 4 and then to Subcondition 1 (3.8 m/s). Take the variation from Subcondition 4 to Subcondition 1, for example, and analyze the control effect of the proposed control strategy. First, the stochastic gradient method is used to identify the system operating condition and then, according to (7.107) to (7.110), the weighting coefficient of each subcontroller is calculated, shown in Table 7.13. It can be seen that the current system operating condition falls in the wind speed interval of Subcondition 1, but it does not completely match with the characteristic model of subcontroller 1. Thus combined control is required (i.e., some of the subcontrollers combined according to the weighting coefficients). In this case, examine the dynamic

Table 7.13 Weighting coefficients of subcontrollers in the 16-machine system after the wind speed varies to 3.8 m/s.

Subcontroller number	Weighting coefficient
Subcontroller 1	0.6593
Subcontroller 2	0.2398
Subcontroller 3	0.0982
Subcontroller 4	0.0015
Subcontroller 5	0.0011
Subcontroller 6	0.0001
Subcontroller 7	0
Subcontroller 8	0
Subcontroller 9	0
Subcontroller 10	0

responses of the G1–G8 relative power angle and the G1–G13 relative power angle when: (1) only subcontroller 1 is used; (2) a combined controller is used; and (3) a nonmatching subcontroller is used. The simulation results are shown in Figure 7.46.It can be seen that, when a non-matching controller or

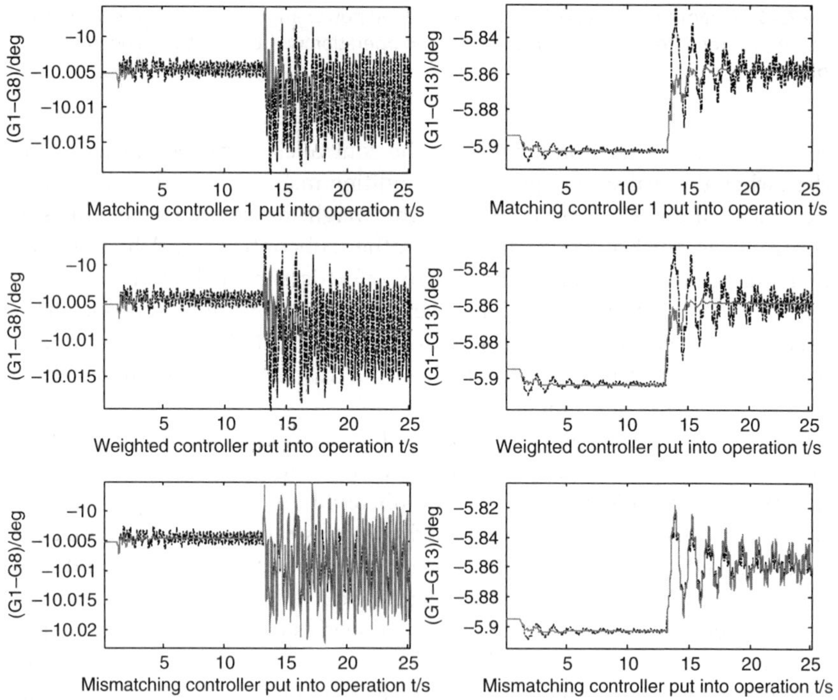

Figure 7.46 Dynamic responses of the G1–G8 relative power angle and the G1–G13 relative power angle after different controllers are put into operation.

Table 7.14 Operating condition information.

Subcondition number	Load interval	Characteristic value
Subcondition 1	21.200+j1.055–22.200+j1.105	21.700+j1.080
Subcondition 2	22.200+j1.105–23.200+j1.155	22.700+j1.130
Subcondition 3	23.200+j1.155–24.200+j1.205	23.700+j1.180
Subcondition 4	24.200+j1.205–25.200+j1.255	24.700+j1.230
Subcondition 5	25.200+j1.255–26.200+j1.305	25.700+j1.280
Subcondition 6	26.200+j1.305–27.200+j1.355	26.700+j1.330
Subcondition 7	27.200+j1.355–28.200+j1.405	27.700+j1.380

subcontroller 1 is used, effective control is not possible and system oscillation even deteriorates. When the combined controller is used, effective control of the system can be achieved.

Time-Varying Power System Adaptive Control Based on Load Variation For the test system, suppose the load connected to bus 52 in Area 3 is time-varying. The system is divided into seven different subconditions according to the load variation. Thus the state set of the continuous Markov time-varying system model can be formed, denoted as $S=\{1,2,\ldots,7\}$. The average value of the load connected to bus 52 is 24.700+j1.230, the variance of active load is 1 (i.e., 4.05% of the active load), and the variance of the reactive load is 0.05 (i.e., 4.07% of the reactive load). The information of different subconditions is shown in Table 7.14.

Similarly, use the modal analysis method to obtain the state matrix of the 16-generator system in each subcondition and then apply the reduced-order state matrix and the following state transfer probability density matrix π to (7.102) (the disturbance attenuation degree γ is 0.1), so that the control feedback response matrix K_i corresponding to each subcondition of the system in (7.102) can be gained. The following two different cases will result according to whether the current system operating condition matches with the characteristic model of any subcondition or not.

$$\pi=\begin{bmatrix} -0.3005 & 0.2420 & 0.0540 & 0.0044 & 0.0001 & 0 & 0 \\ 0.2420 & -0.5425 & 0.2420 & 0.0540 & 0.0044 & 0.0001 & 0 \\ 0.0540 & 0.2420 & -0.5965 & 0.2420 & 0.0540 & 0.0044 & 0.0001 \\ 0.0044 & 0.0540 & 0.2420 & -0.6008 & 0.2420 & 0.0540 & 0.0044 \\ 0.0001 & 0.0044 & 0.0540 & 0.2420 & -0.5965 & 0.2420 & 0.0540 \\ 0 & 0.0001 & 0.0044 & 0.0540 & 0.2420 & -0.5425 & 0.2420 \\ 0 & 0 & 0.0001 & 0.0044 & 0.0540 & 0.2420 & -0.3005 \end{bmatrix}$$

a) The current operating condition matches with the characteristic model of a certain subcondition.

Suppose the 16-machine system changes from Subcondition 4 to Subcondition 2 and then to Subcondition 3 (where the load is 23.700+j1.180). Take the change from Subcondition 2 to Subcondition 3, for example, and analyze the control effect of the proposed control strategy. First, according to the stochastic gradient method, the weighting coefficient of each subcontroller after the system operating condition changes can be gained, shown in Table 7.15. It can be seen that, except for subcontroller 3, the weighting coefficients of all the other subcontrollers are very small and negligible. In this case, examine the dynamic responses of the G1–G10 relative power angle and the G4–G10 relative power angle when: (1) only subcontroller 3 (the matching controller) is used; (2) a combined controller is used (i.e., all the subcontrollers combined according to the weighting coefficients); and (3) a nonmatching subcontroller is used. The simulation results are shown in Figure 7.47, where the black dashed line represents the simulation result when the controller is not applied and the red solid line represents the simulation result after the controller is applied. It can be seen that, when a nonmatching subcontroller is used, effective control is not possible and the system oscillation even deteriorates. However, when only the matching controller is used, the control effect is almost the same as that when using the combined controller.

b) The current operating condition does not match with the characteristic model of a certain subcondition.

Suppose the 16-machine system changes from Subcondition 4 to Subcondition 5, and then to Subcondition 2 (where the load is 22.500+j1.145). Take the change from Subcondition 5 to Subcondition 2, for example, and analyze the control effect of the proposed control strategy. First, the stochastic gradient method is used to identify the system operating condition. The weighting coefficient of each subcontroller after the system operating condition changes is shown in Table 7.16.

It can be seen that the current system operating condition falls in the load interval of Subcondition 2, but it does not completely match with the

Table 7.15 Weighting coefficients of subcontrollers in the 16-machine system after the load varies to 23.700+j1.180.

Subcontroller number	Weighting coefficient
Subcontroller 1	0
Subcontroller 2	0.0026
Subcontroller 3	0.9953
Subcontroller 4	0.0021
Subcontroller 5	0
Subcontroller 6	0
Subcontroller 7	0

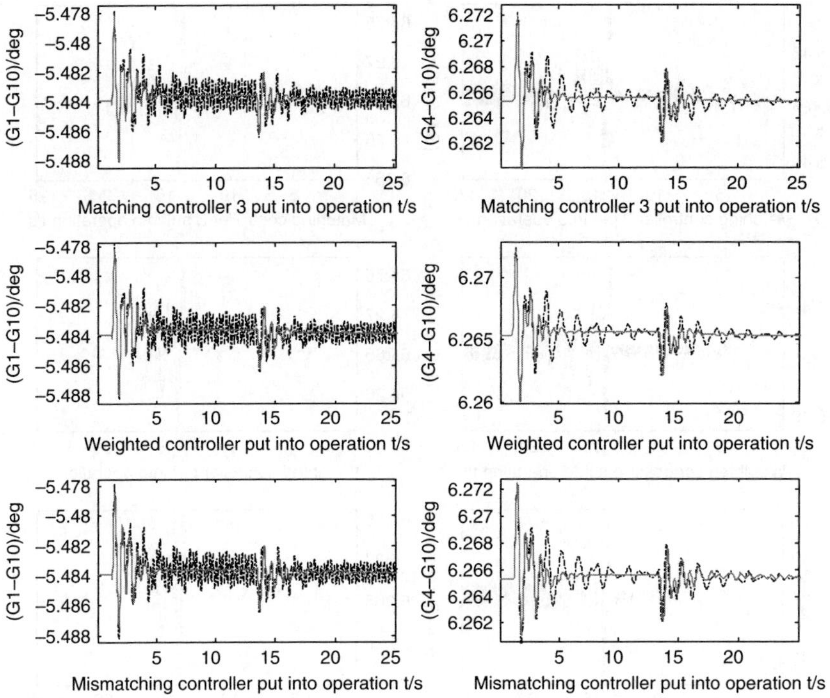

Figure 7.47 Dynamic responses of the G1–G10 relative power angle and the G4–G10 relative power angle after different controllers are put into operation in the new system operating condition.

Table 7.16 Weighting coefficients of subcontrollers in the 16-machine system after the load varies to 22.500+*j*1.145.

Subcontroller number	Weighting coefficient
Subcontroller 1	0.2591
Subcontroller 2	0.5986
Subcontroller 3	0.1339
Subcontroller 4	0.0075
Subcontroller 5	0.0009
Subcontroller 6	0
Subcontroller 7	0

characteristic model of subcontroller 2. Thus combined control is required (i.e. some of the subcontrollers combined according to the weighting coefficients). In this case, examine the dynamic responses of the G1–G10 relative power angle and the G4–G10 relative power angle when: (1) only subcontroller 2 is used; (2) the combined controller is used; and (3) the nonmatching subcontroller is used. The simulation results are shown in Figure 7.48. It can be

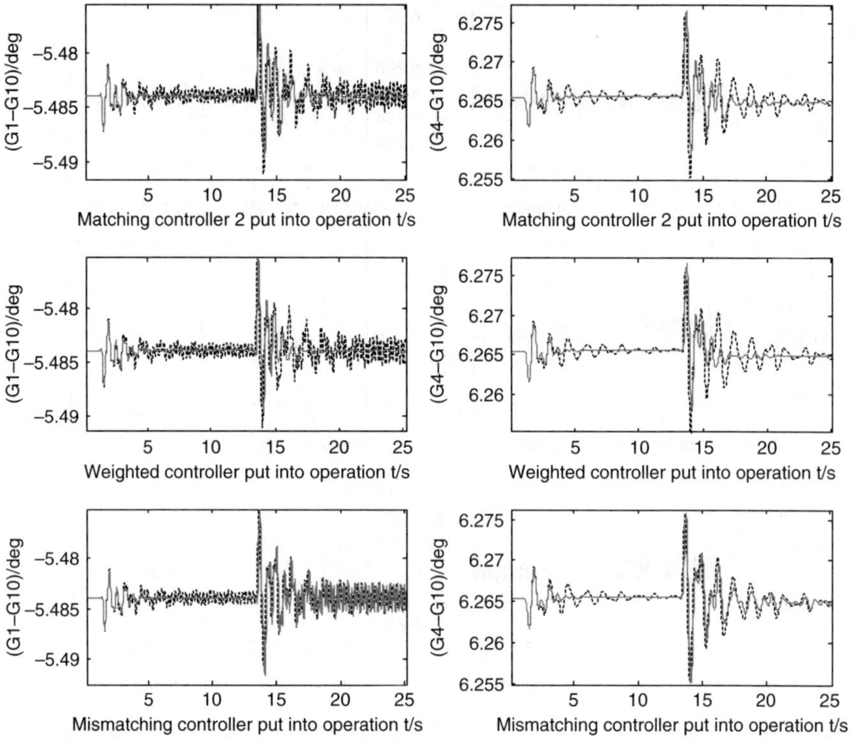

Figure 7.48 Dynamic responses of the G1–G10 relative power angle and the G4–G10 relative power angle after different controllers are put into operation.

seen that, when the combined controller is used, effective control of the system can be achieved. When a nonmatching controller or subcontroller 2 is used, effective control is not possible and system oscillation may even deteriorate.

7.3.3 Discrete Markov Model Based Adaptive Control Strategy of Multiple-Condition Power System

7.3.3.1 Discrete Markov Power System Model with Multiple Operating Conditions

The model of a power system with multiple operating conditions can be expressed using the discrete Markov model as follows:

$$\begin{cases} \mathbf{x}_{k+1} = \left(\boldsymbol{A}(s_k) + \Delta\boldsymbol{A}(s_k)\right)\boldsymbol{x}_k + \boldsymbol{B}(s_k)\boldsymbol{u}_k + \boldsymbol{G}(s_k)\omega_k \\ \boldsymbol{z}_k = \boldsymbol{C}(s_k)\boldsymbol{x}_k + \boldsymbol{D}(s_k)\boldsymbol{u}_k + \boldsymbol{L}(s_k)\omega_k \end{cases} \tag{7.111}$$

where $\boldsymbol{x} \in \boldsymbol{R}^n$ is the state vector, $\boldsymbol{u} \in \boldsymbol{R}^p$ is the control input vector, $\boldsymbol{z} \in \boldsymbol{R}^r$ is the control output vector. The process noise $\boldsymbol{\omega}$ is a zero mean noise series. $\boldsymbol{A}, \boldsymbol{B}, \boldsymbol{G}, \boldsymbol{C}, \boldsymbol{D}, \boldsymbol{L}$ are the coefficient matrices of corresponding vectors, respectively. $\Delta\boldsymbol{A}$ is the uncertain parameter and $\{s_k, k \geq 0\}$ is the Markov chain taking values in limited space $S=\{1,2,\ldots,l\}$. This corresponds with multiple operating conditions of cascading failure and describes the evolution of a system condition during fault; its state probability can be expressed as

$$\Pr\{s_{k+1}=j|s_k=i\}=p_{ij}, \quad \sum_{j=1}p_{ij}=1(p_{ij}\geq 0) \tag{7.112}$$

For each $s_k=i\in S$, denote $A(s_k), B(s_k), G(s_k), C(s_k), D(s_k), L(s_k)$ as $A_i, B_i, G_i, C_i, D_i, L_i$, respectively, and the uncertain parameter in (7.111) meets the matching conditions:

$$\Delta A_k(s_k)=\Delta A_k(i)=H_iF_k(i)M_i \tag{7.113}$$

where H_i and M_i are the known matrix and real matrix $F(i, t)$ reflects the structure information of uncertain parameters and meets

$$F_k^{\mathrm{T}}(i)F_k(i)\leq I \tag{7.114}$$

Formulas (7.111) to (7.114) establish the discrete Markov model for a power system with multiple operating conditions and this model is used to further deduce the stability criterion and develop an adaptive control strategy.

7.3.3.2 Robust Stochastic Stabilization Based on Discrete Markov Model for Power System with Multiple Operating Condition

Definition 7.2 For the system in (7.111), when $u_k=\mathbf{0}$, $\omega_k=\mathbf{0}$, if the following equation is true for all allowable uncertainty ΔA_i:

$$\lim_{N\to\infty} E\left\{\sum_{k=0}^{N}|x_k|^2\Big|x_0,s_0\right\}\leq \tilde{M}(x_0,s_0) \tag{7.115}$$

then the discrete Markov system such as (7.111) is robust stochastic stable.

Lemma 7.2 Denote D, E, F as real matrices with appropriate dimensions and $P>I$, $\|F\|\leq 1$.Then:

1) For all $\varepsilon>0$ and vector $x, y\in R^n$, there exists $2x^{\mathrm{T}}\mathbf{DFEy}\leq\varepsilon^{-1}x^{\mathrm{T}}DD^{\mathrm{T}}x+\varepsilon y^{\mathrm{T}}EE^{\mathrm{T}}y$
2) $DFE+E^{\mathrm{T}}F^{\mathrm{T}}D^{\mathrm{T}}\leq\varepsilon DD^{\mathrm{T}}+\varepsilon^{-1}E^{\mathrm{T}}E$

where I is a unit matrix.

Theorem 7.4 For a given constant $\gamma>0$, if there exists a set of positive definite symmetric matrix $Q_i>0$, $i\in S$, which establishes the following matrix inequality:

$$\begin{bmatrix} -Q_i & \mathbf{0} & Q_iA_i^{\mathrm{T}}\Omega_i & Q_iC_i^{\mathrm{T}} \\ * & -\gamma^2 I & G_i^{\mathrm{T}}\Omega_i & L_i^{\mathrm{T}} \\ * & * & -\psi & \mathbf{0} \\ * & * & * & -I \end{bmatrix}<0 \tag{7.116}$$

$$\Omega_i = \left[\sqrt{p_{i1}}\boldsymbol{I} \quad \sqrt{p_{i2}}\boldsymbol{I} \quad \cdots \quad \sqrt{p_{il}}\boldsymbol{I} \right] \tag{7.117}$$

$$\psi = diag\left(Q_1, Q_2, \cdots, Q_l\right) \tag{7.118}$$

then when $\boldsymbol{u}_k \equiv \boldsymbol{0}$, the system in (7.111) is robust stochastic stable and satisfies interference decay rate:

$$\sum_{k=0}^{N} E\left[\|\boldsymbol{z}_k\|^2\right] \leq \gamma^2 \sum_{k=0}^{N} E\left[\|\omega_k\|^2\right] \tag{7.119}$$

7.3.3.3 Adaptive Control Strategy Considering Discrete Markov Model for Power System with Multiple Operating Conditions

Design of Adaptive Controller Considering Discrete Markov Model for a Power System with Multiple Operating Conditions For a discrete Markov model for a power system with multiple operating conditions in (7.111), if for all allowable uncertainty, the following conditions are satisfied:

1) There exists a feedback control matrix $\boldsymbol{K}_i$ such that the system in (7.111) is robust stochastic stable.
2) For a given scalar $\gamma > 0$, the initial state $\boldsymbol{x}_0 = \boldsymbol{0}$ and all $\omega_k \neq \boldsymbol{0}$, the control output $\boldsymbol{z}_k$ meets the following formula:

$$\sum_{k=0}^{N} E\left[\|\boldsymbol{z}_k\|^2\right] \leq \gamma^2 \sum_{k=0}^{N} E\left[\|\omega_k\|^2\right] \tag{7.120}$$

3) The steady-state variance of each operating condition meets the following constraints:

$$\text{Var}\left[\boldsymbol{x}_k(s_k)\right] := \lim_{k\to\infty} E\left[\boldsymbol{x}_k(s_k)\boldsymbol{x}_k^{\text{T}}(s_k)\right] = \left[\hat{\boldsymbol{Q}}(s_k)\right]_{ii} < \sigma_i^2(s_k) \tag{7.121}$$

$$\hat{\boldsymbol{Q}}(s_k) = \lim_{k\to\infty} \hat{\boldsymbol{Q}}_k(s_k) = \lim_{k\to\infty} E\left[\boldsymbol{x}_k(s_k)\boldsymbol{x}_k^{\text{T}}(s_k)\right] = \sum_{i=1}^{l} \hat{\boldsymbol{Q}}(i)\boldsymbol{P}_i \tag{7.122}$$

where $\sigma_i^2(s_k) > 0(i = 1,2,\ldots,l)$ is the acceptable variance upper bound of an actual requirement.

The following equation can be obtained from (7.113) and (7.116):

$$\begin{bmatrix} -\boldsymbol{Q}_i & \boldsymbol{0} & \boldsymbol{Q}_i\bar{\boldsymbol{A}}_i^{\text{T}}\Omega_i & \boldsymbol{Q}_i\bar{\boldsymbol{C}}_i^{\text{T}} \\ * & -\gamma^2\boldsymbol{I} & \boldsymbol{G}_i^{\text{T}}\Omega_i & \boldsymbol{L}_i^{\text{T}} \\ * & * & -\psi & 0 \\ * & * & * & -\boldsymbol{I} \end{bmatrix} = \begin{bmatrix} -\boldsymbol{Q}_i & \boldsymbol{0} & \varsigma_i & \boldsymbol{Q}_i\left(\boldsymbol{C}_i + \boldsymbol{D}_i\boldsymbol{K}_i\right) \\ * & -\gamma^2\boldsymbol{I} & \boldsymbol{G}_i^{\text{T}}\Omega_i & \boldsymbol{L}_i^{\text{T}} \\ * & * & -\psi & \boldsymbol{0} \\ * & * & * & -\boldsymbol{I} \end{bmatrix} \tag{7.123}$$

$$\varsigma_i = \boldsymbol{Q}_i\left(\boldsymbol{A}_i + \boldsymbol{B}_i\boldsymbol{K}_i + \boldsymbol{H}_i\boldsymbol{F}(i,k)\boldsymbol{M}_i\right)^{\text{T}}\Omega_i \tag{7.124}$$

$$\begin{bmatrix} -Q_i & 0 & Q_i\bar{A}_i^{\mathrm{T}}\Omega_i & Q_i\bar{C}_i^{\mathrm{T}} \\ * & -\gamma^2 I & G_i^{\mathrm{T}}\Omega_i & L_i^{\mathrm{T}} \\ * & * & -\psi & 0 \\ * & * & * & -I \end{bmatrix} = \begin{bmatrix} -Q_i & 0 & Q_i\left(A_i+B_iK_i\right)^{\mathrm{T}}\Omega_i & Q_i\left(C_i+D_iK_i\right) \\ * & -\gamma^2 I & G_i^{\mathrm{T}}\Omega_i & L_i^{\mathrm{T}} \\ * & * & -\psi & 0 \\ * & * & * & -I \end{bmatrix}$$
$$+\begin{bmatrix} 0 \\ 0 \\ \Omega_i^{\mathrm{T}}H_i \\ 0 \end{bmatrix}\begin{bmatrix} F(i,k)M_iQ_i & 0 & 0 & 0 \end{bmatrix}$$
$$+\begin{bmatrix} Q_i\left(F(i,k)M_i\right)^T \\ 0 \\ 0 \\ 0 \end{bmatrix}\begin{bmatrix} 0 & 0 & H_i^{\mathrm{T}}\Omega_i & 0 \end{bmatrix} \tag{7.125}$$

By Lemma 7.2, the following inequality exists in (7.123):

$$\begin{bmatrix} -Q_i & 0 & \varsigma_i & Q_i\left(C_i+D_iK_i\right) \\ * & -\gamma^2 I & G_i^{\mathrm{T}}\Omega_i & L_i^{\mathrm{T}} \\ * & * & -\psi & 0 \\ * & * & * & -I \end{bmatrix}$$
$$\leq \begin{bmatrix} -Q_i & 0 & Q_i\left(A_i+B_iK_i\right)^{\mathrm{T}}\Omega_i & Q_i\left(C_i+D_iK_i\right) \\ * & -\gamma^2 I & G_i^{\mathrm{T}}\Omega_i & L_i^{\mathrm{T}} \\ * & * & -\psi & 0 \\ * & * & * & -I \end{bmatrix}$$
$$+\varepsilon_i\begin{bmatrix} 0 \\ 0 \\ \Omega_i^{\mathrm{T}}H_i \\ 0 \end{bmatrix}\begin{bmatrix} 0 & 0 & H_i^{\mathrm{T}}\Omega_i & 0 \end{bmatrix}+\varepsilon_i^{-1}\begin{bmatrix} Q_iM_i^{\mathrm{T}} \\ 0 \\ 0 \\ 0 \end{bmatrix}\begin{bmatrix} M_iQ_i & 0 & 0 & 0 \end{bmatrix}$$
$$\leq \begin{bmatrix} -Q_i & 0 & Q_i\left(A_i+B_iK_i\right)^{\mathrm{T}}\Omega_i & Q_i\left(C_i+D_iK_i\right) \\ * & -\gamma^2 I & G_i^{\mathrm{T}}\Omega_i & L_i^{\mathrm{T}} \\ * & * & \varepsilon_i\Omega_i^{\mathrm{T}}H_iH_i^{\mathrm{T}}\Omega_i-\psi & 0 \\ * & * & * & -I \end{bmatrix}$$
$$+\varepsilon_i^{-1}\begin{bmatrix} Q_iM_i^T \\ 0 \\ 0 \\ 0 \end{bmatrix}\begin{bmatrix} M_iQ_i & 0 & 0 & 0 \end{bmatrix}<0 \tag{7.126}$$

For controlled system satisfying the above assumptions, set $Y_i = Q_i K_i$ and use the Schur complement lemma to obtain the linear matrix inequality (LMI) of the feedback control matrix when the discrete Markov system with multiple operating conditions is robust stochastic stable. Meanwhile, the method can be used to obtain the robust stochastic stability criterion of a discrete Markov system with multiple operating conditions based on allowable uncertainty condition (7.111) to (7.113), which is similar to Theorem 7.4 proving the system satisfying the interference decay rate γ.

Theorem 7.5 For given positive constants $\gamma > 0$, $\sigma_{ij}^2 > 0 (i = 1,\ldots,l;\ j = 1,\ldots,n)$, if there exists a constant $\varepsilon_i > 0$, positive definite symmetric matrix, $Q_i > 0$, and matrix $Y_i > 0$, which establishes the following matrix inequality:

$$\begin{bmatrix} -Q_i & 0 & \Xi_{1i} & Q_i C_i^{\mathrm{T}} + Y_i^{\mathrm{T}} D_i^{\mathrm{T}} & Q_i M_i^{\mathrm{T}} \\ * & -\gamma^2 I & G_i^{\mathrm{T}} \Omega_i & L_i^{\mathrm{T}} & 0 \\ * & * & \Xi_{2i} & 0 & 0 \\ * & * & * & -I & 0 \\ * & * & * & * & -\varepsilon_i I \end{bmatrix} < 0 \tag{7.127}$$

$$\Xi_{1i} = \left(Q_i A_i^{\mathrm{T}} + Y_i^{\mathrm{T}} B_i^{\mathrm{T}}\right)\Omega_i \tag{7.128}$$

$$\Xi_{2i} = \varepsilon_i \Omega_i^{\mathrm{T}} H_i H_i^{\mathrm{T}} \Omega_i - \psi \tag{7.129}$$

$$\left[Q_i\right]_{ij} \le \sigma_{ij}^2 \left(i = 1,\ldots,l;\ j = 1,\ldots,n\right) \tag{7.130}$$

then there exists a feedback gain $K_i = Y_i Q_i^{-1}$ such that the system in (7.111) is robust stochastic stable.

The inequality (7.127) to (7.130) is linear matrix inequality (LMI) about constants σ_{ij}, ε_i, and matrix variable Q_i, Y_i. A convex set is constituted by the matrix inequality and matrix variable satisfying (7.127) to (7.130). In order to make the calculation results of LMI more suitable for actual needs, the control satisfying specific requirements can be designed by allowable uncertainty conditions (7.111) – (7.113). Then the minimum variance robust controller as shown in Theorem 7.6 can be obtained.

Theorem 7.6 For a given positive constant $\gamma > 0$ and a set of weighed coefficients $\beta_{ij} > 0 (i = 1,\ldots,l;\ j = 1,\ldots,n)$, where $\sum_{i=1}^{l}\sum_{j=1}^{n}\beta_{ij} = 1$, if the following optimization problem has the solution:

$$\min_{Q_i, Y_i, \varepsilon_i} \sum_{i=1}^{l}\sum_{j=1}^{n}\beta_{ij}\sigma_{ij}^2 \quad s.t. (7.117)-(7.118), (7.127)-(7.130) \tag{7.131}$$

Then $u_k = K_i x_k = Y_i Q_i^{-1} x_k$ is the minimum variance robust controller of the system in (7.111).

The solving method of the linear objective function minimization problem is used to solve the optimization problem of (7.131). If there exist positive constants $\varepsilon_i > 0$, $\xi_i > 0$, and a positive definite symmetric matrix $\boldsymbol{Q}_i > 0$ and matrix $Y_i > 0$, then the robust control satisfying the minimum variance constraints can be expressed as $\boldsymbol{u}_t = \boldsymbol{K}_i \boldsymbol{x}_t = \boldsymbol{Y}_i \boldsymbol{P}_i^{-1} \boldsymbol{x}_t$.

Switch Structure In actual operation, there is a need of on-line identification for the current operating condition, and appropriate controller combination should be selected for effective control based on the weighted index.

The multivariable identification model of a system can be expressed as follows:

$$\boldsymbol{y}(t) = \boldsymbol{\Phi}(t)\theta + \boldsymbol{v}(t) \tag{7.132}$$

where $\boldsymbol{y}(t)$ is the output variable matrix, $\boldsymbol{\Phi}(t)$ is the regression information matrix and contains the input and output data, $\boldsymbol{\theta}$ is the parameter matrix needed to be identified, and $\boldsymbol{v}(t)$ is the white noise vector.

The loss function can be defined as

$$J(\theta) = \left\| y(t) - \Phi(t)\theta \right\|^2 \tag{7.133}$$

Denote $\boldsymbol{\mu}(t)$ as the step and $\boldsymbol{\mu}(t) = 1/\boldsymbol{r}(t), \boldsymbol{r}(t) = \boldsymbol{r}(t-1) + \left\|\boldsymbol{\Phi}(t)\right\|^2$, and the negative gradient method is used to search the minimization $J(\boldsymbol{\theta})$, then the stochastic gradient algorithm to estimate the parameters vector $\hat{\boldsymbol{\theta}}$ can be obtained:

$$\begin{aligned} \hat{\boldsymbol{\theta}}(t) &= \hat{\boldsymbol{\theta}}(t-1) - \frac{\boldsymbol{\mu}(t)}{2} grad\left[J\left(\hat{\boldsymbol{\theta}}(t-1)\right)\right] \\ &= \hat{\boldsymbol{\theta}}(t-1) - \frac{\boldsymbol{\Phi}^{\mathrm{T}}(t)}{\boldsymbol{r}(t)}\left[\boldsymbol{y}(t) - \boldsymbol{\Phi}(t)\hat{\boldsymbol{\theta}}(t-1)\right] \end{aligned} \tag{7.134}$$

$$\boldsymbol{r}(t) = \boldsymbol{r}(t-1) + \left\|\boldsymbol{\Phi}(t)\right\|^2, \boldsymbol{r}(0) = 1 \tag{7.135}$$

The controller when the system is actually put into use can be defined as a combination of the matching controller of each operating condition. The combination controller can be formed as

$$\boldsymbol{K} = \alpha_1 \boldsymbol{K}_1 + \alpha_2 \boldsymbol{K}_2 + \cdots + \alpha_l \boldsymbol{K}_l \tag{7.136}$$

where the weighted coefficient α_i at time k can be expressed as

$$J_i(k) = \kappa e_i^2(k) + \beta \int_0^k e^{-\lambda(k-\tau)} e_i^2(\tau) d\tau, \quad \kappa, \beta, \lambda \geq 0 \tag{7.137}$$

$$\alpha_i(k) = \frac{\exp\left(-\left|J_i(k)\right|^2\right)}{\sum_{i=1}^{N} \exp\left(-\left|J_i(k)\right|^2\right)}, \quad \sum_{i=1}^{N} \alpha_i(k) = 1 \tag{7.138}$$

$$\boldsymbol{e}_i = \boldsymbol{y}_i - \hat{\boldsymbol{y}}_i = \boldsymbol{y}_i - \boldsymbol{\Phi}(t)\hat{\boldsymbol{\theta}} + \boldsymbol{v}(t) \tag{7.139}$$

where κ, β are the measurement coefficients and ensure that both the instantaneous measurement and the time interval measurement have an effect, λ is the forgetting factor and can be used to reduce the proportion of past error of the index function, and $\boldsymbol{e}_i$ is the error between the current control model output and the system output. It can be seen from (7.136) to (7.139) that the output error of the model of each operating condition and the controlled object is estimated at each moment; thus the combination of controllers can be corrected and updated adaptively. The higher the matching degree of the operating condition of the controlled object and the ith condition model, the greater the weight of the weighted coefficient α_i will be.

7.3.3.4 Simulation Analysis

Take the IEEE 16-machine 68-bus New England–New York interconnected system, for example. Group 4 contains machines that are a simplified representation of the New England system. Group 5 represents the New York system. The thick lines indicate the tie lines, which are the root cause of the low-frequency inter-area oscillations. The most important connections in this system model are those between groups 4 and 5 (bus 1 to bus 2, bus 1 to 27, and bus 8 to 9). The dashed lines indicate the coherent group boundaries. The load applies the mixed-load model. The active power is 15% constant power, 25% constant current, and 60% constant impedance. The reactive power is 15% constant power, 25% constant current, and 60% constant admittance.

Take the cascading failure leading to multiple operating conditions as an example to verify the correctness and validity of this method. Supposing the initial contingency is a line 1–31 fault and is cut out from service, the load is transferred to the remaining lines and the state transfer probability p_{ij} of the remaining lines is calculated as shown in Table 7.17.

In the case of the same fault path of the previous stage, the larger the transfer probability, the higher is the probability of a line becoming the fault line of the next stage in the cascading failure. Therefore, the system operating conditions corresponding to a fault occurring on the lines with relatively large state transfer probabilities are chosen to form the state set of the discrete Markov system and lines with relatively small state transfer probabilities are taken as nonfault lines. The chosen operating condition set is as shown in Table 7.18, which is denoted by $\boldsymbol{S} = \{1,2,3,4,5\}$, and the probability density matrix is shown as follows:

$$\boldsymbol{p} = \begin{bmatrix} 1 & 0 & 0 & 0 & 0 \\ 0 & 0.0155 & 0.5957 & 0.3784 & 0.0105 \\ 0 & 0 & 1 & 0 & 0 \\ 0 & 0 & 0 & 1 & 0 \\ 0 & 0 & 0 & 0 & 1 \end{bmatrix} \tag{7.140}$$

All the eigenvalues are usually calculated by the QR method, but this method cannot provide phase compensation design information and is not conducive to coordination design of a multimachine system controller. Then SMA (selective modal analysis) is used to reduce the system order and only the state equations

Table 7.17 State transfer probability of some lines after a failure outage of line 1–31.

Line	State transfer probability
1–30	0.5759
30–31	0.3784
31–38	0.0105

Table 7.18 System state sets after failure outage of line 1–31

Operating condition	Operating condition set
Condition 1	Normal operating condition
Condition 2	Line 1–31 outage
Condition 3	Line 1–31 and Line 1–30 outage
Condition 4	Line 1–31 and 30–31 outage
Condition 5	Line 1–31 and Line 31–38 outage

related to the generator electromechanical transient are retained, which contributes to overcoming the above problems and improving the control effects.

Furthermore, since certain conditions must be met when solving inequality using LMI, not all inequalities have a feasible solution. This problem is also taken into account in the simulation process, so the SMA method is used to reduce the order of the 16-machine system corresponding to each operating condition. Only the state variable associated with an electromechanical transient of the generator is retained after order reduction, that is, the power angle and speed of the generator. Then substitute the reduced-order state matrix and the transfer probability matrix of (7.140) into (7.131) and take the interference decay rate γ as 0.1. Thus the control feedback matrix $\boldsymbol{K}_i$ can be obtained corresponding to each model in the system model library. There are two different cases based on whether the system operating condition matches the operating condition model.

Cases When Current Operating Condition Matches a Feature Model of a Certain Operating Condition After a line 1–31 fault and cut-out from service, line 30–31 is disconnected and the system condition switches to condition 4. In order to analyze the control effect of the proposed controller under the condition, the stochastic gradient method is used to perform a state identification. The corresponding weighed coefficient of each controller after an operating condition changes is then calculated using (7.136) to (7.139), as shown in Table 7.19. It can be seen that the system operating condition is consistent with condition 4 and thus the weighed coefficient of other subcontrollers is very small and can be ignored except for subcontroller 4. The dynamic response of the relative power angle of G1–G8 and G1–G13 after subcontroller 4, combination controller, and mismatching controller are put into operation when operating conditions

Table 7.19 Weighted coefficient of the subcontroller after failure outage of line 1–31 of a 16-machine system.

Subcontroller	Weighted coefficient
Subcontroller 1	0
Subcontroller 2	0
Subcontroller 3	0.0132
Subcontroller 4	0.9703
Subcontroller 5	0.0165

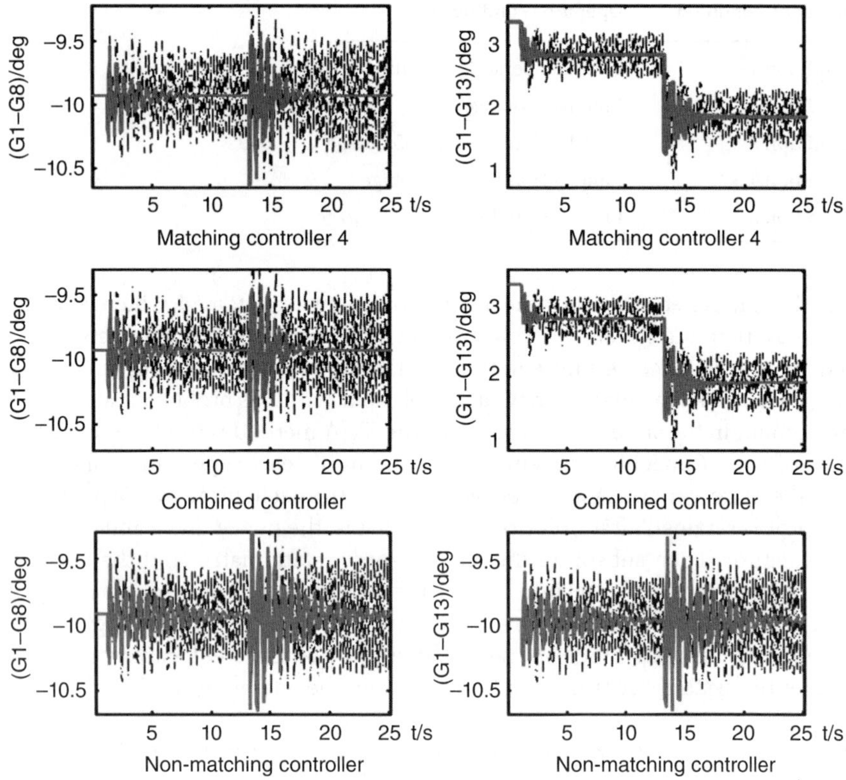

Figure 7.49 Dynamic responses of the relative power angle after different controllers are put into operation when the operating conditions match.

match, as shown in Figure 7.49, where the dotted line represents the case of no controller put into operation and the solid line represents the case after the control is put into operation.

It can be seen from Figure 7.49 that when the system operating condition matches the feature model of a certain condition, the control effect of subcontroller 4 and weighted controller is almost the same. However, when the mismatching controller is put into operation, effective control cannot be realized and may even aggravate system oscillation.

Table 7.20 Weighted coefficient of a subcontroller after failure outage of line 35–34 of a 16-machine system.

Subcontroller	Weighted coefficient
Subcontroller 1	0.0399
Subcontroller 2	0.3280
Subcontroller 3	0.2721
Subcontroller 4	0.0991
Subcontroller 5	0.2609

Cases When Current Operating Condition Mismatches a Feature Model of a Certain Operating Condition After a line 1–31 fault and cut-out from service, line 35–34 is disconnected and the system does not match any certain operating condition. The stochastic gradient method is used to do state identification and the corresponding weighed coefficient of each controller is calculated by (7.136) to (7.139), as shown in Table 7.20 after the operating condition changes. It can be seen that weighted control should be realized using weighed coefficients under this condition. The dynamic response of the relative power angle G1–G8 and G1–G13 after the combination controller and mismatching controller put into operation is shown in Figure 7.50, where the dotted line represents the case of no control put into operation and the solid line represents the case after the control is put into operation. Figure 7.50 illustrates that the combination controller can suppress oscillation effectively when the system does not match any certain operating condition. However, when the mismatching controller is put into operation, effective control cannot be realized and may even aggravate system oscillation.

7.3.4 Adaptive Controller Switching Considering Time Delay

7.3.4.1 Time-Delay Model Considering the Jump Characteristic of a Power System

The closed-loop power system considering time delay can be described as

$$\begin{cases} \dot{\mathbf{x}}(t) = \boldsymbol{A}\mathbf{x}(t) + \boldsymbol{BK}\mathbf{u}(t-d(t)) \\ \mathbf{u}(t-d(t)) = \mathbf{x}(t-d(t)) \end{cases} \tag{7.141}$$

where $\mathbf{x} \in \mathbf{R}^n$ is the state vector, $\mathbf{u} \in \mathbf{R}^n$ is the control input vector, $\boldsymbol{A} \in \mathbf{R}^{n \times n}$ is the system state matrix, $\boldsymbol{B} \in \mathbf{R}^{n \times m}$ is the system control matrix, and $\boldsymbol{K} \in \mathbf{R}^{m \times n}$ is the system controller.

Suppose the time delay matrix $\boldsymbol{Ad} = \boldsymbol{BK}$; then the state equation of the power system is

$$\begin{cases} \dot{\mathbf{x}}(t) = \boldsymbol{A}\mathbf{x}(t) + \boldsymbol{A}_d\mathbf{u}(t-d(t)) \\ \mathbf{u}(t-d(t)) = \mathbf{x}(t-d(t)) \end{cases} \tag{7.142}$$

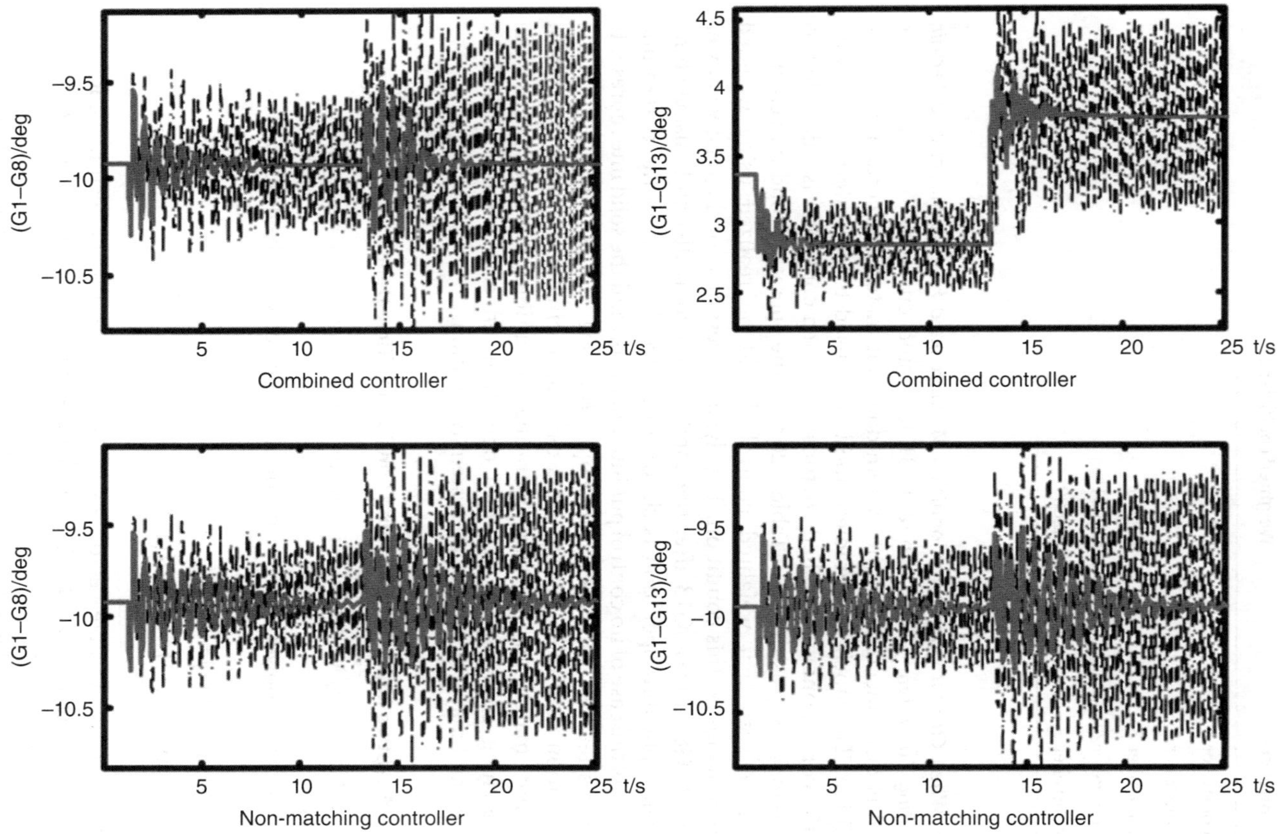

Figure 7.50 Dynamic responses of the relative power angle after different controllers are put into operation when operating conditions mismatch.

When the power system operation state/parameter jumps, the state equation becomes

$$\begin{cases} \dot{\mathbf{x}}(t) = \boldsymbol{A}(\eta)\mathbf{x}(t) + \boldsymbol{A}_d(\eta)\mathbf{u}(t-d(t)) \\ \qquad \mathbf{u}(t-d(t)) = \mathbf{x}(t-d(t)) \end{cases} \tag{7.143}$$

In (7.143), the time delay $d(t)$ meets the following conditions:

$$0 \le d(t) \le h, \dot{d}(t) < \mu \tag{7.144}$$

where $\mathbf{x}(t) \in R^n$ is the state vector and η is a discrete Markov process defined in the finite state space $S = \{1,2,\ldots,N\}$. The state transfer probability of η is

$$p_{ij} = P(\eta_{t+\Delta} = j \mid \eta_t = i) = \begin{cases} \pi_{ij}\Delta + \mathrm{o}(\Delta), & i \ne j \\ 1 + \pi_{ii}\Delta + \mathrm{o}(\Delta), & i = j \end{cases} \tag{7.145}$$

where

$$\lim_{\Delta \to \infty} \frac{\mathrm{o}(\Delta)}{\Delta} = 0 \qquad (\Delta > 0) \tag{7.146}$$

$$\pi_{ii} = -\sum_{i \ne j} \pi_{ij} \quad (\pi_{ij} \ge 0, i \ne j) \tag{7.147}$$

In (7.145), π_{ij} is the element in the Markov transfer probability density matrix $\boldsymbol{\pi}$, which represents the transfer probability density that renders the system modal at state i when time $= t$ and renders the system modal at state j when time $= t + \Delta$.

7.3.4.2 Time-Delay Control Strategy Considering the Jump Characteristic of a Power System

For any appropriately dimensioned matrices $\mathbf{M}_1$ and $\mathbf{M}_2$, according to the Newton–Leibniz formula:

$$2\left[x^T(t)M_1 + x^T(t-h)M_2\right] \times \left[x(t) - \int_{t-h}^{t} \dot{x}(s)ds - x(t-h)\right] = 0 \tag{7.148}$$

Apply (7.143) and (7.148) to the differential equation of the Lyapunov–Krasovskii functional so that the following Theorem and Inference can be gained [14,15].

Theorem 7.7 For given scalar quantities $h > 0$ and μ, if there exist $\mathbf{L}_i = \mathbf{L}_i^{\mathrm{T}} > 0\ (i = 1,2)$, $\mathbf{W}_i = \mathbf{W}_i^{\mathrm{T}} \ge 0\ (i = 1,2)$, $\mathbf{R} = \mathbf{R}^{\mathrm{T}} > 0$,

$$\mathbf{Y} = \begin{bmatrix} \mathbf{Y}_{11} & \mathbf{Y}_{12} & \mathbf{Y}_{13} \\ * & \mathbf{Y}_{22} & \mathbf{Y}_{23} \\ * & * & \mathbf{Y}_{33} \end{bmatrix} \ge 0$$

and any appropriately dimensioned matrices $\mathbf{M}_i(i=1,2,3)$, $\mathbf{S}_i(i=1,2)$, and $\mathbf{V}_i(i=1,2)$, which satisfy the following matrix inequalities:

$$\Lambda = \begin{bmatrix} \Lambda_{11} & \Lambda_{12} & \Lambda_{13} & h\mathbf{S}_1^T \\ * & \Lambda_{22} & \Lambda_{23} & h\mathbf{S}_2^T \\ * & * & \Lambda_{33} & 0 \\ * & * & * & -h\mathbf{R} \end{bmatrix} < 0 \quad (7.149)$$

$$\begin{bmatrix} \mathbf{Y}_{11} & \mathbf{Y}_{12} & \mathbf{Y}_{13} & \mathbf{M}_1 \\ * & \mathbf{Y}_{22} & \mathbf{Y}_{23} & \mathbf{M}_2 \\ * & * & \mathbf{Y}_{33} & \mathbf{M}_3 \\ * & * & * & \mathbf{L}\mathbf{R}^{-1}\mathbf{L} \end{bmatrix} \geq 0 \quad (7.150)$$

then the time-delay power system considering the jump characteristic in (7.131) could damp the inter-area oscillation by obtaining the time-delay controller $\mathbf{K}_{mi}$:

$$\mathbf{K}_{mi} = \mathbf{V}_i\mathbf{L}_i^{-1}\left(i=1,2 \right) \quad (7.151)$$

When i=1, the time-delay controller $\mathbf{K}_{m1}$ is used to damp the inter-area oscillation before the system jump; when i=2, the time-delay controller $\mathbf{K}_{m2}$ is used to damp the inter-area oscillation after the system jump, where

$$\Lambda_{11} = \mathbf{W}_i + \mathbf{M}_1 + \mathbf{M}_1^T + \mathbf{S}_1 + \mathbf{S}_1^T + h\mathbf{Y}_{11} + \sum_{j=1}^{2}\pi_{ij}\mathbf{L}_j$$

$$\Lambda_{12} = \mathbf{A}_i\mathbf{L}_i - \mathbf{S}_1^T + \mathbf{S}_2 + \mathbf{M}_2^T + h\mathbf{Y}_{12}$$

$$\Lambda_{13} = \mathbf{M}_3^T - \mathbf{M}_1 + h\mathbf{Y}_{13}$$

$$\Lambda_{22} = -\mathbf{S}_2 - \mathbf{S}_2^T + h\mathbf{Y}_{22}$$

$$\Lambda_{23} = \mathbf{A}_{di}\mathbf{L}_i + \mathbf{B}_i\mathbf{V}_i - \mathbf{M}_2^T + h\mathbf{Y}_{23}$$

$$\Lambda_{33} = -(1-\mu)\mathbf{W}_i - \mathbf{M}_3 - \mathbf{M}_3^T + h\mathbf{Y}_{33}$$

Consider that (7.150) contains nonlinear items such as $\mathbf{L}\mathbf{R}^{-1}\mathbf{L}$; thus Theorem 7.7 is a nonlinear minimizing problem. In previous literature, this is solved by iteration, which is time-consuming and will greatly lower the solving efficiency of the controller. Therefore, in this paper the Schur complement method is used to decouple the nonlinear item $\mathbf{L}\mathbf{R}^{-1}\mathbf{L}$ and transform the matrix inequalities containing the nonlinear items to standard linear matrix inequalities, so that the low solving efficiency problem caused by iteration can be avoided. Equation (7.150) is equivalent to

$$-\begin{bmatrix} \mathbf{Y}_{11} & \mathbf{Y}_{12} & \mathbf{Y}_{13} & \mathbf{M}_1 \\ * & \mathbf{Y}_{22} & \mathbf{Y}_{23} & \mathbf{M}_2 \\ * & * & \mathbf{Y}_{33} & \mathbf{M}_3 \\ * & * & * & 0 \end{bmatrix} - \begin{bmatrix} \mathbf{0} & \mathbf{0} & \mathbf{0} & \mathbf{L} \end{bmatrix}\mathbf{R}^{-1}\begin{bmatrix} 0 \\ 0 \\ 0 \\ \mathbf{L} \end{bmatrix} \leq 0 \quad (7.152)$$

Using the Schur complement, (7.152) could be transformed to

$$\begin{bmatrix} -\mathbf{Y}_{11} & -\mathbf{Y}_{12} & -\mathbf{Y}_{13} & -\mathbf{M}_1 & \mathbf{0} \\ * & -\mathbf{Y}_{22} & -\mathbf{Y}_{23} & -\mathbf{M}_2 & \mathbf{0} \\ * & * & -\mathbf{Y}_{33} & -\mathbf{M}_3 & \mathbf{0} \\ * & * & * & \mathbf{0} & \mathbf{L} \\ * & * & * & * & \mathbf{R} \end{bmatrix} \le 0 \tag{7.153}$$

which is also equivalent to (7.150); thus (7.150) can be substituted by (7.153). In addition, the following Inference can be gained.

Inference 7.1 For given scalar quantities $h>0$ and μ, if there exist $\mathbf{L}_i=\mathbf{L}_i^{\mathrm{T}}>0\,(i=1,2)$, $\mathbf{W}_i=\mathbf{W}_i^{\mathrm{T}}\ge 0\,(i=1,2)$, $\mathbf{R}=\mathbf{R}^{\mathrm{T}}>0$,

$$\mathbf{Y}=\begin{bmatrix} \mathbf{Y}_{11} & \mathbf{Y}_{12} & \mathbf{Y}_{13} \\ * & \mathbf{Y}_{22} & \mathbf{Y}_{23} \\ * & * & \mathbf{Y}_{33} \end{bmatrix} \ge 0$$

and any appropriately dimensioned matrices $\mathbf{M}_i\,(i=1,2,3)$, $\mathbf{S}_i\,(i=1,2)$, and $\mathbf{V}_i\,(i=1,2)$, which satisfy the following matrix inequalities:

$$\Lambda=\begin{bmatrix} \Lambda_{11} & \Lambda_{12} & \Lambda_{13} & h\mathbf{S}_1^T \\ * & \Lambda_{22} & \Lambda_{23} & h\mathbf{S}_2^T \\ * & * & \Lambda_{33} & 0 \\ * & * & * & -h\mathbf{R} \end{bmatrix} < 0 \tag{7.154}$$

$$\begin{bmatrix} -\mathbf{Y}_{11} & -\mathbf{Y}_{12} & -\mathbf{Y}_{13} & -\mathbf{M}_1 & \mathbf{0} \\ * & -\mathbf{Y}_{22} & -\mathbf{Y}_{23} & -\mathbf{M}_2 & \mathbf{0} \\ * & * & -\mathbf{Y}_{33} & -\mathbf{M}_3 & \mathbf{0} \\ * & * & * & \mathbf{0} & \mathbf{L} \\ * & * & * & * & \mathbf{R} \end{bmatrix} \le 0 \tag{7.155}$$

then matrix inequalities containing nonlinear items can be transformed to standard linear matrix inequalities. Thus the low solving efficiency problem caused by iteration could be avoided. Since the function of the time-delay controller $\mathbf{K}_{m1}$ is equal to that of the time-delay controller without considering the jump characteristic, the following simulation verification is mainly focused on the control effect of the time-delay controller $\mathbf{K}_{m2}$.

7.3.4.3 Simulation Verification

IEEE 16-Machine System Take the IEEE 16-machine 68-bus New England–New York interconnected system, for example.The test system can be divided into five areas. Areas 1, 2, and 3 are equivalent systems, Area 4 is the New York system, and Area 5 is the New England system. There are three connection lines between

Area 4 and Area 5, that is, line 1–2, line 8–9, and line 1–27. Line 1–2 and line 8–9 are both double circuit lines. The generators apply the sixth-order detailed model, the excitation system applies the IEEE-DC1 excitation, and the load applies the WECC load model – 80% constant active power load, 80% constant reactive power impedance load, and 20% dynamic load.

Simulation Tests Taking the IEEE 16-machine 68-bus system, for example, the effectiveness and feasibility of the proposed time-delay controller considering the jump characteristic of power system is verified. Shown in Table 7.21 are nine different operating conditions of the 16-machine system. The time-delay upper bound and time-delay varying rate of different jump cases are shown in Table 7.22.

It can be seen from Table 7.21 that there are eight different cases of system jump from normal operation (Case 1) to other operating conditions (Case 2 to Case 9). When the system jumps from Case 1 to Case 2 and Case 3, the structure

Table 7.21 Case bank of different operating conditions.

Case number	Operating condition
1	Normal operation
2	40% load increase in Area 4
3	40% load increase in Area 5
4	One line breaks in double circuit line 1–2; 40% load increase in Area 4
5	One line breaks in double circuit line 8–9; 40% load increase in Area 4
6	One line breaks in double circuit line 1–27; 40% load increase in Area 4
7	One line breaks in double circuit line 1–2; 40% load increase in Area 5
8	One line breaks in double circuit line 8–9; 40% load increase in Area 5
9	One line breaks in double circuit line 1–27; 40% load increase in Area 5

Table 7.22 Time-delay upper bound and time-delay varying rate of different jump cases for the 16-machine system.

Jump case	Upper bound of the time delay	The varying rate of the time delay
Case1–Case 2	67.1 ms	0.1
Case1–Case 3	68.2 ms	0.1
Case1–Case 4	49.3 ms	0.1
Case1–Case 5	48.7 ms	0.1
Case1–Case 6	43.6 ms	0.1
Case1–Case 7	51.9 ms	0.1
Case1–Case 8	52.5 ms	0.1
Case1–Case 9	47.9 ms	0.1

of the power system remains unchanged during the time-domain simulations. When the system jumps from Case 1 to Case 4 and Case 9, the structure of the power system changes during the time-domain simulations. The load buses are not eliminated to obtain a reduced-order system. In the eight jump cases, time-delay controllers $\mathbf{K}_{m1}$ and $\mathbf{K}_{m2}$ are designed for damping inter-area oscillation before and after the jump, respectively, using the proposed method. Figure 7.50 show the dynamic responses of the generator power angle difference in the eight jump cases, where the solid lines represent the dynamic responses when the time-delay controller is applied and the dashed lines represent the dynamic responses when the time-delay controller is not applied. Suppose the system is in a normal operation before the jump and a three-phase transient fault occurs on line 28–29 at t=0.1s and lasts for 0.2 s. When the controller is not applied, the system has obvious oscillatory features and is not able to restore stability in a short time. However, when the time-delay controller $\mathbf{K}_{m1}$ is applied, the inter-area oscillation is damped within 10–15 s and the system is quickly restored to the normal operation state, which can be seen in Figure 7.51.

Then, at t=20 s, the system jumps to another operating condition from the normal operation state. In order to guarantee that the system after the jump does not become unstable due to the influence of the time-delay controller $\mathbf{K}_{m1}$, the time-delay controller $\mathbf{K}_{m2}$ is applied. It can be seen that, after t=20 s, the power angle difference oscillation without a controller becomes more obvious and the system is more difficult to restore to stability. However, the time-delay controller $\mathbf{K}_{m2}$ is able to use the jump characteristic of the power system and so is adaptable to the switch between different operating conditions. Even when the system operating condition changes drastically, the oscillation can still be damped within 10–15 s using the proposed controller.

Concerning the inter-area oscillation of power system, when taking into account the random jump characteristic of the power system, a pre-jump time-delay controller $\mathbf{K}_{m1}$ and a post-jump time-delay controller $\mathbf{K}_{m2}$ are designed, which are applicable to different jump cases. Figure 7.51 shows that controller $\mathbf{K}_{m1}$ and $\mathbf{K}_{m2}$ are effective in damping the inter-area oscillation of the jump system in different jump cases.

To examine the advantages of the proposed method further, a comparison is made between the time-delay controller put forward by the traditional method and the time-delay controller $\mathbf{K}_{m2}$ designed in this paper concerning the eight jump cases. As shown in Figure 7.52, the solid lines represent the closed-loop responses of the proposed method and the dashed lines represent the closed-loop responses of the traditional method. It can be seen that, in different jump cases, the time-delay controller $\mathbf{K}_{m2}$ designed in this paper is able to damp the inter-area oscillation fast and effectively, with a better control effect than the traditional time-delay controller, which does not consider the jump characteristic of a power system.

A new class of Lyapunov–Krasovskii functional is constructed and then loose items formed with the improved free weighting matrix and the Newton–Leibniz formula are added to the derivative of the functional along the system. On this basis, a set of linear matrix inequalities (LMIs) is formed and then a time-delay controller and time-delay stability criterion of the jump system are gained. The

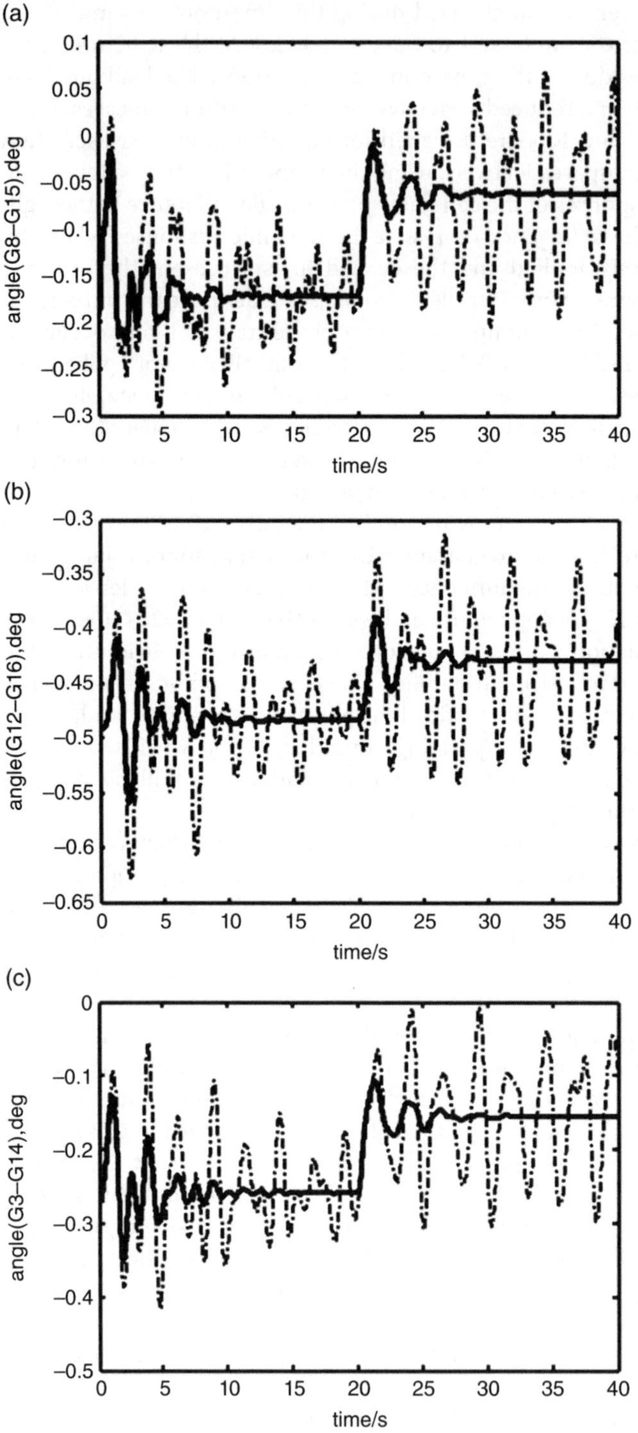

Figure 7.51 Dynamic responses of the power system under different operating conditions: (-) with the jump controller; (- · -)without the controller.

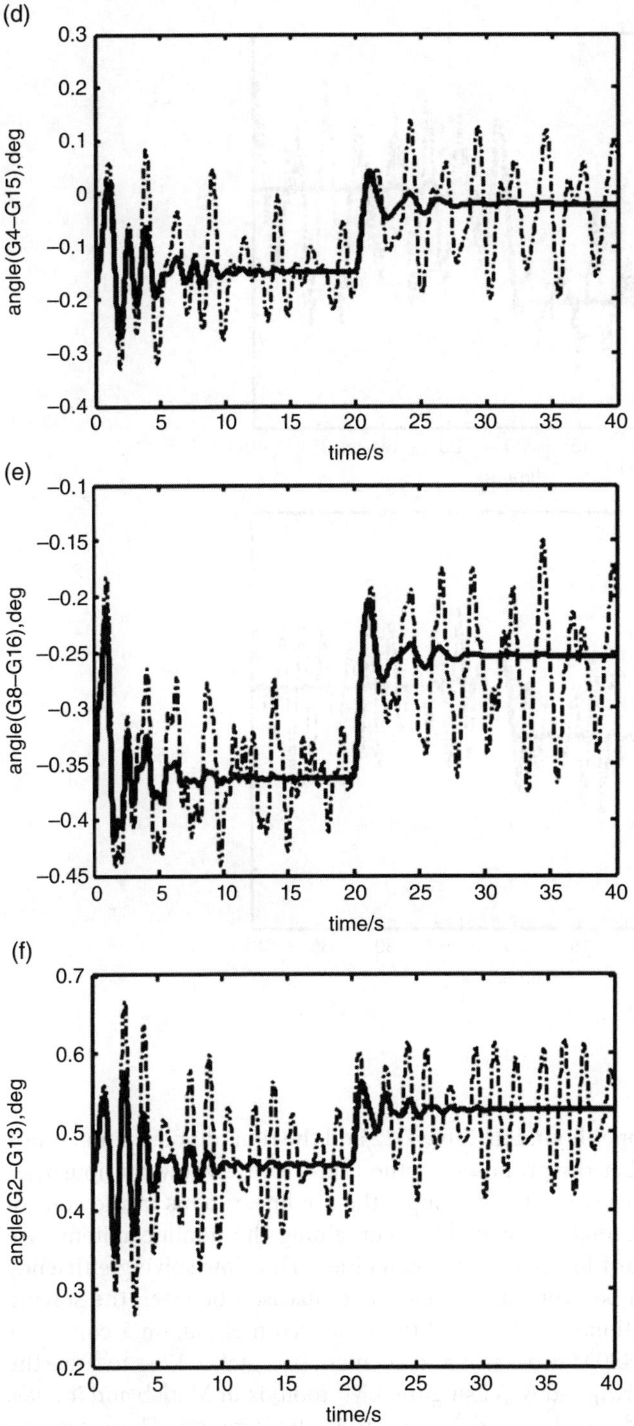

Figure 7.51 (Continued)

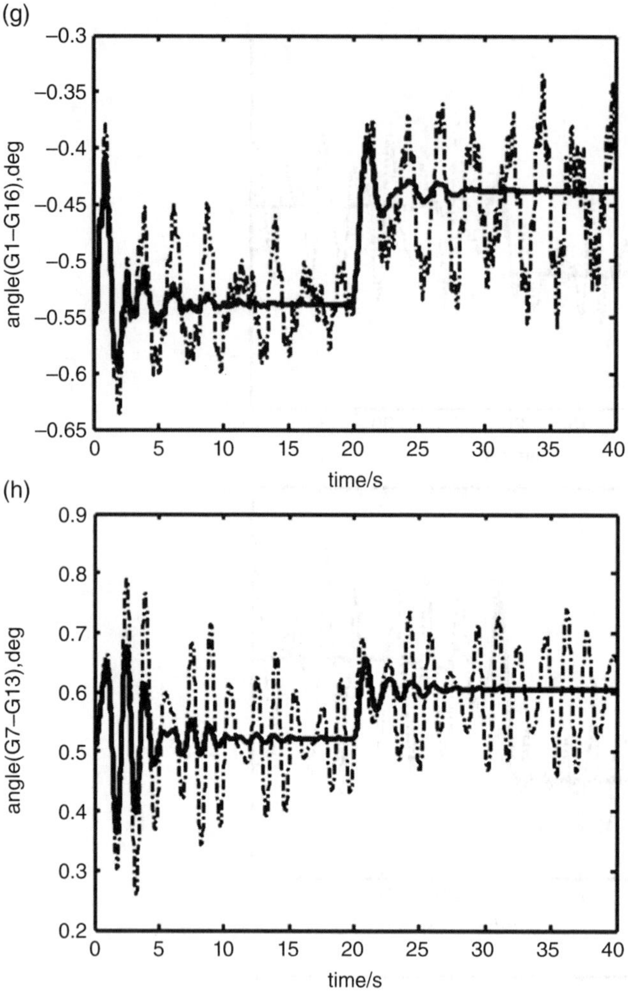

Figure 7.51 (Continued)

time-delay stability upper bound in Table 7.22 and the time-domain simulations in Figure 7.52 show that the proposed method has relatively low cautiousness. The Schur complement is used to decouple the nonlinear items in the control algorithm, so that the matrix inequalities containing the nonlinear items are transformed to standard linear matrix inequalities. Thus low solving efficiency due to iteration could be avoided. To make a comparison between the solving efficiency of the traditional method and the proposed method, on a computer with a Windows 7, i5-430M processor and 2G memory, it takes 178 s to solve the time-delay controller $\mathbf{K}_{m1}$ and $\mathbf{K}_{m2}$ using the LMI toolbox in Matlab and it takes 873 s to solve the traditional time-delay controller by iteration. Therefore, the proposed method has a higher solving efficiency.

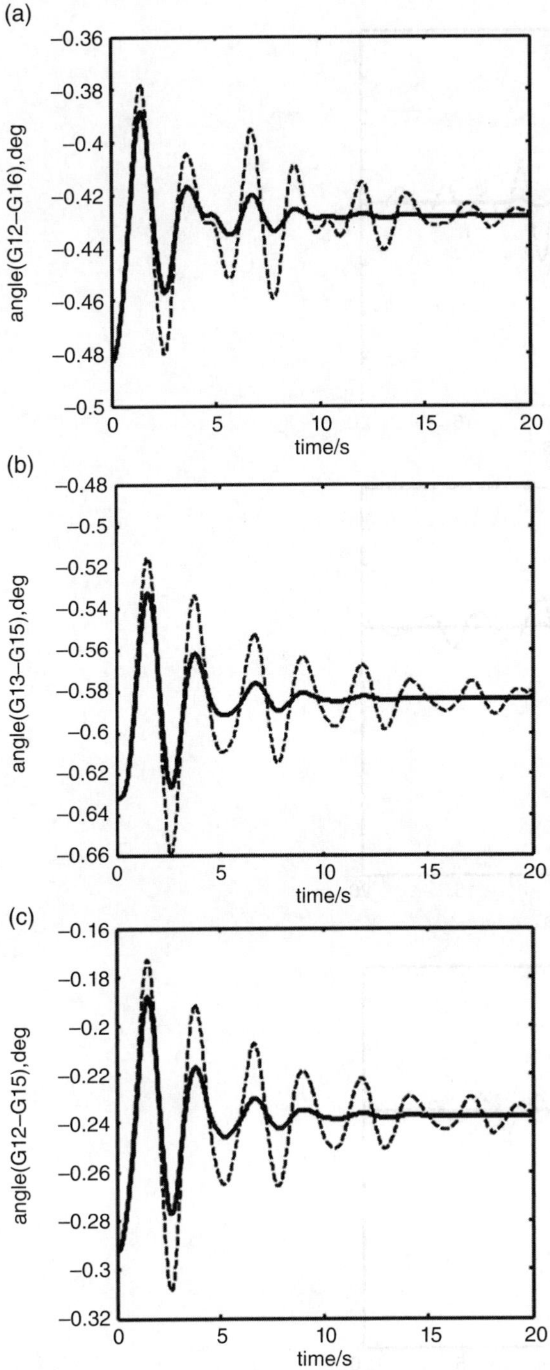

Figure 7.52 Comparison between the traditional method and the proposed method.

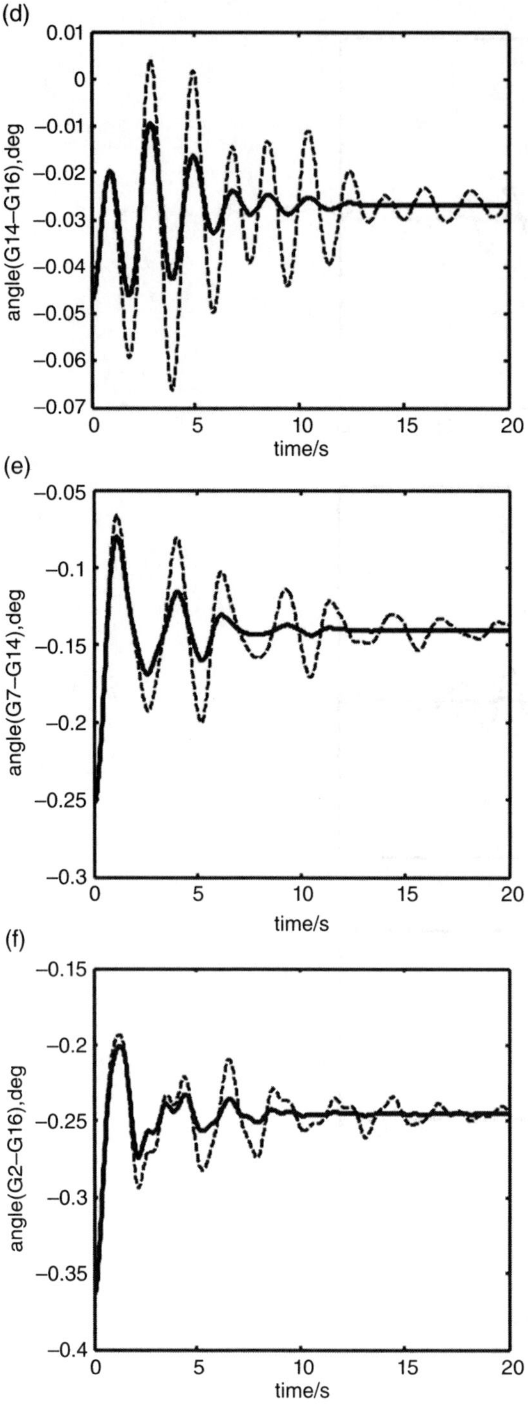

Figure 7.52 (Continued)

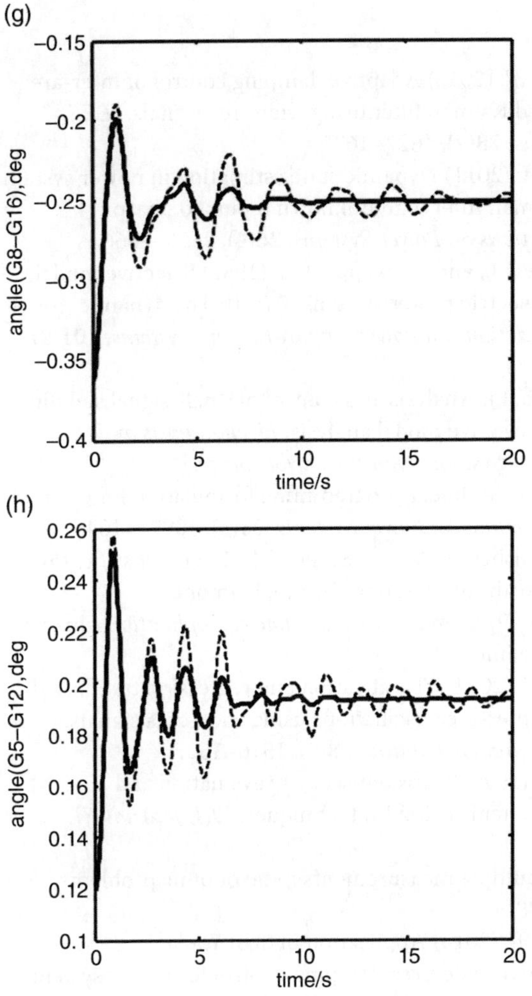

Figure 7.52 (Continued)

7.4 Summary

In order to achieve accurate and efficient adaptive control aiming at a power system operating at a wide range of changing conditions, the system state identification matching and controller switching are analyzed and discussed in this chapter. With regard to operating condition identification matching, the federated Kalman filter and classification and regression tree based wide-area adaptive control are introduced. The two strategies are independent of the system model and can identify the current operating condition quickly and put in the proper controller to damp the oscillation. As for controller switching, dual Youla parameterization and Markov theory based adaptive wide-area damping control are used to achieve a smooth handoff between controllers and thus avoid the impact of controller switching on the power system.

References

1 Ma, J., Wang, T., Wang, Z., *et al.* (2013) Adaptive damping control of inter-area oscillations based on federated Kalman filter using wide area signals. *IEEE Transactions on Power Systems*, **28**(2), 1627–1635.

2 Ghahremani, E. and Kamwa, I. (2011) Dynamic state estimation in power system by applying the extended Kalman filter with unknown inputs to phasor measurements. *IEEE Transactions on Power Systems*, **26**(4), 2556–2566.

3 Verghese, G.C., Perez-Arriage, I.J., and Schweppe, F.C. (1982) Selective modal analysis with applications to electric power systems, Part II: The dynamic stability problem. *IEEE Transactions on Power Apparatus and Systems*, **101**(9), 3126–3134.

4 Pagola, F.L. and Perez-Arriaga, I.J. Analysis and control of small-signal stability in electric power systems by selective modal analysis. *Eigenanalysis and Frequency Domain Methods for System Dynamic Performance.*

5 Gou, B. (2008) Generalized integer linear programming formulation for optimal PMU placement. *IEEE Transactions on Power Systems*, **23**(3), 1099–1104.

6 Breiman, L., Friedman, J.H., Olshen, R.A., and Stone, C.J. (1984) *Classification and Regression Trees*, Wadsworth International Group, Belmont.

7 Ming, Q., Li W., and Li, H. (2009) *Data Mining Technology Application Instance*, Mechanical Industry Press, Beijing.

8 Ma, J., Wang, T., Gao, X., *et al.* (2014) Classification and regression tree-based adaptive damping control of inter-area oscillations using wide-area signals. *IET Generation Transmission and Distribution*, **8**(9), 1516–1527.

9 Arora, C.M. and Surana, S.L. (1996) Transient security evaluation and preventive control of power systems using PR techniques. *EI(I) Journal-EL*, **76**, 199–203.

10 Fisher, R. (1936) The use of multiple measurements in taxonomic problems. *Annals of Eugenics*, **7**, 179–188.

11 Ma, J., Wang, T., Wang, S., *et al.* (2014) Application of dual Youla parameterization based adaptive wide-area damping control for power system oscillations. *IEEE Transactions on Power Systems*, **29**(4), 1602–1610.

12 Skogestad, S. and Postlethwaite, I. (2001) *Multivariable Feedback Control*, John Wiley& Sons, Inc., New York.

13 Tay, T., Mareels, I., and Moore, J.B. (1997) *High Performance Control*, Birkhauser, Boston.

14 Bao, L. (2003) *Modern Control Theory*, Mechanical Industry Press, Beijing.

15 Jie, S. and Chengshan, W. (2008) Design of H_∞controller for damping interarea oscillations with consideration of the delay of remote signal. *Proceedings of the CSEE*, **28**(1), 30–34.

Index

Power System Wide-Area Stability Analysis and Control, First Edition. Jing Ma.

d

t

y